T0331918

Functions of Completely Regular Growth

Mathematics and Its Applications (*Soviet Series*)

Volume 81

Functions of Completely Regular Growth

by

L.I. Ronkin
Institute for Low Temperature Physics and Engineering,
UKR. SSR Academy of Sciences,
Kharkov, Ukraine, C.I.S.

KLUWER ACADEMIC PUBLISHERS
DORDRECHT / BOSTON / LONDON

Library of Congress Cataloging-in-Publication Data

Ronkin, L. I. (Lev Isaakovich)
Functions of completely regular growth / by L.I. Ronkin.
p. cm. -- (Mathematics and its applications. Soviet series ; v. 81)
Includes bibliographical references and index.
ISBN 0-7923-1677-0 (HB : acid free paper)
1. Functions of complex variables. I. Title. II. Series: Mathematics and its applications (Kluwer Academic Publishers). Soviet series ; 81.
QA331.7.R66 1992
515'.98--dc20 92-5987

ISBN 0-7923-1677-0

Published by Kluwer Academic Publishers,
P.O. Box 17, 3300 AA Dordrecht, The Netherlands.

Kluwer Academic Publishers incorporates
the publishing programmes of
D. Reidel, Martinus Nijhoff, Dr W. Junk and MTP Press.

Sold and distributed in the U.S.A. and Canada
by Kluwer Academic Publishers,
101 Philip Drive, Norwell, MA 02061, U.S.A.

In all other countries, sold and distributed
by Kluwer Academic Publishers Group,
P.O. Box 322, 3300 AH Dordrecht, The Netherlands.

Printed on acid-free paper

Printed in the Netherlands

SERIES EDITOR'S PREFACE

'Et moi, ..., si j'avait su comment en revenir, je n'y serais point allé.'

Jules Verne

The series is divergent; therefore we may be able to do something with it.

O. Heaviside

One service mathematics has rendered the human race. It has put common sense back where it belongs, on the topmost shelf next to the dusty canister labelled 'discarded nonsense'.

Eric T. Bell

Mathematics is a tool for thought. A highly necessary tool in a world where both feedback and nonlinearities abound. Similarly, all kinds of parts of mathematics serve as tools for other parts and for other sciences.

Applying a simple rewriting rule to the quote on the right above one finds such statements as: 'One service topology has rendered mathematical physics ...'; 'One service logic has rendered computer science ...'; 'One service category theory has rendered mathematics ...'. All arguably true. And all statements obtainable this way form part of the raison d'être of this series.

This series, *Mathematics and Its Applications*, started in 1977. Now that over one hundred volumes have appeared it seems opportune to reexamine its scope. At the time I wrote

> "Growing specialization and diversification have brought a host of monographs and textbooks on increasingly specialized topics. However, the 'tree' of knowledge of mathematics and related fields does not grow only by putting forth new branches. It also happens, quite often in fact, that branches which were thought to be completely disparate are suddenly seen to be related. Further, the kind and level of sophistication of mathematics applied in various sciences has changed drastically in recent years: measure theory is used (non-trivially) in regional and theoretical economics; algebraic geometry interacts with physics; the Minkowsky lemma, coding theory and the structure of water meet one another in packing and covering theory; quantum fields, crystal defects and mathematical programming profit from homotopy theory; Lie algebras are relevant to filtering; and prediction and electrical engineering can use Stein spaces. And in addition to this there are such new emerging subdisciplines as 'experimental mathematics', 'CFD', 'completely integrable systems', 'chaos, synergetics and large-scale order', which are almost impossible to fit into the existing classification schemes. They draw upon widely different sections of mathematics."

By and large, all this still applies today. It is still true that at first sight mathematics seems rather fragmented and that to find, see, and exploit the deeper underlying interrelations more effort is needed and so are books that can help mathematicians and scientists do so. Accordingly MIA will continue to try to make such books available.

If anything, the description I gave in 1977 is now an understatement. To the examples of interaction areas one should add string theory where Riemann surfaces, algebraic geometry, modular functions, knots, quantum field theory, Kac-Moody algebras, monstrous moonshine (and more) all come together. And to the examples of things which can be usefully applied let me add the topic 'finite geometry'; a combination of words which sounds like it might not even exist, let alone be applicable. And yet it is being applied: to statistics via designs, to radar/sonar detection arrays (via finite projective planes), and to bus connections of VLSI chips (via difference sets). There seems to be no part of (so-called pure) mathematics that is not in immediate danger of being applied. And, accordingly, the applied mathematician needs to be aware of much more. Besides analysis and numerics, the traditional workhorses, he may need all kinds of combinatorics, algebra, probability, and so on.

In addition, the applied scientist needs to cope increasingly with the nonlinear world and the extra mathematical sophistication that this requires. For that is where the rewards are. Linear models are honest and a bit sad and depressing: proportional efforts and results. It is in the nonlinear world that infinitesimal inputs may result in macroscopic outputs (or vice versa). To appreciate what I am hinting at: if electronics were linear we would have no fun with transistors and computers; we would have no TV; in fact you would not be reading these lines.

There is also no safety in ignoring such outlandish things as nonstandard analysis, superspace and anticommuting integration, *p*-adic and ultrametric space. All three have applications in both electrical engineering and physics. Once, complex numbers were equally outlandish, but they frequently proved the shortest path between 'real' results. Similarly, the first two topics named have already provided a number of 'wormhole' paths. There is no telling where all this is leading - fortunately.

Thus the original scope of the series, which for various (sound) reasons now comprises five subseries: white (Japan), yellow (China), red (USSR), blue (Eastern Europe), and green (everything else), still applies. It has been enlarged a bit to include books treating of the tools from one subdiscipline which are used in others. Thus the series still aims at books dealing with:

- a central concept which plays an important role in several different mathematical and/or scientific specialization areas;
- new applications of the results and ideas from one area of scientific endeavour into another;
- influences which the results, problems and concepts of one field of enquiry have, and have had, on the development of another.

Just how a function grows as the norm of its argument goes to infinity is a difficult and important matter to analyze and characterize. Early successes date from the thirties for entire functions of one variable and quickly found a considerable array of applications.

Since then much has happened and in particular the author and a number of colleagues found the right generalizations for functions of several variables. A great deal of the literature is in the Russian language and not well enough known in the West. The author has worked in this field for over 25 years and here presents the complete theory for the cases of holomorphic functions of completely regular growth of one or several complex variables and of subharmonic functions of completely regular growth on $\mathbf{R}^n$ or a cone. It is a pleasure to welcome such a unique and complete volume in this series.

The shortest path between two truths in the real domain passes through the complex domain.

J. Hadamard

La physique ne nous donne pas seulement l'occasion de résoudre des problèmes ... elle nous fait pressentir la solution.

H. Poincaré

Never lend books, for no one ever returns them; the only books I have in my library are books that other folk have lent me.

Anatole France

The function of an expert is not to be more right than other people, but to be wrong for more sophisticated reasons.

David Butler

Bussum, 10 February 1992

Michiel Hazewinkel

TABLE OF CONTENTS

PREFACE

The theory of functions of completely regular growth[1] (f.c.r.g) of one variable was constructed by B.Levin and A.Pfluger in the nineteen-thirties. In short, these functions are defined as function such that the logarithm of their moduli is asymptotically close to a positively homogeneous function (an indicator) out of some exceptional set. The matter of the theory is as follows: c.r.g. of an entire function is equivalent to the existence of a certain regularity in its zero distribution . Originally this theory was perceived as an exotic one. But soon enough it was applied in various branches of mathematics (differential operators, convolution equations, interpolation, functional analysis, etc.) and in some branches of physics (such as radio physics). In the nineteen sixties the theory of f.c.r.g. attracted the attention of many investigators, and was considerably developed. It was extended to subharmonic functions in $\mathbb{R}^n$ (by V.Azarin) and to holomorphic functions in the half-plane $\mathbb{C}^+ = \{z \in \mathbb{C}: Im z \geq 0\}$ (by N.Govorov and others). In the nineteen seventies a new approach to this theory was suggested by V.Azarin, P.Agranovič and L.Ronkin. Their method was based on the study and application of weak convergence of a certain family of functions generated by the initial function. It enables one to give a comparatively simple and clear description of the classical theory, and to describe the theories of f.c.r.g. that are subharmonic in $\mathbb{R}^n$ or holomorphic in the half-plane. However, the main result is the extension of the theory of f.c.r.g. to entire and plurisubharmonic functions of several variables.

Following the above-mentioned method, in the first four chapters we give a uniform account of the whole theory of holomorphic f.c.r.g of one (in $\mathbb{C}$ or $\mathbb{C}^+$) or several variables (in $\mathbb{C}^n$), as well as of

---

[1] For brevity we shall sometimes the word "completely" in the phrase "completely regular growth".

subharmonic f.c.r.g. (in $\mathbb{R}^n$ or in a cone).

In the nineteen-seventies and -eighties, simultaneously with the constructing of the global theory great attention was paid to special classes of f.c.r.g in $\mathbb{C}^n$ (or in similar domains) which especially often occur in applications. One of these classes is the set of entire functions of several variables that are Fourier transforms of distributions with compact (resp. finite) support. A systematic description of the properties of the functions from these classes is given in **Chapters 5** and **6**. The last **Chapter 7** is devoted to holomorphic mappings.

We now sketch in some detail the contents of this monograph.

In **Chapter 1** we present of Levin-Pfluger theory of entire f.c.r.g. of one variable in accordance with the above-mentioned approach. In particular, the relationship between growth regularity of the function and its zero distribution, the properties of rays of c.r.g., and the theorem on addition of indicators are given. We use standard, fairly simple and well-known properties of entire functions and distributions. The case of functions of one variable is distinguished because, first, these occur often used in applications, and, secondly, in this case there are not so many additional restrictions. Therefore **Chapter 1** is self-contained and simple enough. On the other hand it gives a model for further constructions.

Chapter 2 is devoted to the study of subharmonic f.c.r.g. in $\mathbb{R}^n$. A considerable part of this chapter consists of auxiliary knowledge on $\mathcal{D}'$-convergence of families of subharmonic functions and on the properties of their derivatives. Subsequently we determine different criteria for c.r.g., introduce the concept of a limit set (from Azarin) and of a ray of c.r.g. (from L.Gruman), and study their properties.

Entire functions in $\mathbb{C}^n$ are considered in **Chapter 3**. If f(z) is an entire function, then $ln|f(z)|$ is plurisubharmonic, and hence subharmonic. Therefore the simple statements on growth regularity of entire functions follow from the corresponding statements on subharmonic functions (**Chapter 2**). At the same time there are special problems for entire functions. For example, at the beginning of this chapter we consider the relation between growth regularity of the restrictions of an entire function f(z), $z \in \mathbb{C}^n$, on complex lines $z = \lambda w$,

$w \in \mathbb{C}^n$ and growth regularity of $f(z)$ in all of $\mathbb{C}^n$. Then we prove the converse theorem on addition of indicators (Favorov's theorem). Unlike the direct theorem, the converse theorem does not follow from the corresponding statement on subharmonic functions. **Chapter 3** is completed with the recent Sigurdsson's result on the possibility of approximating in $\mathbb{C}^n$ a plurisubharmonic function by the logarithm of the modulus of an entire function. In particular, the theorem on existence of f.c.r.g. with prescribed theorem may be extracted from Sigurdsson's result. This theorem strengthens the corresponding theorem of Martineau and Kiselman.

In **Chapter 4** we study f.c.r.g. on a cone in $\mathbb{R}^n$. For the same reason as the case of entire function of one variable has been distinguished in **Chapter 1**, here the case of functions holomorphic in the upper half-plane is distinguished. As in the previous chapters, the exposition is based on the study and application of weak convergence of the corresponding family of functions. In this way we obtain Govorov's theory of f.c.r.g. in $\mathbb{C}^+$ and $\overline{\mathbb{C}}^+$, as well as some other results. Subsequently all results are transferred to subharmonic functions on a cone. The proofs are omitted because they are too long. Although their conceptual basis is similar to the case of functions in the half-plane, the technical difference is rather essential , and the apparatus used is less simple.

In **Chapter 5** we consider entire functions of exponential type that are bounded for real values of independent the variables. The study of Fourier transforms of distributions with compact support can be easily reduced to the study of such functions of exponential type which are bounded on $\mathbb{R}^n$, $n \geq 2$. Note that these functions are not necessarily of c.r.g., unlike the case of one variable. In **Chapter 6** we give the corresponding "Vauthier's" example, and determine conditions on the spectrum support of Fourier transforms such that if they are satisfied, then it is impossible to construct a similar example. Subsequently we study sufficient conditions for the following statement to hold: a discrete set in $\mathbb{R}^n$ is a set of uniqueness for the functions from the class under consideration. Besides we study conditions under which the following statement holds: if a function is bounded on a set $E \subset \mathbb{R}^n$, then it is bounded by a proportional constant on al of $\mathbb{R}^n$.

In **Chapter 6** we consider quasipolynomials, i.e. finite sums of exponentials with coefficients which are entire functions of zero degree. Note that Fourier transforms of distributions with finite support (consisting of a finite number of points) are quasipolynomials with constant or polynomial coefficients. We study the asymptotic behaviour and zero distribution of quasipolynomials, as well as the structure of functions which are quasipolynomials with respect to every variable. We also consider problems on the divisibility of quasipolynomials.

The theory of holomorphic mappings $\mathbb{C}^n \to \mathbb{C}^m$, $n \geq m \geq 2$, of completely regular growth is in a initial stage of development now. Moreover, it is not clear what definition of these is reasonable. **Chapter 7** is devoted to holomorphic mappings $\mathbb{C}^n \to \mathbb{C}^m$. Here the notions of plurisubharmonic function and holomorphic mapping of d-regular growth are introduced, and it is shown that regularity of a function $u(z)$ involves the convergence of the currents $(dd^c(t^{-\rho}u(tz)))^q$, $t \to \infty$. From this we derive a number of statements on the asymptotic behaviour of functions which are known from Nevanlinna's theory of value distribution of holomorphic. Then we consider holomorphic almost periodic mappings. Such mappings are not mapping of d-regular growth, but our main result is obtained by means of weak convergence, as well as the basic results of the theory of f.c.r.g. In **Chapter 7** we also prove analogues of Jessen's classical theorem on the existence of a density for the zero set in a strip for holomorphic almost-periodic functions of one variable. For this the notions of Jessen function and current are introduced and studied.

We assume that the reader is acquainted with standard results on holomorphic and subharmonic functions and distributions. We recall a great deal of this information here.

This book was originally written in Russian. A.Ronkin and I.Yedvabnik translated it into English. A.Ronkin, P.Agranovič and A.Rashkovskiĭ did a lot to eliminate mathematical inaccuracy and slips of the pen that were in the text. I express my deep gratitude to all of them.

CHAPTER 1

ENTIRE FUNCTIONS OF COMPLETELY REGULAR GROWTH OF ONE VARIABLE

§1 Preliminaries

Here we introduce some notation and recall some facts on entire functions of one variable (for example, see B.Levin [1], A.Gol'dberg-I.Ostrovskiǐ [1], L.Ronkin [1]).

Let $U(a,R)$ be the disk of radius R with the centre a: $U(a,R) = \{z \in \mathbb{C}: |z-a| < R\}$. For brevity $U(R) = U(0,R)$, $U = U(1)$. As usual, we let $C(E)$ be the space of continuous functions on a set E, $C^{\infty}(E)$ the space of infinitely differentiable functions, and $C^{k}(E)$ the space of k-smooth functions on E[2]. Let G be a domain in $\mathbb{C}$ (or in $\mathbb{C}^n$, or in $\mathbb{R}^n$, $n \geq 1$). Then $D(G)$ and $\mathcal{D}'(G)$ are, respectively the spaces of test functions and distributions on G (for example, see V.Vladimirov [1], L.Hörmander [1]). Let us denote by $\xrightarrow{\mathcal{D}'}$ convergence in $\mathcal{D}'$, and by $\mathcal{D}'$-lim the corresponding limit. We denote by $C_0^k(G)$ the set of k-smooth functions with support in the domain G. Therefore $\mathcal{D}(G) = C_0^{\infty}(G)$.

Let $\nu(t)$ be a monotone non-decreasing function on $\mathbb{R}^+ = \{t \in \mathbb{R}: t \geq 0\}$. Then the number

$$\rho(\nu) = \overline{\lim_{t \to \infty}} \frac{\ln^{+}\nu^{+}(t)}{\ln t},$$

where $\nu^{+} = max\{0,\nu\}$, is referred to as the order of growth (or simply the order) of $\nu(t)$. The type of the function $\nu(t)$ with respect to the order ρ, $0 < \rho < \infty$, is denoted by $\sigma(\nu,\rho)$ and is defined by the equality

[2] I.e. k times continuously differentiable function.

$$\sigma(\nu,\rho) = \overline{\lim_{t\to\infty}} \frac{\nu(t)}{t^{\rho}} .$$

Note that the following equality is valid

$$\rho(\nu) = \gamma(\nu) \overset{\text{def}}{=} \inf\left\{\alpha: \int^{\infty}\nu(t)t^{-\alpha-1}dt < \infty\right\}.$$

Besides, if $\int^{\infty}\nu(t)t^{-\rho-1}dt < \infty$ then $\sigma(\nu,\rho) = 0$. The function $\nu(t)$ is sade to be of minimal, normal, or maximal type if $\sigma(\nu,\rho) = 0$, $0 < \sigma(\nu,\rho) < \infty$, or $\sigma(\nu,\rho) = \infty$, respectively.

Let $A = \{a_j\}_{j=1}^{\infty}$ be a sequence of points from $\mathbb{C}$ such that $|a_j| \le |a_{j+1}|$, $\forall j$, and $|a_j| \to \infty$ if $j \to \infty$. Put $|A| = \bigcup_{j=1}^{\infty} a_j$. Let $\tilde{n}_A(a)$ be the number of elements in the set $\{j: a_j = a\}$. Associate to the sequence A with the measure $\bar{n}_A$ with the support $|A|$. Let $n_A(t) = \int_{\bar{U}(t)} d\bar{n}_A$. In other words, $n_A(t)$ is a number of points of A in the close disc $|z| \le t$. Note that $n_A(t)$ is usually called the counting function of sequence A. The convergence exponent of A is the greatest lower bound γ_A of all numbers α for which $\sum |a_j|^{-\alpha} < \infty$. Its value is equal to $\gamma(n_A)$, and therefore to $\rho(n_A)$.

We take H(G) to be the space of functions which are holomorphic on $G \subset \mathbb{C}^n$. Let $f(z) \in H(\mathbb{C})$. In other words, f(z) is an entire function of one complex variable. Set

$$M_f(R) = \max_{|z|=R} |f(z)|.$$

Note that $M_f(R)$ is a monotone increasing function (if $f(z) \not\equiv$ const) and $\ln M_f(e^t)$ is a convex function. The order of growth of the function $\ln M_f(R)$ is called the order of growth (or simply the order) ρ_f of the entire function f(z). Thus,

$$\rho_f = \overline{\lim_{R\to\infty}} \frac{\ln^+\ln^+ M_f(R)}{\ln R}.$$

Similarly, the type of f(z) with respect to the order $\rho > 0$ is defined as

$$\sigma_f = \sigma(lnM_f,\rho) = \lim_{R\to\infty} \frac{lnM_f^+(R)}{R^\rho}.$$

Below we will generally study entire functions of finite type only, i.e. functions $f(z)$ such that $\sigma(M_f,\rho) < \infty$. We denote by $H(\mathbb{C},\rho]$ the class of such functions.

Let us enumerate the zeros $\{a_j\}$ of an entire function $f(z)$ in such a way that $|a_j| \le |a_{j+1}|$, $\forall j$, while the number of occurrences of a zero a_j is equal to its multiplicity. We denote this sequence by Z_f (Z_f is called a divisor) and let $\bar{n}_f$, $n_f(t)$, γ_f be the corresponding $\bar{n}_A$, $n_A(t)$, γ_A. So $n(t)$ is the number of zeros of $f(z)$ in the close disc $|z| \le t$, taken with multiplicities. Note that the measure $\bar{n}_f$ is equal to the Riesz associated measure of the subharmonic function $ln|f(z)|$.

We recall that a function $u(z)$ is called subharmonic in $G \subset \mathbb{C}$ if it is upper semicontinuous and if for any point $a \in G$ and every $r > 0$ which is small enough the following inequality is true :

$$u(a) \le \frac{1}{2\pi}\int_0^{2\pi} u(a+re^{i\theta})d\theta.$$

The Laplacian of such a function, as an element of $\mathcal{D}'$, is a positive distribution, and therefore it is identical to a positive measure. This measure, divided by 2π, is called the Riesz associated measure of the function $u(z)$. Subharmonic functions will be described more completely, but in general without proofs, in **Chapter 2**, which is dedicated to subharmonic f.c.r.g. A more detailed description of the theory of subharmonic functions can be found in, for example, L.Ronkin [1].

If $f(0) \ne 0$, then the following equality (Jensen's formula) takes place

$$\frac{1}{2\pi}\int_0^{2\pi} ln|f(re^{i\theta})|d\theta = \int_0^r \frac{n_f(t)}{t}dt + ln|f(0)|. \qquad (1.1.1)$$

In particular, it follows from this equality that the maximum modulus of a function modulus is an upper bound for the number of its zeros. More precisely,

$$n_f(r) \le \frac{1}{\ln k}\{\ln M_f(kr) - \ln|f(0)|\},\ \forall k > 1. \qquad (1.1.2)$$

Therefore $\gamma_f \le \rho_f$, $\forall f\in H(\mathbb{C},\rho]$.

Let $A = \{a_j\}_{j=1}^{\infty}$ be a sequence of the above-mentioned type (in particular, $A = Z_f$). Assuming $\gamma_A < \infty$, define

$$q = q_A = \min\left\{k\in\mathbb{Z}: \int^{\infty} t^{-k-2} n_A(t)dt < \infty\right\}$$

and

$$G_q(u) = (1-u)\exp\left\{u + \frac{u^2}{2} + \ldots + \frac{u^q}{q}\right\}.$$

G_q is called a Weierstrass primary factor. The following estimates hold for it:

$$\ln|G_q(u)| \le C\frac{|u|^{q+1}}{1+|u|},\ \forall u\in\mathbb{C}, \qquad (1.1.3)$$

and

$$|\ln G_q(u)| \le C_\delta|u|^{q+1},\ |u| \le \delta < 1. \qquad (1.1.3')$$

Using these inequalities it is easy to prove that if $a_j \ne 0$, $\forall j$, then the product

$$\Pi(z) = \Pi_A(z) = \prod_{j=1}^{\omega} G_q\left(\frac{z}{a_j}\right),\ \omega \le \infty,$$

called a Weierstrass canonical product, converges in $\mathbb{C}$. This product is an entire function with $Z_\Pi = A$. It is true that

$$\ln M_\Pi(R) \le C(q)K_q(R, n_A(t)), \qquad (1.1.4)$$

where $C(q) < \infty$ and

$$K_q(R, n_A(t)) = R^q\left\{\int_0^R \frac{n_A(t)}{t^{q+1}}dt + R\int_R^\infty \frac{n_A(t)}{t^{q+2}}dt\right\}$$

From this estimate, in view of the inequality $\gamma_A \le \rho_\Pi$ it follows that $\gamma_A = \rho_\Pi$; if $\gamma_A \notin \mathbb{Z}$, then $n_A(R)$ and $\ln M_\Pi(R)$ are functions of the same type (i.e. $\sigma(n_A, \rho_\Pi)$, $\sigma(\ln M_\Pi, \rho_\Pi)$ are either simultaneously zero, finite and non-zero, or equal to ∞). If $\gamma_A = p\in\mathbb{Z}$, then it is more complicated to deal with the type of $\Pi(z)$. In this case the sufficient statement concerning the type of $\Pi(z)$ holds good after the replacement of $\sigma(n_A,\rho_\Pi)$ by $max\{\sigma(n_A,\rho_\Pi),\ \delta_A\}$, where

$$\delta_A = \overline{\lim_{R\to\infty}} \left| \sum_{|a_j|=R} a_j^{-p} \right|.$$

Every function of finite order can be factored by means of a canonical product as follows (Hadamard's theorem)

$$f(z) = z^k\, \Pi(z)\, e^{g(z)},$$

where $k \ge 0$ is the order of vanishing of zero of $f(z)$ at the point $z = 0$, $\Pi(z)$ is a canonical product constructed for the sequence $A = Z_f\setminus\{0\}$, and $g(z)$ is a polynomial of degree at most q.

More refined (than σ_f) growth characteristics are considered for functions $f(z)$ that have at most normal type with respect to the order ρ; namely, the indicators

$$h_f(\theta) = \overline{\lim_{r\to\infty}} \frac{ln|f(re^{i\theta})|}{r^\rho}$$

and

$$\mathfrak{L}_f(z) = \overline{\lim_{t\to\infty}} \frac{ln|f(tz)|}{t^\rho},\ z \in \mathbb{C}.$$

It is obvious that

$$\mathfrak{L}_f(re^{i\theta}) = r^\rho h_f(\theta). \tag{1.1.5}$$

Thus $\mathfrak{L}_f(z)$ satisfies the condition of being positively homogeneous of order ρ, i.e. $\mathfrak{L}_f(tz) = t^\rho\mathfrak{L}_f(z)$, $\forall t > 0$. Moreover, $\mathfrak{L}_f(z)$ is a subharmonic function. This is equivalent to ρ-trigonometrical convexity of the function $h_f(\theta)$, i.e. in this case the following inequality takes place

$$h_f(\theta)\sin\rho(\theta_1-\theta_2) + h_f(\theta_2)\sin\rho(\theta-\theta_1) + h_f(\theta_1)\sin\rho(\theta_2-\theta) \ge 0$$

for all θ_1, θ_2, θ such that $\theta_1 < \theta < \theta_2$, $\theta_2 - \theta_1 \le \frac{\pi}{\rho}$. If $\mathfrak{L}(z)$ is positively homogeneous and subharmonic, then $\mathfrak{L}(z)$ is the indicator $\mathfrak{L}_f(z)$ of some entire function $f\in H(\mathbb{C},\rho]$ (this is sufficient condition). In the case $\rho = 1$ this condition means that $\mathfrak{L}_f(z)$ is an ordinary convex function. Note also that ρ-trigonometrical convexity of this function implies the continuity of $h_f(\theta)$ and the inequality $h_f(\theta) + h_f(\theta + \frac{\pi}{\rho}) \geq 0$, $\forall\theta\in\mathbb{R}$.

By means of its indicator the function f(z) can be estimated not only on any individual ray $z = te^{i\theta}$, $t > 0$, but also on the whole plane $\mathbb{C}$, namely: $\forall\varepsilon > 0\ \exists C_\varepsilon < \infty$ such that

$$ln|f(z)| \leq C_\varepsilon + \varepsilon|z|^\rho + \mathcal{L}_f(z). \tag{1.1.6}$$

All these facts and definitions concerning to the function $ln|f(z)|$ can be extended almost unchanged to any subharmonic function u in $\mathbb{C}$. In this case $\mu_u(\overline{U}(t))$ is the analogue of $n_f(t)$, where μ_u is the Riesz associated measure of the subharmonic function u(z) under consideration. Below we will also write μ_u instead of $\mu_u(\overline{U}(t))$.

§2 Regularity of growth, $\mathcal{D}'$-convergence and right distribution of zeros

1. **Definition, statement of the main theorem.** Let $E \subset \mathbb{R}$ be a measurable set, $[E]_r = \{x\in E: |x| < r\}$, and let $mes[E]_r$ be its Lebesgue measure. E is called a set of relative measure zero if $\overline{\lim_{r\to\infty}}\, r^{-1} mes[E]_r = 0$.

An entire function f(z) is called a function of completely regular growth (f.c.r.g.) if it has at most normal type with respect to the order $\rho > 0$ and $r^{-\rho} ln|f(re^{i\theta})| \rightrightarrows h_f(\theta)$, as $r \to \infty$ through a set of numbers outside some set E of relative measure zero.

The subject matter of the Levin-Pfluger theory is as follows: regularity of growth of a function is equivalent to the existence of a certain regularity in the distribution of its zero set. Another criterion for c.r.g. is formulated in terms of the weak convergence of the functions $u_t = t^{-\rho} ln|f(te^{i\theta})|$. Note that the characterization of the exceptional set in the definition of f.c.r.g. can be improved upon. This requires some new definitions and notation.

Let $A = \{a_j\}_{j=1}^{\infty}$ be a sequence of points in $\mathbb{C}$ such that $|a_j| \leq |a_{j+1}|$, $\forall j$, $|a_j| \to \infty$ as $j \to \infty$, while the corresponding counting function $n_A(t)$ has at most normal type with respect to the order ρ, $0 < \rho < \infty$. Let $Y(R,\theta_1,\theta_2) = \{z\in\mathbb{C}: |z| < R,\ \theta_1 < \arg z < \theta_2\}$, and let $n_A(R,\theta_1,\theta_2)$ be the number of points a_j (counted with their multiplicities) inside $Y(R,\theta_1,\theta_2)$. We say that A has an angular density (with respect to the order ρ) if for all but a denumerable set $\mathfrak{U}$ of values θ_1, θ_2 the limit $\lim\limits_{R\to\infty} R^{-\rho} n_A(R,\theta_1,\theta_2)$ exists. If $\rho\notin\mathbb{Z}$, then a sequence A which has an angular density is called regularly distributed. For $\rho = p\in\mathbb{Z}$, A is called regularly distributed if A has an angular density and, moreover, the limit

$$\lim_{R\to\infty} \sum_{|a_j|<R} a_j^{-p} \overset{\text{def}}{=} \kappa_A \tag{1.2.1}$$

exists.

Note that this condition is automatically fulfilled if $p = q + 1$, where $q = q_A$ was introduced in §1. In the case $p = q$ the existence of this limit is equivalent to certain symmetry in the distribution of the points a_j.

A set $E \subset \mathbb{C}$ is called a C_0-set if it can be covered by a system of discs $U_j = U(a_j,r_j)$ such that

$$\lim_{R\to\infty} \frac{1}{R} \sum_{\{j:\,|a_j|<R\}} r_j = 0.$$

It is easy to see that the following definition is equivalent to the previous one.

A set $E \subset \mathbb{C}$ is called a C_0-set if it is possible to cover the intersection $E\cap U(R)$ by a system of discs $U_j = U(a_j,r_j)$, $j = 1, 2, \ldots$, $a_j = a_j(R)$, $r_j = r_j(R)$, such that

$$\lim_{R\to\infty} \frac{1}{R} \sum_j r_j = 0.$$

Now we have everything required to formulate of the following theorem, which, in particular, contains a statement on the relationship between c.r.g. and the distribution of zeros. This relationship, as noted above, is fundamental in the theory of f.c.r.g. in $\mathbb{C}$.

Theorem 1.2.1. Let $f\in H(\mathbb{C},\rho]$. Then the following assertions are equivalent:

(a) $f(z)$ is a function of c.r.g. ;

(b) there exists a C_0-set E such that

$$ln|f(z)| = \mathfrak{L}_f(z) + o(|z|^\rho)$$

as $z \to \infty$, $z \notin E$;

(c) the sequence Z_f of zeros of $f(z)$ is regularly distributed

(d) the limit $\mathcal{D}' - \lim_{t \to \infty} \dfrac{ln|f(tz)|}{t^\rho}$ exist.

Note that (b) ⇒ (a) is obvious.

2. Proof of (a) ⇒ (d). According to the definition of f.c.r.g., the logarithm of its modulus is asymptotically close to the indicator outside an exceptional set E. The function belongs to a class $H(\mathbb{C},\rho]$ and E is a "lean" set. That is why it can be expected that integrals of the logarithm of the modulus of the function over a bounded part of E are sufficient small and, thus, integrals over bounded regions of $\mathbb{C}$ are asymptotically close to the corresponding integrals of the indicator. Thus one can easily obtain that (d) is valid, and that the limit is equal to the indicator. Let us outline the main reasons for this.

Let f be a f.c.r.g. and E a set of relative measure zero corresponding to $f(z)$ by the definition of c.r.g. Put $E^* = \{z: |z| \in E\}$, $[E^*]_R = E^* \cap U(R)$.

The boundedness of the indicator $h_f(z)$ and (1.1.5) imply that

$$\lim_{R \to \infty} R^{-\rho-2} \int_{[E^*]_R} |\mathfrak{L}_f(z)|d\omega_z = 0, \quad \left(d\omega_z = \frac{i}{2} dz \wedge d\bar{z}\right). \tag{1.2.2}$$

Let us show that

$$\lim_{R \to \infty} R^{-\rho-2} \int_{[E^*]_R} |ln|f(z)||d\omega_z = 0 \tag{1.2.3}$$

as well. Without loss of generality we may assume $f(0) \neq 0$, and moreover $f(0) = 1$. Then, using Jensen's formula (1.1.1), we conclude that

$$\frac{1}{2\pi}\int_0^{2\pi}[-ln|f(re^{i\theta})|]^+d\theta \le \frac{1}{2\pi}\int_0^{2\pi}[ln|f(re^{i\theta})|]^+d\theta,$$

and hence

$$\frac{1}{2\pi}\int_0^{2\pi}|ln|f(re^{i\theta})||d\theta \le \frac{2}{2\pi}\int_0^{2\pi}[ln|f(re^{i\theta})|]^+d\theta \le 2ln^+M_f(r).$$

Thus

$$\int_{[E^*]_R}|ln|f(z)||d\omega_z \le 2\int_{[E]_R}ln^+M_f(r)rdr \le 2Rln^+M_f(R)\cdot mes[E]_R.$$

It follows that

$$\lim_{R\to\infty} R^{-\rho-2}\int_{[E^*]_R}|ln|f(z)||d\omega_z \le 2\lim_{R\to\infty}\frac{ln^+M_f(R)}{R^\rho}\cdot\lim_{R\to\infty}\frac{mes[E]_R}{R} = 0.$$

Hence equality (1.2.3) is valid.

Let $\varphi\in D(\mathbb{C})$. Without loss of generality we may assume that $supp\varphi \subset U$ and $\|\varphi\|_\infty \overset{def}{=} max\{|\varphi(z)|: z\in U\} < 1$. Then

$$\left|\int_{\mathbb{C}}(R^{-\rho}ln|f(Rz)|-\mathcal{L}_f(z))\varphi(z)d\omega_z\right| =$$

$$= \left|\int_U(R^{-\rho}ln|f(Rz)|-\mathcal{L}_f(z))\varphi(z)d\omega_z\right| \le$$

$$\le \frac{1}{R^{\rho+2}}\int_{U(R)}|ln|f(\zeta)|-\mathcal{L}_f(\zeta)|d\omega_\zeta.$$

Now, by (1.2.2), (1.2.3) and the fact that

$$|ln|f(\zeta)| - \mathcal{L}_f(\zeta)| = o(|\zeta|^\rho)$$

as $\zeta\to\infty$, $\zeta\notin E^*$, we obtain

$$\left|\int_{\mathbb{C}}(R^{-\rho}ln|f(Rz)| - \mathcal{L}_f(z))\varphi(z)d\omega_z\right| \le$$

$$\le \frac{1}{R^{\rho+2}} \int_{U(R)\setminus[E^*]_R} |\ln|f(\zeta)| - \mathfrak{L}_f(\zeta)|d\omega_\zeta +$$

$$+ \frac{1}{R^{\rho+2}} \int_{[E^*]_R} |\ln|f(\zeta)||d\omega_\zeta + \frac{1}{R^{\rho+2}} \int_{[E^*]_R} |\mathfrak{L}_f(\zeta)|d\omega_\zeta = o(1).$$

Hence the following equality holds

$$\mathcal{D}'\text{-}\lim_{R\to\infty} \frac{\ln|f(Rz)|}{R^\rho} = \mathfrak{L}_f(z).$$

So (a) ⇒ (d) is proved.

Remark. In the proof only the fact that $\mathfrak{L}_f(z)$ is positively homogeneous and $h_f(\theta)$ is bounded are used, but not the fact that these functions are indicators of f. That is why the implication (a) ⇒ (d) can be formally stated in a more general form, namely:

If $sup\{|h(\theta|: 0 \le \theta \le 2\pi\} < \infty$ and $t^{-\rho}\ln|f(te^{i\theta})| \overset{\to}{\to} h(\theta)$, as $t \to \infty$ outside some set of relative measure zero, then

$$\mathcal{D}'\text{-}\lim_{t\to\infty} t^{-\rho}\ln|f(tz)| = |z|^\rho h(\arg z).$$

But one can easily see that $\mathcal{D}'\text{-}\lim_{t\to\infty} t^{-\rho}\ln|f(tz)|$ has to coincide with the indicator $\mathfrak{L}_f(z)$ if the limit exists. In fact, let $\mathfrak{L}(z)\in\mathcal{D}'(\mathbb{C})$ and $\mathcal{D}'\text{-}\lim_{t\to\infty} t^{-\rho}\ln|f(tz)| = \mathfrak{L}(z)$. Since $\Delta\mathfrak{L} = \mathcal{D}'\text{-}\lim_{t\to\infty} t^{-\rho}\Delta\ln|f(tz)|$, we have $\Delta\mathfrak{L} \ge 0$. This implies (see, for example, L.Ronkin [1]) that $\mathfrak{L}(z)$ is (or may be considered as) an ordinary subharmonic function. It is also obvious that $\mathfrak{L}(tz) = t^\rho\mathfrak{L}(z)$. Let us show that $\mathfrak{L}(z) = \mathfrak{L}_f(z)$. By estimate (1.1.6) it follows that $\mathfrak{L}(z) \le \mathfrak{L}_f(z)$. Let $\alpha_\varepsilon\in C^\infty(\mathbb{R})$ be such that $\alpha_\varepsilon \ge 0$, $supp\alpha_\varepsilon \subset (0,\varepsilon)$ and $\int_0^\varepsilon \alpha_\varepsilon(t)dt = 2\pi$. Now, if we set $\varphi_\varepsilon(z) = \alpha_\varepsilon(|z|)$, then the definition of $\mathcal{D}'$-lim and the properties of subharmonic function $\ln|f(z)|$ imply that

$$\int_{\mathbb{C}} \mathfrak{L}(z)\varphi_\varepsilon(z-z_0)d\omega_z = \lim_{t\to\infty} \frac{1}{t^\rho} \int_{\mathbb{C}} \ln|f(z)|\varphi_\varepsilon(z-z_0)d\omega_z \ge$$

$$\geq \overline{\lim_{t \to \infty}} \frac{1}{t^{\rho}} \ln|f(tz_0)| = \mathfrak{L}_f(z_0).$$

Letting $\varepsilon \to \infty$, we obtain $\mathfrak{L}(z) \geq \mathfrak{L}_f(z)$. Therefore $\mathfrak{L}(z) = \mathfrak{L}_f(z)$.

3. **Proof of (d) ⇒ (c).** First of all we show that the existence of an angular density follows from the existence of $\mathcal{D}'\text{-}\lim_{t \to \infty} t^{-\rho}\ln|f(tz)|$. Note that, as mentioned above, the measure $\tilde{n}_f$ constructed from the sequence (divisor) Z_f of zeros of an entire function $f(z)$ is equal to the Riesz mass of $\ln|f(z)|$. Hence it may be identified with the distribution $\frac{1}{2\pi}\Delta\ln|f(z)|$, where Δ is the Laplacian. That is why the proof of the existence of the angular density can be reduced to proving the convergence of the Laplacians $\Delta(t^{-\rho}\ln|f(tz)|)$ convergence in $\mathcal{D}'$. Then it is sufficient to use the well-known fact that $\mathcal{D}'$-convergence of functions implies $\mathcal{D}'$-convergence of their derivatives.

Now let the limit exist:

$$\mathfrak{L}(Z) = \mathcal{D}'\text{-} \lim_{t \to \infty} t^{-\rho}\ln|f(tz)|. \tag{1.2.4}$$

Then, as mentioned at the end of §1.2, $\mathfrak{L}(z) = \mathfrak{L}_f(z)$. We denote the measures that are Riesz associated to the functions $t^{-\rho}\ln|f(tz)|$ and $\mathfrak{L}_f(z)$ by $\tilde{n}^t$ and μ_f, respectively.

It is obvious that $\tilde{n}^t(Y(1,\theta_1,\theta_2)) = t^{-\rho}\tilde{n}^t(Y(t,\theta_1,\theta_2)$, and $\mu_f(Y(t,\theta_1,\theta_2)) = t^{\rho}\nu((\theta_1,\theta_2))$, where ν is the measure on $S = \partial U$ defined, as an element of $\mathcal{D}'(S)$, by the equality $\nu = h_f''(\theta) + \rho^2 h_f(\theta)$.

From (1.2.4) it follows that

$$\mathcal{D}'\text{-}\lim_{\to \infty} \tilde{n}^t = \frac{1}{2\pi}\mathcal{D}'\text{-}\lim_{\to \infty}\Delta(t^{-\rho}\ln|f(tz)|) = \frac{1}{2\pi}\Delta\mathfrak{L}_f(z) = \mu_f \quad .$$

Now, using a well-known property[3)] of $\mathcal{D}'$-convergent families of positive measures, we conclude that

---

[3)] This property is as follows.

If positive measures μ_t, $t > 0$, are such that $\mathcal{D}'\text{-}\lim_{\to \infty} \mu_t = \mu$ exists, then $\lim_{t \to \infty} \mu_t(G) = \mu(G)$ for any domain G satisfying the condition $\mu(\partial G) = 0$ (see, for example, L.Ronkin [1]).

$$\lim_{t\to\infty} \int_G \varphi(z)d\tilde{n}^t = \int_G \varphi(z)d\mu_f$$

for any $\varphi\in C(\bar{G})$ and any domain $G \subset\subset \mathbb{C}$ satisfying the condition $\mu_f(\partial G)=0$. In particular, for such domains

$$\lim \tilde{n}^t(G) = \mu_f(G).$$

Considering a sector $Y(t,\theta_1,\theta_2)$ as the domain G and taking into account the above-mentioned properties of $\tilde{n}^t$ and μ_t, we find that $\lim_{t\to\infty} t^{-\rho}\tilde{n}^t(Y(t,\theta_1,\theta_2)) = \nu((\theta_1,\theta_2))$, $\forall\theta_1,\theta_2$, $\nu(\{\theta_1\}) = \nu(\{\theta_2\}) = 0$. It is obvious that the set $\{\ \theta:\ \nu(\{\theta\})\neq 0\ \}$ is at most countable. This proves that if (d) is valid, then the sequence Z_f has an angular density.

To prove that Z_f is regularly distributed it is also necessary to show that if $\rho = p\in\mathbb{N}$, then

$$\lim_{t\to\infty} \sum_{|a_j|<t} a_j^{-p} \overset{\text{def}}{=} \chi_f \tag{1.2.5}$$

exists. For this we use Nevanlinna's formula (see, for example, A.Gol'dberg, I.Ostrovskiĭ [1]).

$$\frac{1}{p!}\left.\frac{d^p \ln f}{dz^p}\right|_{z=0} = \frac{1}{\pi t^p}\int_0^{2\pi} \ln|f(te^{i\theta})|e^{-ip\theta}d\theta + \frac{1}{p}\sum_{|a_j|<t}\left(\frac{\bar{a}_j^p}{t^{2p}} - \frac{1}{a_j^p}\right).$$

Using the notations given above, this formula can be transformed to

$$\frac{1}{p}\sum_{|a_j|<t} a_j^{-p} =$$

$$= \frac{1}{\pi t^p}\int_0^{2\pi} \ln|f(te^{i\theta})|e^{-ip\theta}d\theta + \frac{1}{p}\int_U \bar{\zeta}^p d\tilde{n}^t(\zeta) + \frac{1}{p!}\left.\frac{d^p \ln f}{dz^p}\right|_{z=0}.$$

Since $\tilde{n}^t \to \mu_f$ and $\mu_f(\partial U) = 0$, we find

$$\exists \lim_{t\to\infty} \int_U \bar{\zeta}^p\, d\tilde{n}^t(\zeta) = \int_U \bar{\zeta}^p\, d\mu_f(\zeta).$$

Therefore, to prove the existence of the limit (1.2.5) it is sufficient to assure of the existence of

$$\lim_{t \to \infty} \frac{1}{t^p} \int_0^{2\pi} ln|f(te^{i\theta})|e^{-ip\theta}d\theta,$$

or, equivalently, the existence of

$$\lim_{t \to \infty} \frac{1}{t^p} \int_0^{2\pi} ln|f(tre^{i\theta})|e^{-ip\theta}d\theta, \ r > 0. \qquad (1.2.6)$$

For this we represent the function $f(tz)$ in the circle U as

$$f(tz) = \left(\prod_{|a_j|<t} \frac{tz-a_j}{t-\bar{a}_j z} \right) F_t(z). \qquad (1.2.7)$$

It is clear that $F_t(z)$ does not vanish in U, $|F_t(z)| = |f(tz)|$ on ∂U, and the family of functions $t^{-p}ln|F_t(z)|$, harmonic in U, is uniformly bounded above. Besides, by (1.2.7) it follows that

$$\frac{1}{t^p} \int_U ln|F_t(z)|\psi(z)d\omega_z =$$

$$= \frac{1}{t^p} \int_U ln|f(tz)|\psi(z)d\omega_z + \int_U \left(\int_U ln\left|\frac{1-z\zeta}{z-\zeta}\right|\psi(z)d\omega_z \right) d\tilde{n}^t(\zeta).$$

Since the function $\int_U ln\left|\frac{1-z\zeta}{z-\zeta}\right|\psi(z)d\omega_z$ is continuous for all $\psi \in D(\mathbb{C})$ and $\mu_f(\partial U) = 0$, the following limit exists:

$$\lim_{t \to \infty} \int_U \left(\int_U ln\left|\frac{1-z\zeta}{z-\zeta}\right|\psi(z)d\omega_z \right) d\tilde{n}^t(\zeta), \ \forall \psi \in D(\mathbb{C}).$$

It is also clear that

$$\exists \lim_{t \to \infty} \frac{1}{t^p} \int_U ln|f(tz)|\psi(z)d\omega_z, \ \forall \psi \in D(\mathbb{C}).$$

Hence

$$\exists \lim_{t \to \infty} \frac{1}{t^p} \int_U ln|F_t(z)|\psi(z)d\omega_z, \ \forall \psi \in D(\mathbb{C}),$$

or, equivalently, $\exists\mathcal{D}'\text{-}\lim\, t^{-p} ln|F_t(z)|$. This and the above-mentioned properties of $F_t(z)$ implies (see **Lemma 2.1.2**) that as $t \to \infty$ the functions $t^{-p} ln|F_t(z)|$ converge uniformly on each compact set $K \subset U$ to some function $v(z)$, harmonic in U. In particular, for fixed r, $0 < r < 1$, as $t \to \infty$,

$$\frac{1}{t^p} ln|F_t(re^{i\theta})| \to v(re^{i\theta})$$

uniformly in θ, and hence (see (1.2.7)) the existence of the limit (1.2.6) reduced to the existence of the limit

$$\lim_{t \to \infty} \int_U \left(\int_0^{2\pi} e^{-ip\theta} ln\left|\frac{re^{i\theta}-\zeta}{1-\bar{\zeta}re^{i\theta}}\right| d\theta \right) d\tilde{n}^t(\zeta).$$

It is easy to see[4] that the function $\int_0^{2\pi} e^{-ip\theta} ln\left|\frac{re^{i\theta}-\zeta}{1-\bar{\zeta}re^{i\theta}}\right| d\theta$ is continuous in U. Therefore, using the properties of the measures $\tilde{n}^t$, we obtain that the latter limit exists.

Thus, the existence of limit (1.2.5) is proved and hence the sequence Z_f is regularly distributed.

4. Proof of (c) ⇒ (b) in the case of a nonintegral order. Since for the factor $z^k e^{g(z)}$ in the Hadamard factorization of an entire function (see §1), assertion (b) is valid for any ρ, (c) ⇒ (b) has to be proved for $f(z) = \Pi(z)$ only, i.e. for a Weierstrass canonical product. We sketch this part of the proof of **Theorem 1.2.1**, which is the most laborious part.

First of all we will show that the existence of the angular density of Z_f implies $\mathcal{D}'$-convergence of the measures $\tilde{n}^t$, defined in the previous section, to some positively homogeneous measure $\tilde{\mu}$. Further we shall estimate the quantity

---

[4] $\int_0^{2\pi} e^{-ip\theta} ln|re^{i\theta}-\zeta| d\theta$ is equal to $\bar{\zeta} r^{-2p}$ when $|\zeta| \le r$, and to ζ^{-p} when $|\zeta| \ge r$.

$$\left| \ln|f(z)| - \int_{\mathbb{C}} \ln\left|G_q(\frac{z}{\zeta})\right| d\tilde{\mu} \right| = \left| \int_{\mathbb{C}} \ln\left|G_q(\frac{z}{\zeta})\right| d\tilde{n}^t - \int_{\mathbb{C}} \ln\left|G_q(\frac{z}{\zeta})\right| d\tilde{\mu} \right|.$$

The domain of integration, i.e. the plane $\mathbb{C}$, will be divided into parts. The manner of estimation in each of these parts will vary. If we are in asymptotically (when $r \to \infty$) small neighbourhoods $|\zeta| < \varepsilon r$ and $|\zeta| > kr$ of zero and infinity, then each of the mentioned integrals will be asymptotically small, because the neighbourhoods are small and f belongs to $H(\mathbb{C},\rho]$; $n_f(t)$ and $\tilde{\mu}(t)$can then be estimated correspondingly.

In the "interval" $\varepsilon r < |z| < kr$ between these neighbourhoods we will distinguish two domains, namely: a neighbourhood of the point z such that $\ln\left|G_q\left(\frac{z}{\zeta}\right)\right|$ is equal to $-\infty$ at z, and the complement of this neighbourhood. In the complement the difference between the integrals will be small if $\tilde{n}_f$ and $\tilde{\mu}$ are asymptotically near. In the neighbourhood the difference can be large. Using Hayman's ε-normal points method we will prove that the last is possible only when the neighbourhood contains many (in some sense) points from Z_f. Since there is an estimate for the number of such points in all of $\mathbb{C}$ (i.e. for the function $\tilde{n}_f(t)$), there can not be too many accumulation points of Z_f. It will be shown that they form a C_0-set. This will be used to obtain the equation

$$\int_{\mathbb{C}} \left| \ln G_q\left(\frac{z}{\zeta}\right)\right| d\tilde{\mu} = \mathfrak{L}_f(z),$$

which completes the proof.

Now set

$$f(z) = \prod_k G_q\left(\frac{z}{a_k}\right),$$

where q and $\prod_k G_q\left(\frac{z}{a_k}\right)$ are as before, and let the sequence Z_f be regularly distributed. We consider the function

$$\nu(\theta_1,\theta_2) = \lim_{t\to\infty} \tilde{n}_f(Y(t,\theta_1,\theta_2)). \tag{1.2.8}$$

If the zeros a_k are regularly distributed and, therefore, have an angular density, then the function $\nu(\theta_1,\theta_2)$ is correctly defined by

(1.2.8) for any $\theta_1 \notin \mathfrak{U}$, $\theta_2 \notin \mathfrak{U}$, where the set $\mathfrak{U}$ is at most countable. If $\theta_1 = \theta_1^0 \notin \mathfrak{U}$ is fixed, then $\nu(\theta_1^0, \theta_2)$ is monotone in $\theta_2 \in [\theta_1^0, \theta_1^0 + 2\pi) \setminus \mathfrak{U}$. Thus it defines the measure ν on $[\theta_1^0, \theta_1^0 + 2\pi]$ or, equivalently, on ∂U, namely,

$$\nu(\Gamma) = \int_{\Gamma} d\nu(\theta_1^0, \theta_2), \quad \forall \Gamma \subset [\theta_1^0, \theta_1^0 + 2\pi).$$

Then we define a positively homogeneous measure $\tilde{\mu}$ on $\mathbb{C}$ by setting $d\tilde{\mu}(z) = \rho |z|^{\rho-1} \otimes d\nu(\theta)$, $\theta = \arg z$. So $\tilde{\mu}(Y(r, \theta_1, \theta_2)) = r^{\rho} \tilde{\mu}(Y(1, \theta_1, \theta_2))$. The measure ν is naturally called the angular density of the zeros a_k, $k = 1, 2, \dots$.

We show that if the measures $\tilde{n}^t$ are constructed as in §1.2, i.e. $\tilde{n}^t(E) = t^{-\rho} \tilde{n}_f(tE)$, then they $\mathcal{D}'$-converge to the measure $\tilde{\mu}$. Moreover we shall prove that

$$\lim_{t \to \infty} \int_U \varphi d\tilde{n}^t = \int_{\bar{U}} \varphi d\tilde{\mu}, \quad \forall \varphi \in C(\bar{U}).$$

For this we take arbitrary $\varphi \in C(\bar{U})$ and $\varepsilon > 0$ and divide the disc U into a finite number N of "collars"

$$Y'_j = Y(r_{2,j}, \theta_{1,j}, \theta_{2,j}) \setminus Y(r_{1,j}, \theta_{1,j}, \theta_{2,j}), \quad \theta_{1,j} \notin \mathfrak{U}, \ \theta_{2,j} \notin \mathfrak{U},$$

which are so small that

$$\sup\{ |\varphi(z_1) - \varphi(z_2)| : z_1 \in Y'_j, \ z_2 \in Y'_j, \ j = 1, \dots, N\} < \varepsilon.$$

Then

$$\left| \int_{\bar{U}} \varphi d\tilde{n}^t - \int_{\bar{U}} \varphi d\tilde{\mu} \right| \le \sum_{j=1}^{N} \left| \int_{Y'_j} \varphi d\tilde{n}^t - \int_{Y'_j} \varphi d\tilde{\mu} \right| \le$$

$$\le \sum_{j=1}^{N} \varepsilon \int_{Y'_j} d\tilde{n}^t + \sum_{j=1}^{N} \varepsilon \int_{Y'_j} d\tilde{\mu} + \|\varphi\|_{\infty} \sum_{j=1}^{N} \left| \int_{Y'_j} d\tilde{n}^t - \int_{Y'_j} d\tilde{\mu} \right| =$$

$$= \varepsilon \tilde{n}^t(\bar{U}) + \varepsilon \tilde{\mu}(\bar{U}) + \|\varphi\|_{\infty} \sum_{j=1}^{N} | \tilde{n}^t(Y'_j) - \tilde{\mu}(Y'_j) |$$

From the construction of $\tilde{\mu}$ and the partition of U into Y'_j it follows that

$$\lim_{t \to \infty} \tilde{n}^t(Y'_j) = \tilde{\mu}(Y'_j), \ \forall j = 1,\dots,N.$$

Moreover, it is obvious that $\lim_{t \to \infty} \tilde{n}^t(\bar{U}) = \tilde{\mu}(\bar{U})$. Therefore

$$\lim_{t \to \infty} \left| \int_U \varphi \, d\tilde{n}^t - \int_U \varphi \, d\tilde{\mu} \right| \le 2\varepsilon\tilde{\mu}(U),$$

and since $\varepsilon > 0$ is arbitrary, we have

$$\lim_{t \to \infty} \int_U \varphi d\tilde{n}^t = \int_U \varphi d\tilde{\mu}.$$

Note also that a simple analysis of this reasoning gives that if $\varphi = \varphi(z,w) \in C(\bar{U} \times K)$, where K is a compact set in $\mathbb{C}$, then the functions

$$\psi_t(w) = \int_{\bar{U}} \varphi(z,w) d\tilde{n}^t(z)$$

converge (as $t \to \infty$) uniformly to $\psi(z) = \int_{\bar{U}} \varphi(z,w) d\tilde{\mu}(z)$ on K.

In this reasoning the order ρ was arbitrary. Let now $q < \rho < q + 1$, where $q \in \mathbb{Z}^+$, and consider the function

$$J(z) = \int ln \left| G_q\left(\frac{z}{\zeta}\right) \right| d\tilde{\mu}(\zeta). \tag{1.2.9}$$

By (1.1.3) and the properties of the measure μ and the corresponding function $\tilde{\mu}(t) = \tilde{\mu}(\bar{U}(t)) = t^\rho \tilde{\mu}(1)$, integral (1.2.9) converges.

It is obvious that J(z) is subharmonic in $\mathbb{C}$ and positively homogeneous of degree ρ. Let us prove that outside some C_0-set E the equation $ln|f(z)| - J(z) = \gamma(z)|z|^\rho$ is valid, where $\gamma(z) \to 0$ as $z \to \infty$, $z \notin E$. It suffices to show that $\forall \varepsilon > 0 \ \forall R > 0$ the set $E_{\varepsilon,R} = \{z: |z| < R, \ |ln|f(z)| - J(z)| > \varepsilon|z|^\rho\}$ can be covered by an at most countable system $\mathfrak{A}(R,\varepsilon)$ of discs $U_j = U(z_j, r_j)$, $z_j = z_j(R,\varepsilon)$, $r_j = r_j(R,\varepsilon)$, such that

$$\lim_{\varepsilon \to 0} \overline{\lim_{R \to \infty}} \frac{1}{R} \sum_j r_j(R,\varepsilon) = 0. \tag{1.2.10}$$

In fact, if such a system $\mathfrak{A}(R,\varepsilon)$ exists, then one can choose $\varepsilon_j \downarrow 0$ and

$R\uparrow\infty$ in such a manner that

$$\frac{1}{R}\sum_k r_k(R,\varepsilon_j) < \frac{1}{j}, \qquad \forall R \ge R_{j-1}.$$

Then, assuming $E = \bigcup_{j=1}^{\infty}\{E_{\varepsilon_{j+1}},R_{j+1}\setminus U(R_j)\}$ we obtain that outside E the function $\gamma(z) = |z|^{-\rho}\ln|f(z)| - J(z)$ tends to zero as $|z| \to \infty$, and only the fact that E is a C_0-set is to be verified. But it follows from the construction of E that for $R_{j-1} \le R < R_j$ the set $E \cap U(r)$ is contained in $R_{\varepsilon,R}$. Thus it can be covered by a system $\mathfrak{A}(R,\varepsilon_j)$ of discs U_j such that the sum of the radii is less than $\frac{1}{j}R$. Therefore E is a C_0-set.

To obtain the required characterization of the set $E_{\varepsilon,R}$ we represent the difference $\ln|f(z)| - J(z)$ as:

$$\ln|f(z)| - J(z) =$$

$$= f_{t,1}(z) + f_{t,2}(z) + f_{t,3}(z) - J_{t,1}(z) - J_{t,2}(z) - J_{t,3}(z),$$

where

$$f_{t,1}(z) = \int_{|\zeta|\ge t|z|} \ln\left|G_q\left(\frac{z}{\zeta}\right)\right| d\tilde{n}_f(\zeta),$$

$$f_{t,2}(z) = \int_{\frac{1}{t}|z|<|\zeta|<t|z|} \ln\left|G_q\left(\frac{z}{\zeta}\right)\right| d\tilde{n}_f(\zeta),$$

$$f_{t,3}(z) = \int_{|\zeta|<\frac{1}{t}|z|} \ln\left|G_q\left(\frac{z}{\zeta}\right)\right| d\tilde{n}_f(\zeta),$$

$$J_{t,1}(z) = \int_{|\zeta|\ge t|z|} \ln\left|G_q\left(\frac{z}{\zeta}\right)\right| d\tilde{\mu}(\zeta),$$

$$J_{t,2}(z) = \int_{\frac{1}{t}|z|<|\zeta|<t|z|} \ln\left|G_q\left(\frac{z}{\zeta}\right)\right| d\tilde{\mu}(\zeta),$$

$$J_{t,3}(z) = \int_{|\zeta|<\frac{1}{t}|z|} \ln\left|G_q\left(\frac{z}{\zeta}\right)\right| d\tilde{\mu}(\zeta),$$

It follows from the estimate (1.1.3') that

$$|f_{t,1}(z)| \le C_1|z|^{q+1} \int_{t|z|}^{\infty} \frac{dn_f(s)}{s^{q+1}} \le C_1 \frac{|z|^\rho}{t^{q+1-\rho}}.$$

Therefore, $\forall\varepsilon > 0\ \exists t_0(\varepsilon,R)$:

$$|f_{t,1}(z)| \le \varepsilon|z|^\rho,\ \forall t \ge t_0,\ |z| \ge 1.$$

The function $J_{t,1}(z)$ can be similarly.

Now we estimate $f_{t,3}(z)$ and $J_{t,3}(z)$. Since

$$\ln|t-1| - C\left|\frac{z}{\zeta}\right|^q \le \ln\left|G_q\left(\frac{z}{\zeta}\right)\right| \le C\left|\frac{z}{\zeta}\right|^q$$

for some constant C as $|\zeta| < \frac{1}{t}|z|$, we obtain

$$|f_{t,3}(z)| < C|z|^q \int_{|\zeta|<\frac{1}{t}|z|} |\zeta|^{-q}d\tilde{n}_f(\zeta) + \ln|t-1| \int_{|\zeta|<\frac{1}{t}|z|} d\tilde{n}_f(\zeta) =$$

$$= C|z|^q \int_0^{\frac{1}{t}|z|} t^{-q}dn_f(t) + \ln|t-1|dn_f\left(\frac{1}{t}|z|\right) =$$

$$= C|z|^q \left\{ \frac{n_f\left(\frac{1}{t}|z|\right)}{\left(\frac{1}{t}|z|\right)^q} + q\int_0^{\frac{1}{t}|z|} \frac{dn_f(s)}{s^{q+1}} \right\} + \ln|t-1|dn_f\left(\frac{1}{t}|z|\right) \le$$

$$\le C_1|z|^q \left\{\left(\frac{1}{t}|z|\right)^{\rho-q} + q\left(\frac{1}{t}|z|\right)^{\rho-q}\right\} + C\left(\frac{1}{t}|z|\right)^\rho \ln|t-1|.$$

Therefore the following is valid for sufficient large t:

$$|f_{t,3}(z)| \le \varepsilon|z|^\rho.$$

A similar estimate for $J_{t,3}(z)$ is obvious.

Then we estimate the difference $\Phi(z) = J_{t,2}(z) - f_{t,2}(z)$. For this we consider the kernel

$$\mathfrak{K}_q^\delta\left(\frac{z}{\zeta}\right) = \max\left\{ln\left|1-\frac{\zeta}{z}\right|,\ ln\delta\right\} + ln\frac{z}{\zeta} + Re\sum_{k=1}^{q}\frac{z^k}{k\zeta^k}$$

Now $\Phi(z)$ can be represented as

$$\Phi(z) = \int_{\frac{1}{t}|z|<|\zeta|<t|z|} \mathfrak{K}_q^\delta\left(\frac{z}{\zeta}\right)d\tilde{n}_f(\zeta) + \int_{|\zeta-z|<\delta|z|}\{ln|z-\zeta|-ln|\zeta|\}d\tilde{n}_f(\zeta) +$$

$$- \int_{\frac{1}{t}|z|<|\zeta|<t|z|} \mathfrak{K}_q^\delta\left(\frac{z}{\zeta}\right)d\tilde{\mu}(\zeta) - \int_{|\zeta-z|<\delta|z|}\{ln|z-\zeta|-ln|\zeta|\}d\tilde{\mu}(\zeta) =$$

$$= |z|^\rho\left\{\int_{\frac{1}{t}<|w|<t} \mathfrak{K}_q^\delta\left(\frac{e^{iargz}}{w}\right)d\tilde{n}_f^{|z|}(w) - \int_{\frac{1}{t}<|w|<t} \mathfrak{K}_q^\delta\left(\frac{e^{iargz}}{w}\right)d\tilde{\mu}(w)\right\} +$$

$$+ \int_{|\zeta-z|<\delta|z|}\{ln|z-\zeta| - ln|\zeta|\}d\tilde{n}_f - \int_{|\zeta-z|<\delta|z|}\{ln|z-\zeta| - ln|\zeta|\}d\tilde{\mu} =$$

$$= \Phi_1 + \Phi_2 + \Phi_3.$$

Since $\mathfrak{K}_q^\delta(e^{-i\theta}w)$ is jointly continuous in the variables $\theta = argz$ and w, for $\frac{1}{t} < |w| < t$ and $\tilde{n}^{|z|} \xrightarrow{\mathcal{D}} \tilde{\mu}$, we have $\Phi_1(z) \to 0$ as $|z| \to \infty$. Thus, for any $\varepsilon > 0$ and $t > 1$ there is R" such that

$$|\Phi_1(z)| \le \varepsilon|z|^\rho,\ |z| > R".$$

Now we estimate $|\Phi_3|$. Taking into account that $d\tilde{\mu}(re^{i\varphi}) = \rho r^{\rho-1}dr\otimes d\nu$, we obtain

$$|\Phi_3(z)| = |z|^\rho\left|\int_{|1-w|<\delta}(ln|1-w|-ln\delta)d\tilde{\mu}(e^{i\theta}w)\right| = |z|^\rho\int_0^\delta(ln\delta-lns)d\mu_1(s),$$

where $\mu_1(s) = \tilde{\mu}(U(e^{i\theta},s))$, $\theta = argz$, $\mu_1(s)$ is estimated as follows

$$\mu_1(s) = \rho \int_{-\arcsin \delta}^{\arcsin \delta} d\nu(\theta+\psi) \int_{\cos\psi - \sqrt{s^2-\sin^2\psi}}^{\cos\psi + \sqrt{s^2-\sin^2\psi}} t^{\rho-1} dt \le$$

$$\le \rho 2s\nu([0,2\pi))(1+s)^{\rho-1} \le \rho 2^{\rho} s\nu([0,2\pi)).$$

Therefore

$$|\Phi_3(z)| \le |z|^{\rho} \int_0^{\delta} (ln\delta - lns)d\mu_1(s) \le |z|^{\rho} \int_0^{\delta} s^{-1} d\mu_1(s) \le 2^{\rho}\rho\delta\left(1+ln\frac{1}{\delta}\right)|z|^{\rho}.$$

So for $\delta > 0$ sufficiently small,

$$|\Phi_3(z)| \le \varepsilon|z|^{\rho}.$$

Now it remains to estimate $|\Phi_2(z)|$. We will obtain an estimate in points z for which the following condition is satisfied

$$\int_{U(z,\lambda)} |\zeta|^{-\rho} d\tilde{n}_f(\zeta) \le \frac{\lambda}{\sqrt{\delta}\,|z|}, \quad \forall\lambda < \delta|z|.$$

Such points are called normal points of the measure $\tilde{n}_f$ (with respect to the given ρ and δ). Let $E_{\rho,\delta}(\tilde{n}_f)$ be the set of all normal points of the measure $\tilde{n}_f$. For $z \in E_{\rho,\delta}(\tilde{n}_f)$ we have

$$|\Phi_2(z)| = -\int_{U(z,\delta|z|)} ln\left|\frac{\zeta-z}{\delta z}\right| d\tilde{n}_f(\zeta) = \int_{U(z,\delta|z|)} |\zeta|^{\rho} ln\left|\frac{\delta z}{\zeta-z}\right| \frac{d\tilde{n}_f(\zeta)}{|\zeta|^{\rho}} \le$$

$$\le |z|^{\rho}(1+\delta)^{\rho} \int_{U(z,\delta|z|)} ln\frac{\delta|z|}{|\zeta-z|} \frac{d\tilde{n}_f(\zeta)}{|\zeta|^{\rho}} \le$$

$$\le |z|^{\rho}(1+\delta)^{\rho} \int_0^{\delta|z|} ln\frac{\delta|z|}{s} d\left(\int_{U(z,s)} \frac{d\tilde{n}_f(\zeta)}{|\zeta|^{\rho}}\right) =$$

$$= |z|^{\rho}(1+\delta)^{\rho}\int_0^{\delta|z|}\frac{ds}{s}\left(\int_{U(z,s)}\frac{d\tilde{n}_f(\zeta)}{|\zeta|^{\rho}}\right) \le$$

$$\le |z|^{\rho}(1+\delta)^{\rho}\sqrt{\delta} \le \sqrt{\delta}\, 2^{\rho}|z|^{\rho}.$$

Combining the estimates of $f_{t,1}$, $J_{t,1}$, $f_{t,3}$, $J_{t,3}$, Φ_1, Φ_2, Φ_3, we conclude that if $\varepsilon > 0$ is fixed, then we can choose $t > 1$, $\delta > 0$, and R_0 such that

$$|ln|f(z)|-J(z)| = |f_{t,1} - J_{t,1} + f_{t,3} - J_{t,3} + \Phi_1 + \Phi_2 + \Phi_3| < \varepsilon|z|^{\rho},$$

$\forall z\in E_{\rho,\delta}(\tilde{n}_f)$, $|z| > R_0$. Hence the inclusion $E_{\varepsilon,R} \subset (\mathbb{C}\setminus E_{\rho,\delta}(\tilde{n}_f))\cup U(R_0))$ hold. Now, to complete the proof of the fact that outside a C_0-set

$$ln|f(z)| - J(z) = o(1)\varepsilon|z|^{\rho},$$

it is suffices to choose a system $\mathfrak{A}(R,\varepsilon)$ which satisfies the condition (1.2.10) and covers the set $\mathbb{C}\setminus E_{\rho,\delta}(\tilde{n}_f)$. Such a system can be constructed as follows.

Let $z\in\hat{E} = \mathbb{C}\setminus E_{\rho,\delta}(\tilde{n}_f)$. Then $\exists\lambda = \lambda_z$, $0 < \lambda < \delta|z|$,

$$\int_{U(z,\lambda_z)}|\zeta|^{-\rho}d\tilde{n}_f(\zeta) > \frac{\lambda}{\sqrt{\delta}\ |z|}.$$

Let us consider the set $\mathfrak{A}$ of all discs $U(z,\lambda_z)$ with centres $z\in(\hat{E}\cap U(R))$. Using the theorem on selection of a cover of a finite multiplicity[5)] we choose an at most countable family of discs U_j =

---

5) This theorem is due to L.Ahlfors in the case n = 2 and A.Besicovitch in the case of a space of arbitrary dimension (see, for example, M.Guzman [1]). It can be formulated as follows.

Let a set $A \subset \mathbb{R}^n$ be covered with balls in such a way that every point $x\in A$ is the centre of some ball $B(x)$ with radius $r(x)$ and $sup\{r(x)\} < \infty$. Then there is an at most countable subsystem $\{B(x_j)\}$ of the system $\{B(x)\}$ such that $\{B(x_j)\}$ is a cover of A and each point $x\in A$ belongs to

$U(z,\lambda_{z_j})$ covering the set $\hat{E} \cap U(R)$ with multiplicity ≤ 7. Now we estimate $\sum \lambda_{z_j}$. We have

$$\sum \lambda_{z_j} \le \sqrt{\delta} \sum |z_j| \int_{U_j} |\zeta|^{-\rho} d\tilde{n}_f(\zeta) \le$$

$$\le \sqrt{\delta} \sum_{k=-\infty}^{0} \sum_{2^{k-1}R \le |z_j| \le 2^k R} |z_j| \int_{U_j} |\zeta|^{-\rho} d\tilde{n}_f(\zeta) \le$$

$$\le 7\sqrt{\delta}\, R \sum_{k=-\infty}^{0} 2^{k+1} \int_{2^{k-1}R \le |\zeta| \le 2^k R} |\zeta|^{-\rho} d\tilde{n}_f(\zeta) = 7\sqrt{\delta}\, R \sum_{k=-\infty}^{0} 2^{k+1} \int_{2^{k-1}R}^{2^k R} \frac{dn_f(s)}{s^\rho} \le$$

$$\le 7\sqrt{\delta}\, R \sum_{k=-\infty}^{0} 2^{k+1} \left\{ (R2^k)^{-\rho} n_f(R2^k) - (R2^{k-1})^{-\rho} n_f(R2^{k-1}) + \rho \int_{2^{k-1}R}^{2^k R} \frac{dn_f(s)}{s^{\rho+1}} \right\} \le$$

$$\le 28\sqrt{\delta}\, R^{\rho-1} n_f(R) + 7\rho\sqrt{\delta}\, R \sum_{k=-\infty}^{0} 2^{k+1} \int_{2^{k-1}R}^{2^k R} \frac{dn_f(s)}{s^{\rho+1}}. \qquad (1.2.11)$$

Since $f(z)$ is a function of at most normal type with respect to the order ρ and $f(z) \ne 0$, we have $n_f(s) \le Cs^\rho$. Therefore we can simplify the estimates (1.2.11) such that

$$\sum \lambda_{z_j} \le 28\sqrt{\delta}\, RC + 28\sqrt{\delta}\, RC\, ln 2 \le \text{const } R\sqrt{\delta}.$$

Thus the condition (1.2.10) is satisfied for the system $\mathfrak{A}(R,\varepsilon) = \{U_j\}$. So it is proved that outside some C_0-set

$$ln|f(z)| - J(z) = o(1)|z|^\rho. \qquad (1.2.12)$$

Moreover, $ln|f(e^{i\theta})| \rightrightarrows h(\theta) = |z|^{-\rho} J(z)$, as $r \to \infty$ outside a set of relative measure zero. Now, according to the **Remark** in §1.2, it follows that $J(z) = \mathcal{D}'\text{-}lim\ t^{-\rho} ln|f(tz)|$, $J(z) = \mathfrak{L}_f(z)$, and hence (1.2.12) coincides with condition (c) of the theorem to be proved. Therefore (c) ⇒ (b) is proved in the case of a noninteger ρ.

---

dimension of the space and does not depend on A. For $\mathbb{C}$, i.e. $n = 2$, this constant equals 7.

5. Proof of (c) ⇒ (b) in case of integral order. Let the sequence Z_f be regularly distributed and $\rho = p \in \mathbb{N}$. Hence either $q = p$ or $q = p-1$. Trying to prove the implication in this case similarly to the previous one, we encounter the fact that the integral defining $J(z)$ diverges. It will be shown below, however, that a minor change in the definition of $J(z)$ not only avoids this difficulty, but also allows us to prove (b) ⇒ (c) for $\rho = p \in \mathbb{N}$ similarly to the case $\rho \notin \mathbb{N}$.

This change in the definition of $J(z)$ is based on the fact that for $\rho = p \in \mathbb{N}$ the condition for regular distribution of the sequence Z_f includes not only the existence of the angular, but also the existence of

$$\kappa_f = \lim_{R \to \infty} \sum_{|a_j|<R} a_j^{-p}.$$

From this requirement it follows that the measure ν defined above (that is, the angular density) is not quite arbitrary when $\rho = p \in \mathbb{N}$. It has to satisfy the condition

$$\int_0^{2\pi} e^{-ip\theta} \, d\nu(\theta) = 0. \qquad (1.2.13)$$

In fact, since the limit defining κ_f exists, we have

$$0 = \lim_{R \to \infty} \sum_{\frac{1}{2}R<|a_j|<R} a_j^{-p} = \lim_{R \to \infty} \int_{\frac{1}{2}R<|\zeta|<R} \zeta^{-p} d\tilde{n}_f(\zeta) =$$

$$= \lim_{R \to \infty} \int_{\frac{1}{2}<|z|<1} z^{-p} d\tilde{n}_f^R(z) = \lim_{R \to \infty} \int_{\frac{1}{2}<|z|<1} z^{-p} d\tilde{\mu}(z) =$$

$$= \int_{\frac{1}{2}}^{1} pr^{p-1} dr \int_0^{2\pi} e^{-ip\theta} d\nu(\theta).$$

Hence the condition (1.2.13) is satisfied.

Now we define $J(z)$ as follows:

$$J(z) = \int_{|\zeta|\le k} ln\left|G_{p-1}\left(\frac{z}{\zeta}\right)\right| d\tilde{\mu}(\zeta) +$$

$$+ \int_{|\zeta|>k} ln\left|G_p\left(\frac{z}{\zeta}\right)\right| d\tilde{\mu}(\zeta) + \frac{1}{p} Re(\kappa_f z^p), \quad (1.2.14)$$

where $d\tilde{\mu}(z) = p|z|^{p-1} d|z| \otimes d\nu(\theta)$ similar, to §1.4.

Note that J(z) does not depend on k, since

$$- \int_{k\le|\zeta|\le k_1} ln\left|G_{p-1}\left(\frac{z}{\zeta}\right)\right| d\tilde{\mu}(\zeta) + \int_{k\le|\zeta|\le k_1} ln\left|G_p\left(\frac{z}{\zeta}\right)\right| d\tilde{\mu}(\zeta) =$$

$$= \frac{1}{p} \int_{k\le|\zeta|\le k_1} \left(\frac{z}{\zeta}\right)^p d\tilde{\mu}(\zeta) = z^p \int_k^{k_1} \frac{dr}{r} \int_0^{2\pi} e^{-ip\theta} d\nu(\theta) = 0.$$

Let us consider the difference $ln|f(z)| - J(z)$. Similarly to §1.4 we represent it as

$$ln|f(z)| - J(z) = \sum_{m=1}^{3} f_{t,m}(z) - \sum_{m=1}^{3} J_{t,m}(z),$$

where the $f_{t,m}(z)$ are the same as in §1.4 and the $J_{t,m}(z)$ are defined, according to the change in the definition of J(z) as follows:

$$J_{t,1}(z) = \int_{|\zeta|>t|z|} ln\left|G_p\left(\frac{z}{\zeta}\right)\right| d\tilde{\mu}(\zeta),$$

$$J_{t,2}(z) = \int_{\frac{1}{t}|z|<|\zeta|<t|z|} ln\left|G_p\left(\frac{z}{\zeta}\right)\right| d\tilde{\mu}(\zeta),$$

and

$$J_{t,3}(z) = \int_{|\zeta|<\frac{1}{t}|z|} ln\left|G_{q-1}\left(\frac{z}{\zeta}\right)\right| d\tilde{\mu}(\zeta) + \frac{1}{q} Re(\kappa_f z^q)$$

when p = q and

$$J_{t,3}(z) = \int_{|\zeta|<\frac{1}{t}|z|} ln\left|G_{q-1}\left(\frac{z}{\zeta}\right)\right| d\tilde{\mu}(\zeta)$$

when p = q+1.

In §2.4, in estimating the difference $|J_{t,2}(z) - f_{t,2}(z)|$ we did not use the assumption $\rho\notin\mathbb{Z}$. Thus all the conclusions of §1.4 related to this difference remain valid in the present case $\rho = p\in\mathbb{N}$. Therefore, to prove our statement similar to §2.4, it is sufficient to show the following: for any fixed $\varepsilon > 0$ we can find numbers $t = t(\varepsilon)$ and $R_0 = R_0(\varepsilon)$ such that when $|z| \geq R_0$,

$$|f_{t,1}(z) + f_{t,3}(z) - J_{t,1}(z) - J_{t,3}(z)| < \varepsilon|z|^{\rho}. \qquad (1.2.15)$$

The estimation of $f_{t,1}(z)$ and $J_{t,1}(z)$ in the case $\rho = p = q$ can be literally repeated in the case $\rho \notin \mathbf{Z}$. Therefore there is a $t_0 = t_0(\varepsilon)$ such that

$$|f_{t,1}(z)| + |J_{t,1}(z)| < \varepsilon|z|^{\rho}, \ \forall t > t_0, \ |z| > 1.$$

For $\rho = p = q+1$ we write $f_{t,1}$ as follows

$$f_{t,1}(z) = \int_{|\zeta|>t|z|} \ln\left|G_p\left(\frac{z}{\zeta}\right)\right| d\tilde{n}_f(\zeta) +$$

$$- \frac{1}{p} Re\left\{z^p \lim_{t'\to\infty} \sum_{t|z|<|a_j|<t'} a_j^{-p}\right\}. \qquad (1.2.16)$$

Since $\lim_{R\to\infty} \sum_{|a_j|<R} a_j^{-p}$ exists (because Z_f is assumed to be regularly distributed), the following inequality is valid for all t larger than some $t'_0 = t'_0(\varepsilon, f)$:

$$\frac{1}{p}\left| \lim_{t'\to\infty} \sum_{t|z|<|a_j|<t'} a_j^{-p} \right| < \varepsilon. \qquad (1.2.17)$$

Further, estimating the integral $\int_{|\zeta|>t|z|} \ln\left|G_p\left(\frac{z}{\zeta}\right)\right| d\tilde{n}_f(\zeta)$ similarly to $f_{t,1}(z)$ when $\rho < q+1$, we obtain for all sufficiently large t and all z, $|z| > 1$,

$$\left| \int_{|\zeta|>t|z|} \ln\left|G_p\left(\frac{z}{\zeta}\right)\right| d\tilde{n}_f(\zeta) \right| \leq \varepsilon|z|^{\rho}.$$

Now we conclude from this inequality and from (1.2.16) and (1.2.17) that, when $p = q + 1$, for all sufficiently large t the estimate

$$|f_{t,1}(z)| \leq \varepsilon|z|^{\rho}, \ |z| > 1,$$

is true. The similar estimate for $J_{t,1}(z)$ is obvious.

In the case $\rho = p = q+1$, $f_{t,3}$ and $J_{t,3}$ can be estimated similarly to the case $\rho \notin \mathbb{Z}$. Thus, the following is true when $\rho = p = q + 1$:

$$|f_{t,3}(z)| + |J_{t,3}(z)| < \varepsilon|z|^{\rho}, \ \forall t > t_1, \ |z| > 1.$$

For $p = q$ we estimate the differences $f_{t,3}(z) - \frac{1}{p}Re\{\kappa_f z^p\}$ and $J_{t,3}(z) - \frac{1}{p}Re\{\kappa_f z^p\}$ instead of $f_{t,3}$ and $J_{t,3}$. We have

$$\left|f_{t,3}(z) - \frac{1}{p}Re\{\kappa_f z^p\}\right| = \left| \int_{|\zeta|<\frac{1}{t}|z|} ln\left|G_q\left(\frac{z}{\zeta}\right)\right| d\tilde{n}_f(\zeta) - \frac{1}{q}Re\{\kappa_f z^q\}\right| =$$

$$= \left| \int_{|\zeta|<\frac{1}{t}|z|} ln\left|G_{q-1}\left(\frac{z}{\zeta}\right)\right| d\tilde{n}_f(\zeta) - \frac{1}{q}\, Re\left\{\left(\sum_{|a_j|<\frac{1}{t}|z|} a_j^{-q} - \kappa_f\right) z^q\right\}\right|.$$

The integral can be estimated similarly to the corresponding one in the case $\rho > q$. At the same time it follows from the definition of κ_f that

$$-\kappa_f + \lim_{|z|\to\infty} \sum_{|a_j|<\frac{1}{t}|z|} a_j^{-q} = 0.$$

Hence the inequality

$$\left|f_{t,3}(z) - \frac{1}{p}Re\{\kappa_f z^p\}\right| \le \varepsilon|z|^{\rho}. \tag{1.2.18}$$

is valid for all z, $|z| > R'(t,\varepsilon)$. It is also clear that for all sufficient large t,

$$\left|J_{t,3}(z) - \frac{1}{p}Re\{\kappa_f z^p\}\right| = \left| \int_{|\zeta|\le t|z|} ln\left|G_{q-1}\left(\frac{z}{\zeta}\right)\right| d\tilde{\mu}(\zeta)\right| < \varepsilon|z|^{\rho}.$$

This inequality and (1.2.18) imply that in the situation under consideration $|f_{t,3}(z) - J_{t,3}(z)| \le \varepsilon|z|^{\rho}$ for all sufficient large t and for all z, $|z| > R'(t,\varepsilon)$.

It is clear that inequality (1.2.14) follows from these estimates. According to the above, the proof of (b) ⇒ (c) in the case $\rho = p$ is finished, and so is the proof of **Theorem 1.2.1**.

6. Remarks.

1) The equality $J(z) = \mathfrak{L}_f(z)$ has been obtained in the proof of

Theorem 1.2.1. It gives the integral representation with respect to the measure $\tilde{\mu}$ of the indicator of the Weierstrass canonical product of an entire f.c.r.g. Recall that $\tilde{\mu}$ was introduced with the aid of an angular density. Since $\tilde{\mu}$ is positively homogeneous, it is easy to turn this representation into one for the indicator $\mathfrak{L}_f(e^{i\theta}) = h_f(\theta)$, as an integral with respect to the measure ν. Also, it is easy to show that

$$\int_0^\infty t^{\rho-1} \ln\left|G_q\left(\frac{z}{t}e^{i\theta}\right)\right| dt = r^\rho \sum_{-\infty}^{\infty} \frac{\cos k\theta}{\rho^2-k^2} = \frac{\pi}{\rho}\frac{\cos\rho(\pi-|\theta|)}{\sin\rho\pi} r^\rho, \quad |\theta| < \pi, \ \rho\notin\mathbb{Z},$$

and

$$\int_0^r t^{\rho-1} \ln\left|G_{p-1}\left(\frac{z}{t}e^{i\theta}\right)\right| dt + \int_r^\infty t^{\rho-1} \ln\left|G_p\left(\frac{z}{t}e^{i\theta}\right)\right| dt =$$

$$= r^\rho\left\{-\frac{\cos p\theta}{2p^2} + \sum_{\substack{k=-\infty \\ k\neq\pm p}}^{\infty} \frac{\cos k\theta}{p^2-k^2}\right\} =$$

$$= r^\rho\left\{-\frac{\cos p\theta}{2p^2} + \frac{\pi-\theta}{p}\sin p\theta\right\}, \quad 0 < \theta < 2\pi, \ p\in\mathbb{N}.$$

Hence

$$h_f(\theta) = \frac{\pi}{\sin\pi\rho} \int_{[\theta-\pi,\theta+\pi]} \cos\rho(\pi-|\theta-\varphi|)\, d\nu(\varphi) \text{ for } \rho\notin\mathbb{Z},$$

and

$$h_f(\theta) = \int_{[\theta-2\pi,\theta)} (\varphi-\theta)\sin p(\theta-\varphi)\, d\nu(\varphi) + \frac{1}{p}Re(\kappa_f e^{i\theta}) \text{ for } \rho = p \in \mathbb{N}.$$

(Recall that in the latter case $\int_0^{2\pi} e^{ip\theta}\, d\nu(\theta) = 0$.)

2) Analysis of the reasoning in §1.3 to prove the existence of the limit (1.2.5), i.e. of κ_f, shows that the following representation for κ_f is valid:

$$\kappa_f = -\frac{1}{(p-1)!}\frac{d^p \ln f}{dz^p}\bigg|_{z=0} + \frac{p}{\pi}\int_0^{2\pi} h_f(\theta)e^{-ip\theta}\, d\theta.$$

Besides it is obvious that the constructions in §1.3 remain valid, with a minor change, also in the case when there is $\mathcal{D}'$-convergence of

$t^{-\rho} ln|f(tz)|$ as $t \to \infty$ through some sequence. More exactly, if $f \in H[\mathbb{C},\rho)$, $\rho = p \in \mathbb{Z}$ and if the sequence $t_j \to \infty$ is such that

$$\exists\ \mathcal{D}'\text{-}\lim_{j \to \infty} t_j^{-p}\ ln|f(t_j z)| = \mathfrak{L}(z),$$

then for almost all $\lambda > 0$,

$$\exists \lim_{j \to \infty} \sum_{|a_j| < t_j \lambda} a_j^{-p} = -\frac{\lambda^p}{(p-1)!} \frac{d^p ln f}{dz^p}\Big|_{z=0} +$$

$$+ \frac{p}{\pi} \int_0^{2\pi} \mathfrak{L}(\lambda e^{i\theta}) e^{-ip\theta}\, d\theta + \int_{U(\lambda)} \bar{\zeta}^p\, d\mu_{\mathfrak{L}}(\zeta).$$

§3 Rays of completely regular growth. Addition of indicators

1. **Equivalence of different definitions of "ray of c.r.g."** The ray $l_\theta = \{ z = re^{i\theta} : 0 \le r < \infty \}$ is called a **ray of completely regular growth** of an entire function $f(z)$, $z \in \mathbb{C}$, which has at most normal type with respect to the order $\rho > 0$, if there is a set $E(\theta)$ in $\mathbb{R}^+$ of respective measure zero such that

$$\lim_{\substack{r \to \infty \\ r \notin E}} \frac{1}{r^\rho} ln|f(re^{i\theta})| = h_f(\theta)$$

It is clear that this holds if and only if every set $E_\varepsilon(\theta) = \{r: ln|f(re^{i\theta}| < (h_f(\theta)-\varepsilon)r^\rho\}$ is of relative measure zero.

Similar to regularity of growth in the whole plane, regularity of growth on a ray can be characterized in terms of weak convergence.

Theorem 1.3.1. Let $f \in H(\mathbb{C},\rho]$. Then the following statements are equivalent:

(a) the ray l_θ is a ray of c.r.g. of $f(z)$;

(b) the functions $\varphi_r(s) = r^{-\rho} ln|f(rse^{i\theta})|$ converge to $\mathfrak{L}_f(se^{i\theta})$, as $r \to \infty$, considered as functionals on the space $C_0(\mathbb{R}^+)$ of functions from $C(\mathbb{R}^+)$ with bounded support;

(c) the following equality holds:

(c) the following equality holds:

$$\lim_{r \to \infty} \frac{1}{r^{\rho+1}} \int_0^r \ln|f(te^{i\theta})|dt = \frac{1}{\rho+1} h_f(\theta).$$

Proof. First of all we prove (a) ⇒ (b). For this we need the following lemma and some of its corollaries.

Lemma 1.3.1. Let $\varphi(z)$ be a holomorphic function in a disc $U(kR)$, $k > 1$, $\varphi(0) \neq 0$, and let $E \subseteq [-R,R]$ be a measurable set. Then

$$\int_E |\ln|\varphi(x)||dx \le$$

$$\le C_1 mesE \ln\frac{2R}{mesE} \ln M_\varphi(kR) + C_2 mesE \ln\frac{2R}{mesE} |\ln|\varphi(0)||+$$

$$+ C_3 mesE \ln M_\varphi(kR) + C_4 mesE |\ln|\varphi(0)||, \qquad (1.3.1)$$

where *mes* is Lebesgue measure on $\mathbb{R}$ and the constants C_j depend on k and depend neither on R nor on the choice of $\varphi(z)$.

Proof. Without loss of generality we may assume $\varphi(z) \neq 0$ when $|z| = R\sqrt{k}$. In this situation we represent $\varphi(z)$ in the form $\varphi(z) = \varphi_1(z)\varphi_2(z)$, where

$$\varphi_1(z) = \prod_{j=1}^{N} \frac{R\sqrt{k}(z-a_j)}{R^2k-\bar{a}_jz},$$

and a_j, $j = 1,\dots, N = n_\varphi(R\sqrt{k})$, are zeros of $\varphi(z)$ in $U(R\sqrt{k})$, counted with multiplicities. Then $\varphi_2(z)$ is holomorphic in $\bar{U}(R\sqrt{k})$ and does not vanish. Therefore $\ln|\varphi_2(z)|$ is harmonic in $U(R\sqrt{k})$, and can thus be represented by a Poisson integral:

$$\ln|\varphi_2(se^{i\theta})| = \frac{1}{2\pi} \int_0^{2\pi} \frac{(R^2k-s^2)\ln|\varphi_2(R\sqrt{k}\, e^{i\psi})|\, d\psi}{R^2k+s^2-2R\sqrt{k}s\cos(\theta-\psi)}.$$

For $0 < s < R$ this implies

$$|\ln|\varphi_2(se^{i\theta})|| \le \frac{1}{2\pi} \frac{\sqrt{k}+1}{\sqrt{k}-1} \int_0^{2\pi} (\ln^+|\varphi_2(R\sqrt{k}e^{i\psi})| + \ln^-|\varphi_2(R\sqrt{k}e^{i\psi})|)d\psi,$$

($\ln^-(\cdot) = \max\{0, -\ln(\cdot)\}$). Since

$$ln|\varphi_2(0)| = \frac{1}{2\pi}\int_0^{2\pi} ln|\varphi_2(R\sqrt{k}e^{i\psi})|d\psi =$$

$$= \frac{1}{2\pi}\int_0^{2\pi} ln^+|\varphi_2(R\sqrt{k}e^{i\psi})|d\psi - \frac{1}{2\pi}\int_0^{2\pi} ln^-|\varphi_2(R\sqrt{k}e^{i\psi})|d\psi,$$

we have

$$|ln|\varphi_2(se^{i\theta})|| \le \frac{\sqrt{k}+1}{\sqrt{k}-1}\left\{\frac{1}{\pi}\int_0^{2\pi} ln^+|\varphi_2(R\sqrt{k}e^{i\psi})|d\psi - ln|\varphi_2(0)|\right\}.$$

Taking into account that $|\varphi_2(R\sqrt{k}e^{i\psi})| = |\varphi(R\sqrt{k}e^{i\psi})|$, $\forall\psi$, and $ln|\varphi_2(0)| \ge ln|\varphi(0)|$, we find that for $0 \le s \le R$,

$$ln|\varphi_2(se^{i\theta})| \le \frac{\sqrt{k}+1}{\sqrt{k}-1}\{2lnM_f(R\sqrt{k}) - ln|\varphi(0)|\},\ \theta\in[0,2\pi].$$

Hence

$$\int_E ln|\varphi_2(x)|dx \le mesE\,\frac{\sqrt{k}+1}{\sqrt{k}-1}\{2lnM_f(R\sqrt{k}) - ln|\varphi(0)|\}. \qquad (1.3.2)$$

Now we estimate $\int_E |ln|\varphi_1(x)||dx$. We have

$$\int_E |ln|\varphi_1(x)||dx = -\int_E ln|\varphi_1(x)|dx = \sum_{j=1}^{N}\int_E ln\left|\frac{R^2k-\bar{a}_jx}{R\sqrt{k}\,(x-a_j)}\right|dx \le$$

$$\le \sum_{j=1}^{N}\left(mesE\left(ln2R + ln\frac{k+\sqrt{k}}{2}\right) - \int_E ln|x-a_j|dx\right). \qquad (1.3.3)$$

We estimate $\int_E ln|x-a_j|dx$. Using the monotonicity of $ln(s)$ we find that

$$\int_E ln|x-a_j|dx \ge \int_E ln|x-Rea_j|dx \ge \int_{-\frac{1}{2}mesE}^{\frac{1}{2}mesE} ln|s|ds = mesE\cdot(lnmesE-1-ln2).$$

This this and (1.3.3) imply that

$$\int_E |ln|\varphi_1(x)||dx \le N mesE ln\frac{2R}{mesE} + N\cdot C_1'\cdot mesE, \qquad (1.3.4)$$

where $C_1' = ln\frac{k+\sqrt{k}}{2} + 1 + ln2$. Recall that here N is the number of zeros of $\varphi(z)$ in the disc $U(R\sqrt{k})$. It is estimated in a standard manner by Jensen's formula. Namely,

$$N = n_\varphi(R\sqrt{k}) = \frac{1}{ln\sqrt{k}} \int_{R\sqrt{k}}^{Rk} \frac{n_\varphi(R\sqrt{k})}{s} ds \le \frac{1}{ln\sqrt{k}} \int_0^{Rk} \frac{n_\varphi(s)}{s} ds =$$

$$= \frac{1}{ln\sqrt{k}}\left\{\frac{1}{2\pi}\int_0^{2\pi} ln|\varphi(Rke^{i\theta})|d\theta - ln|\varphi(0)|\right\} \le \frac{1}{ln\sqrt{k}}\left\{lnM_\varphi(Rk) - ln|\varphi(0)|\right\}.$$

From this and (1.3.4) we conclude that

$$\int_E |ln|\varphi_1(x)||dx \le \frac{1}{ln\sqrt{k}}\{lnM_\varphi(Rk) - ln|\varphi(0)|\} mesE\cdot ln\frac{2R}{mesE} +$$

$$+ \frac{C_1'}{ln\sqrt{k}}\{lnM_\varphi(Rk) - ln|\varphi(0)|\} mesE.$$

Comparing this inequality with (1.3.2) we obtain (1.3.1).

Thus lemma is proved.

Corollary 1. If $f(z)\in H(\mathbb{C},\rho]$ then

$$\sup_{0<\theta<2\pi}\ \sup_{1<r<\infty} \frac{1}{\rho+1}\int_0^r |\ln|f(te^{i\theta})||dt < \infty.$$

Corollary 2. If $f(z)\in H(\mathbb{C},\rho]$, and $E \subset \mathbb{R}^+$ is a set of relative measure zero, then

$$\frac{1}{r^{\rho+1}} \int_{E\cap[0,r]} |ln|f(te^{i\theta})||dt \to 0$$

uniformly in $\theta \in [0,2\pi]$ as $r \to \infty$.

Now suppose that $r^{-\rho}ln|f(re^{i\theta})| \to h_f(\theta)$ as $r \to \infty$ outside some set $E(\theta)$ of relative measure zero (i.e. the condition (a) is satisfied). Let us consider the expression

$$r^{-\rho}\int_0^\infty \psi(s)\ln|f(rse^{i\theta})|ds,$$

where $\psi(s)\in C_0(\mathbb{R}^+)$.

For simplicity of exposition we assume $supp\psi \subset [0,1]$. From the estimate (1.1.6) it follows that

$$\overline{\lim_{r\to\infty}}\ r^{-\rho}\int_0^\infty\psi(s)\ln|f(rse^{i\theta})|ds \le$$

$$\le \int_0^\infty\psi(s)\mathcal{L}_f(se^{i\theta})ds = h_f(\theta)\int_0^1\psi(s)s^\rho ds. \tag{1.3.5}$$

On the other hand,

$$\underline{\lim_{r\to\infty}}\ r^{-\rho}\int_0^\infty\psi(s)\ln|f(rse^{i\theta})|ds = \underline{\lim_{r\to\infty}}\ r^{-\rho-1}\int_0^r\psi\left(\frac{s}{r}\right)\ln|f(se^{i\theta})|ds =$$

$$= \underline{\lim_{r\to\infty}}\ r^{-\rho-1}\left\{\int_{E(\theta)\cap[0,r]}\psi\left(\frac{s}{r}\right)\ln|f(se^{i\theta})|ds + \int_{[0,r]\setminus E(\theta)}\psi\left(\frac{s}{r}\right)\ln|f(se^{i\theta})|ds\right\} \ge$$

$$\ge -\overline{\lim_{r\to\infty}}\ \frac{1}{r^{\rho+1}}\int_{E(\theta)\cap[0,r]}\left|\psi\left(\frac{s}{r}\right)\right|\,|\ln|f(se^{i\theta})||ds +$$

$$+\underline{\lim_{r\to\infty}}\ \frac{1}{r^{\rho+1}}\int_{[0,r]\setminus E(\theta)}\psi\left(\frac{s}{r}\right)\ln|f(se^{i\theta})|ds. \tag{1.3.6}$$

According to **Corollary 2 of Theorem 1.3.1**

$$\overline{\lim_{r\to\infty}}\ \frac{1}{r^{\rho+1}}\int_{E(\theta)\cap[0,r]}\left|\psi\left(\frac{s}{r}\right)\right|\,|\ln|f(se^{i\theta})||ds \le$$

$$\le \|\psi\|_\infty\ \overline{\lim_{r\to\infty}}\ \frac{1}{r^{\rho+1}}\int_{E(\theta)\cap[0,r]}|\ln|f(se^{i\theta})||ds = 0, \tag{1.3.7}$$

and from the behaviour of $\ln|f(se^{i\theta})|$ outside $E(\theta)$ it follows that $\forall\varepsilon > 0$,

$$\underline{\lim_{r\to\infty}}\ \frac{1}{r^{\rho+1}}\int_{[0,r]\setminus E(\theta)}\psi\left(\frac{s}{r}\right)\ln|f(se^{i\theta})|ds \ge$$

$$\geq \underline{\lim}_{r\to\infty} \frac{1}{r^{\rho+1}} \int_{[0,r]\setminus E(\theta)} \psi\left(\frac{s}{r}\right)(\mathfrak{L}_f(se^{i\theta}) - \varepsilon s^{\rho})ds =$$

$$= \underline{\lim}_{r\to\infty} \frac{1}{r^{\rho+1}} \left\{ \int_0^r \psi\left(\frac{s}{r}\right)(\mathfrak{L}_f(se^{i\theta}) - \varepsilon s^{\rho})ds - \int_{E(\theta)} \psi\left(\frac{s}{r}\right)(\mathfrak{L}_f(se^{i\theta}) - \varepsilon s^{\rho})ds \right\} \geq$$

$$\geq \int_0^1 \psi(s)(\mathfrak{L}_f(se^{i\theta})-\varepsilon s^{\rho})ds = (h_f(\theta)-\varepsilon) \int_0^1 \psi(s)ds.$$

Since $\varepsilon > 0$ is arbitrary, it follows from this inequality and (1.3.6), (1.3.7) that

$$\underline{\lim}_{r\to\infty} r^{-\rho} \int_0^{\infty}\psi(s)\ln|f(rse^{i\theta})|ds \geq h_f(\theta) \int_0^{\infty} \psi(s)s^{\rho}ds.$$

Comparison of this inequality with (1.3.5) shows that $\forall\psi\in C_0(\mathbb{R}^+)$,

$$\exists \lim_{r\to\infty} r^{-\rho} \int_0^{\infty}\psi(s)\ln|f(rse^{i\theta})|ds = h_f(\theta) \int_0^{\infty} \psi(s)s^{\rho}ds = \int_0^{\infty} \psi(s)\mathfrak{L}_f(se^{i\theta})ds.$$

Thus the proof of (a) $\Rightarrow$ (b) is finished.

Now we assume that condition (b) is satisfied, i.e. the limit of the functions $\varphi_r(s) = r^{-\rho}\ln|f(rse^{i\theta})|$, $r \to \infty$, regarded as functionals on $C_0(\mathbb{R}^+)$, exists and is equal to $\mathfrak{L}_f(se^{i\theta})$. In as far as this limit (which is in general, a measure) is an ordinary function, the $\varphi_r(s)$ converge to $\mathfrak{L}_f(rs^{i\theta})$ as functionals on every space $C([a,b])$, $0 \leq a < b < \infty$, as well. In particular, this convergence takes place also on the characteristic functions of intervals too. So

$$\lim_{r\to\infty} \frac{1}{r^{\rho+1}} \int_0^r \ln|f(se^{i\theta})|ds = \lim_{r\to\infty} \frac{1}{r^{\rho}} \int_0^1 \ln|f(rse^{i\theta})|ds =$$

$$= \int_0^1 \mathfrak{L}_f(se^{i\theta})ds = \frac{h_f(\theta)}{\rho+1}.$$

Thus (b) $\Rightarrow$ (c) is proved.

It remains to show that (c) $\Rightarrow$ (a). Let us assume the opposite, i.e. (c) $\not\Rightarrow$ (b). We put, for brevity, $\nu(r) = mes\{E_{\varepsilon}(\theta) \cap [0,r]\}$. Then for some $\varepsilon > 0$ we have

$$\overline{\lim_{r\to\infty}} \frac{\nu(r)}{r} = \tau > 0,$$

and therefore $\exists\{r_j\}_{j=1}^{\infty}$, $r_j\uparrow\infty$, $\lim_{j\to\infty} r_j^{-1}\nu(r_j) = \tau$. It follows from this and estimate (1.1.6) that

$$\overline{\lim_{j\to\infty}} \frac{1}{r_j^{\rho+1}} \int_0^{r_j} ln|f(se^{i\theta})|ds \le$$

$$\le \overline{\lim_{j\to\infty}} \frac{1}{r_j^{\rho+1}} \left\{ \int_{[0,r_j]\setminus E_\varepsilon(\theta)} (h_f(\theta)+o(1))s^\rho ds + \int_{E_\varepsilon(\theta)\cap[0,r_j]} (h_f(\theta)-\varepsilon)s^\rho ds \right\} \le$$

$$\le \overline{\lim_{j\to\infty}} \frac{1}{r_j^{\rho+1}} \left\{ \int_0^{r_j} h_f(\theta)s^\rho ds + \int_0^{r_j} o(1)s^\rho ds - \varepsilon \int_{E_\varepsilon(\theta)\cap[0,r_j]} s^\rho ds \right\} \le$$

$$\le \frac{h_f(\theta)}{\rho+1} - \varepsilon \underline{\lim_{j\to\infty}}\, r_j^{-\rho-1} \int_{E_\varepsilon(\theta)\cap[0,r_j]} s^\rho ds \le$$

$$\le \frac{h_f(\theta)}{\rho+1} - \varepsilon \underline{\lim_{j\to\infty}}\, r_j^{-\rho-1} \int_0^{\nu(r_j)} s^\rho ds \le$$

$$\le \frac{h_f(\theta)}{\rho+1} - \varepsilon \lim_{j\to\infty} \frac{(\nu(r_j))^{\rho+1}}{(\rho+1)r_j^{\rho+1}} = \frac{h_f(\theta)}{\rho+1} - \varepsilon \frac{\tau^{\rho+1}}{\rho+1} < \frac{h_f(\theta)}{\rho+1}.$$

This contradicts (c). Hence (c) $\Rightarrow$ (a).

The proof of the theorem is finished.

2 Regularity of growth on a ray and in a plane. It follows from the definition of an entire function of c.r.g. in $\mathbb{C}$ and on a ray that if $f\in H(\mathbb{C},\rho]$ is of c.r.g. in $\mathbb{C}$, then it is of c.r.g. on every ray l_θ. The validity of the converse statement is not obvious, because the existence of a set $E(\theta)$, $\forall\theta\in[0,2\pi)$, for which: if $r\to\infty$, $r\notin E(\theta)$, then

$$r^{-\rho} ln|f(re^{i\theta}) \to h_f(\theta), \tag{1.3.8}$$

does not a *priori* imply the existence of a set E which is independent of θ and outside which there is convergence as $r \to \infty$, $r \notin E$, for all $\theta \in [0,2\pi)$. Moreover it is not clear that

$$r^{-\rho} \ln|f(re^{i\theta})| \rightrightarrows h_f(\theta)$$

as $r \to \infty$. However, the following statement is true.

Theorem 1.3.2. If $f \in H(\mathbb{C},\rho]$ is of c.r.g. on every ray l_θ, then f is a function of c.r.g. in $\mathbb{C}$.

Proof. Let $\psi(z) \in D(\mathbb{C})$. Then

$$t^{-\rho} \int_{\mathbb{C}} \ln|f(tz)|\psi(z)d\omega_z = t^{-\rho} \int_0^{2\pi} d\theta \int_0^{\infty} \ln|f(tre^{i\theta})|\psi(re^{i\theta})rdr. \qquad (1.3.9)$$

Since f(z) is of c.r.g. on every ray l_θ, it follows from **Theorem 1.3.1** that

$$\exists \lim_{t \to \infty} \int_0^{\infty} \ln|f(tre^{i\theta})|\psi(re^{i\theta})rdr = h_f(\theta)\int_0^{\infty} \psi(re^{i\theta})r^{\rho+1}dr.$$

At the same time it follows from **Corollary 1** of **Lemma 1.3.1** that

$$\sup_{\theta} \sup_{1<t<\infty} \left| \int_0^{\infty} \ln|f(tre^{i\theta})|\psi(re^{i\theta})rdr \right| < \infty .$$

Therefore limit transition under the integral sign is possible in (1.3.9). Thus

$$\exists \lim_{t \to \infty} t^{-\rho} \int_{\mathbb{C}} \ln|f(tz)|\psi(z)d\omega_z =$$

$$= \int_0^{2\pi} h_f(\theta)d\theta\int_0^{\infty} \psi(re^{i\theta})r^{\rho+1}dr = \int_{\mathbb{C}} \mathfrak{L}_f(z)\psi(z)d\omega_z, \quad \forall\psi \in \mathcal{D}'(\mathbb{C}).$$

or, equivalently,

$$\exists\, \mathcal{D}'\text{-}\lim_{t \to \infty} \frac{\ln|f(tz)|}{t^{\rho}} = \mathfrak{L}_f(z).$$

According to **Theorem 1.2.1** it follows that f(z) is an entire function of c.r.g.

The proof of the theorem is finished.

Theorem 1.3.2 proved above deals with a function of c.r.g. on every ray l_θ. There arises natural question about the properties of an entire function of c.r.g. on a set which differs from the set of all rays. We shall state without proof two theorems concerning this case. The proofs can be found in B.Levin [1], V.Azarin [1].

Theorem 1.3.3. Let $f \in H(\mathbb{C},\rho]$ be a function of c.r.g. on the rays l_θ, $\theta \in Q \subset [0,2\pi)$. Then there is a set $E \subset \mathbb{R}^+$ of relative measure zero such that as $r \to \infty$, $r \notin E$, the functions $r^{-\rho} \ln|f(re^{i\theta})|$, $\theta \in Q$, converge to $h_f(\theta)$ uniformly with respect to $\theta \in Q$.

Theorem 1.3.4. A set $\{l_\theta: \theta \in Q \subset [0,2\pi)\}$ consists of all rays of c.r.g. for some function $f \in H(\mathbb{C},\rho]$ if and only if Q is closed (in $[0,2\pi)$).

When $Q \neq [0,2\pi)$, the case $Q = (\alpha,\beta)$ is of the greatest interest. In this case the function is of c.r.g. in the sector $\alpha < \arg z < \beta$. The same results concerning this case will be obtained below on in **Chapters 3** and **4** as corollaries of more general results on subharmonic functions in $\mathbb{R}^n$ and functions holomorphic in the half-plane.

3. **Addition of indicators.** At the end of this section we state an important property of c.r.g. on a ray.

Theorem 1.3.5. Let $f \in H(\mathbb{C},\rho]$ be of c.r.g. on the ray $l_{\theta'}$. Then

$$h_{fg}(\theta') = h_f(\theta') + h_g(\theta'), \ \forall g \in H(\mathbb{C},\rho].$$

Proof. The inequality $h_{fg} \le h_f + h_g$ is obvious. To prove the opposite inequality we give upper and lower bounds for the average

$$\mathfrak{M}_{\ln|fg|}(re^{i\theta'},\delta r) = \frac{1}{\pi(r\delta)^2} \int_{U(re^{i\theta'},\delta r)} \ln|f(z)g(z)|d\omega_z.$$

Taking into account the continuity of the indicator h_{fg}, it follows by (1.1.6) that

$$\mathfrak{M}_{\ln|fg|}(re^{i\theta'},\delta r) \le$$

$$\le \frac{1}{\pi(r\delta)^2} \int_{U(re^{i\theta'},\delta r)} (h_{fg}(\theta) + \varepsilon(r))t^{\rho+1} dt d\theta \le$$

$$\le (h_{fg}(\theta') + \varepsilon(r) + \varepsilon_1(\delta))(1+\delta)r^\rho.$$

where $\varepsilon(r) \to 0$ as $r \to \infty$ and $\varepsilon_1(\delta) \to 0$ as $\delta \to 0$. Therefore

$$\overline{\lim_{\delta\to 0}}\ \overline{\lim_{r\to\infty}} \frac{1}{r^{\rho}} \mathfrak{M}_{ln|fg|}(re^{i\theta'},\delta r) \le h_{fg}(\theta') \tag{1.3.10}$$

To give a lower bound for of the same value we choose $r_j\uparrow\infty$ in such a way that

$$\lim_{j\to\infty} \frac{ln|g(r_j e^{i\theta'})|}{r_j^{\rho}} = h_g(\theta'). \tag{1.3.11}$$

Then we choose some numbers $\eta\in(0,1)$, $\varepsilon>0$ and a point r'_j in each interval $(r_j,(1+r\delta)r_j)$, from some such interval onwards, in such a manner that $ln|f(r'_j e^{i\theta'})| \ge r'^{\rho}_j(h_f(\theta') - \varepsilon)$. This is possible because f is of c.r.g. on the ray $l_{\theta'}$ so that the set $E_\varepsilon(\theta')$ is of relative measure zero. We also assume that $ln|f(z)| \le 0$ when $|argz - \theta'| < arcsin\delta$, $|z| > R_0$. Then the following is true for all sufficiently large j:

$$\mathfrak{M}_{ln|f|}(r_j e^{i\theta'},\delta r_j) \ge \frac{1}{\pi(r_j\delta)^2} \int_{U(r'_j e^{i\theta'_j},(1+\eta)\delta r_j)} ln|f(z)|d\omega_z =$$

$$= (\eta+1)^2\, \mathfrak{M}_{ln|f|}(r'_j e^{i\theta'},(1+\eta)\delta r_j) \ge$$

$$\ge (\eta+1)^2(1+\delta\eta)^{\rho} r_j^{\rho}(h_f(\theta')-\varepsilon). \tag{1.3.12}$$

Since $\eta > 0$ is arbitrary, it follows that

$$\underline{\lim}_{r_j\to\infty} \frac{1}{r_j^{\rho}} \mathfrak{M}_{ln|f|}(r'_j e^{i\theta'},\delta r_j) \ge h_f(\theta') - \varepsilon. \tag{1.3.13}$$

Now using this inequality and taking into account (1.3.11) we find that

$$\overline{\lim_{r\to\infty}} \frac{1}{r^{\rho}} \mathfrak{M}_{ln|fg|}(re^{i\theta'},\delta r) \ge \overline{\lim_{j\to\infty}} \frac{1}{r_j^{\rho}} \mathfrak{M}_{ln|fg|}(r_j e^{i\theta'_j},\delta r_j) \ge$$

$$\geq \underline{\lim}_{j \to \infty} \frac{1}{r_j^{\rho}} \mathfrak{N}_{ln|f|}(r_j e^{i\theta'}, \delta r_j) + \overline{\lim}_{j \to \infty} \frac{1}{r_j^{\rho}} \mathfrak{N}_{ln|g|}(r_j e^{i\theta'_j}, \delta r_j) \geq$$

$$\geq h_f(\theta') - \varepsilon + h_g(\theta').$$

Since $\varepsilon > 0$ is arbitrary, we have

$$\overline{\lim}_{r \to \infty} \frac{1}{r^{\rho}} \mathfrak{N}_{ln|fg|}(re^{i\theta'}, \delta r) \geq h_f(\theta') + h_g(\theta').$$

Hence

$$\lim_{\delta \to 0} \overline{\lim}_{r \to \infty} \frac{1}{r^{\rho}} \mathfrak{N}_{ln|fg|}(re^{i\theta'}, \delta r) \geq h_f(\theta') + h_g(\theta').$$

Comparing this inequality with (1.3.10) we conclude that

$$h_{fg}(\theta') \geq h_f(\theta') + h_g(\theta').$$

We have obtained this inequality under the assumption that $ln|f(z)| \leq 0$ when $|arg\ z - \theta'| < arcsin\delta$, $|z| > R_0$. To get rid of this restriction it is sufficient to estimate $ln|f(z)| - Re((a+bi)z^{\rho})$ where a and b are chosen in such a way that $ln|f(z)| - Re((a+bi)z^{\rho}) < 0$ when $|argz - \theta'| < arcsin\delta$, $|z| > R_0$. This is possible because of the estimate (1.1.6). Here, instead of (1.3.12) the following inequality is obtained:

$$\frac{1}{r_j^{\rho}} \mathfrak{N}_{ln|f|}(r_j e^{i\theta'}, \delta r_j) \geq$$

$$\geq (\eta+1)^2 (1+\delta\eta)^{\rho} r_j^{\rho} (h_f(\theta') - \varepsilon - Re[(a+bi)e^{i\theta'}]) + Re[(a+bi)e^{i\theta'}]).$$

This also leads to (1.3.13).

Corollary. If an entire function $f \in H(\mathbb{C}, \rho]$ is of c.r.g. in $\mathbb{C}$, then the following equality is valid for all $\theta \in [0, 2\pi]$ and all $g \in H(\mathbb{C}, \rho]$:

$$h_{fg}(\theta) = h_f(\theta) + h_g(\theta).$$

As pointed out in **Theorem 1.3.5** and in its **Corollary**, the property of addition of indicators is characteristic for functions of c.r.g.

Theorem 1.3.6. If $f \in H(\mathbb{C}, \rho]$ is such that for fixed $\theta = \theta'$

$$h_{fg}(\theta') = h_f(\theta') + h_g(\theta'), \ \forall g \in H(\mathbb{C},\rho],$$

then f is of c.r.g. on the ray $l_{\theta'}$.

Theorem 1.3.7. If $f \in H(\mathbb{C},\rho]$ is such that $h_{fg} = h_f + h_g$, $\forall g \in H(\mathbb{C},\rho]$, then f is of c.r.g. (in $\mathbb{C}$).

Here we do not give the proofs of these theorems. They will be proved in **Chapter 3** in the more general case for for entire functions in $\mathbb{C}^n$, $n \geq 1$.

Notes

Theorem 1.2.1 (without the assertion (d)) was obtained by B.Levin [2], [3] and A.Pfluger [1], [2], [3], independently and simultaneously. **Theorem 1.3.1** (with the same reservation) was obtained by A.Pfluger as were **Theorems 1.3.2** and **1.3.3**. The possibility of addition to the above results of B.Levin and A.Pfluger of statements on the relationship between completely regular growth and $\mathcal{D}'$-convergence (weak convergence) became clear due to papers by Agranovič-Ronkin [1], V.Azarin [2], [3], P.Agranovič [1]. Note that in these papers a more general situation was considered. The necessity part of **Theorem 1.3.4** was obtained by A.Pfluger [1], while sufficiency part is due to B.Levin. The converse **Theorems 1.3.6** and **1.3.7** were proved by V.Azarin [4]. Note also that besides B.Levin's and A.Pfluger's papers quoted above, the main results of **Chapter 1** were represented earlier in **Chapters II** and **III** of the monograph by B.Levin [4]. Our presentation differs from at in the above-mentioned works of B.Levin and A.Pfluger.

Note also that polynomial asymptotics for entire and subharmonic functions in **C** and for corresponding counting functions were considered by V.Logvinenko [1], Agranovič-Logvinenko [1], [2].

CHAPTER 2

SUBHARMONIC FUNCTIONS OF COMPLETELY REGULAR GROWTH IN $\mathbb{R}^n$

§1 General information on subharmonic functions. $\mathcal{D}'$-convergence

Here we give some facts from the general theory of subharmonic functions, in general without proof. For a detailed account see, for example, L.Ronkin [1], Kennedy-Hayman [1]. Besides, here we give some lemmas concerning sequences of subharmonic functions that converge in the topology of the space of distributions.

1. Elementary information. We denote by $B_r^n(x)$, or simply $B_r(x)$ the ball in the space $\mathbb{R}^n$ with the centre $x\in\mathbb{R}^n$ and radius r. By $S_r^n(x) = S_r(x)$ we denote the corresponding sphere. Thus, $B_r(x) = \{y\in\mathbb{R}^n: |y-x| < r\}$, $S_r(x) = \partial B_r(x) = \{y\in\mathbb{R}^n: |y - x| = r\}$. For brevity we set $B_r(0) = B_r$ and $S_r(0) = S_r$. The Euclidean volume of the ball B_r and the "area" of the sphere S_r are denoted by V_n and σ_{n-1} respectively.

Let a function u(x), $-\infty \le u(x) < \infty$, be measurable and bounded above in the ball $B_r(x^0)$. We denote by $\mathfrak{N}_u(x^0,r)$ the average of u(x) in $B_r(x^0)$. In other words,

$$\mathfrak{N}_u(x^0,r) = \frac{1}{V_n r^n}\int_{B_r(x^0)} u(x)dx,$$

where $dx = d\omega_x = dx_1\wedge\ldots\wedge dx_n$ is the Euclidean volume element in $\mathbb{R}^n$. Similarly, $\mathfrak{M}_u(x^0,t)$ denotes the average of the function u(x) over the sphere $S_t(x^0)$, $0 < t < r$, i.e.

$$\mathfrak{M}_u(x^0,t) = \frac{1}{\sigma_{n-1}t^{n-1}}\int_{S_t(x^0)} u(x)d\sigma,$$

where $d\sigma = d\sigma(x)$ is the $(n-1)$-dimensional volume ("area") element of the sphere $S_t(x^0)$. Let us set

$$M_u(r,x^0) = sup\{u(x):\ x\in B_r(x^0)\},$$

$$M_u(r,) = M_u(r,0).$$

Let G be a domain in $\mathbb{R}^n$, $n \geq 2$. A function u(x) defined in G and taking values from the interval $[-\infty,\infty)$ is called upper semicontinuous in G if

$$\lim_{\varepsilon \to 0} \sup_{y\in B_\varepsilon(x)} u(y) = u(x),\ \forall x\in G.$$

We will denote by $u^*(x)$, or $regu(x)$, the least upper semicontinuous majorant of the function u(x), $x\in G$. Thus,

$$u^*(x) = regu(x) = \lim_{\varepsilon \to 0} \sup_{y\in B_\varepsilon(x)} u(y),\ x\in G.$$

A function u(x) is called subharmonic in a domain G if it is upper semicontinuous in G and for any point $x\in G$ there exists $r_0 = r_0(x)$ such that $u(x) \leq \mathfrak{M}_u(x,r)\ \forall r < r_0$.

Note that the functions $\mathfrak{M}_u(x,r)$ and $\mathfrak{N}_u(x,r)$ are subharmonic with respect to x, monotone nondecreasing with respect to r, continuous with respect to both r and x, and for all admissible values of x and r the following inequalities holds:

$$\mathfrak{N}_u(x,r) \leq \mathfrak{M}_u(x,r) \leq M_u(r,x).$$

Besides, if $0 < r' < r$, then

$$\mathfrak{M}_{|u|}(x,r) \leq 2\mathfrak{M}_{u^+}(x,r) - \mathfrak{M}_u(x,r') \leq 2\mathfrak{M}_{u^+}(x,r) - u(x),$$

$$\mathfrak{N}_{|u|}(x,r) \leq 2\mathfrak{N}_{u^+}(x,r) - \mathfrak{N}_u(x,r') \leq 2\mathfrak{N}_{u^+}(x,r) - u(x), \tag{2.1.1}$$

and hence

$$\mathfrak{M}_{|u|}(x,r) \le 2M_{u^+}(x,r) - \mathfrak{M}_u(x,r') \le 2M_{u^+}(x,r) - u(x),$$
$$\mathfrak{N}_{|u|}(x,r) \le 2M_{u^+}(x,r) - \mathfrak{N}_u(x,r') \le 2M_{u^+}(x,r) - u(x). \tag{2.1.2}$$

Note that also

$$\lim_{r\to 0} \mathfrak{M}_u(x,r) = \lim_{r\to 0} \mathfrak{N}_u(x,r) = u(x), \ \forall x\in G.$$

Let us denote by SH(G) the set of all subharmonic functions in G. Note that for functions $u\in SH(G)$ the maximum principle is valid, and such function can be represented as follows:

$$u(x) = \lim_{j\to\infty} u_j(x), \ x\in G,$$

where $u_j(x)\in(C^\infty(G) \cap SH(G))$ and $u_{j+1} \le u_j$, $j=1,2,\ldots$.

The following lemma is of great importance for the estimation of subharmonic functions.

Hartogs' Lemma. Let uniformly bounded above family of functions $u_t\in SH(G)$, $t > 1$, satisfy for some $A\in\mathbb{R}$ the following condition:

$$\overline{\lim_{t\to\infty}}\, u_t(x) \le A, \ \forall x\in G.$$

Then for any compact set $K \subset G$ and any $\varepsilon > 0$ there exists $t_0 = t_0(K,\varepsilon)$ such that for $t \ge t_0$,

$$u_t \le A + \varepsilon, \ \forall x\in K.$$

Note that the statement of the lemma remains is valid if we replace the number A by a continuous function A(x).

Let $u\in SH(G)$, $u \not\equiv -\infty$. A measure μ_u is called the Riesz associated measure of the function u if it is defined, as an element of $\mathcal{D}'(G)$, by the equality

$$\mu_u = \frac{1}{\theta_n} \Delta u,$$

where $\theta_n = (n-2)\sigma_{n-1}$ for $n > 2$, $\theta_2 = 2\pi$, $\Delta = \sum_{j=1}^{n} \frac{\partial^2}{\partial x_j^2}$. As in the case n = 2, this measure is positive for arbitrary n. The converse statement is also true. More exactly, if for $u\in\mathcal{D}'(G)$ the inequality $\Delta u \ge 0$ is valid, then there is a subharmonic function $\tilde{u}$ which coincides with u as an element of $\mathcal{D}'(G)$. Therefore, if $u = \mathcal{D}'-\lim_{t\to\infty} u_t$, where $u_t \in SH(G)$, $\forall t$, then we may assume (and we shall always do this, without specially saying so) that $u \in SH(G)$[6].

For brevity we denote $\mu_u(\bar{B}_t)$, $u \in SH(G)$, by $\mu_u(t)$. If the function u(x) is harmonic in some neighbourhood of the origin, then the following equality, usually called Jensen's formula (or Jensen-Privalov formula) for subharmonic functions, is valid:

$$\mathfrak{M}_u(0,r) = (n-2)\int_0^r \frac{\mu_u(t)}{t^{n-1}}\,dt + u(0),\ 0 < r < R,\ \text{if } n > 2,$$

and

$$\mathfrak{M}_u(0,r) = \int_0^r \frac{\mu_u(t)}{t}\,dt + u(0), \qquad 0 < r < R,\ \text{if } n = 2.$$

Since $\mu_u(t)$ is monotone, it follows from these equalities that for $k > 1$, $kr < R$,

$$\begin{aligned}\mu_u(r) &\le C(k)r^{n-2}\{M_u(kr) - u(0)\},\ C(k) = \frac{k^{n-2}}{k^{n-2}-1},\ \text{if } n > 2,\\ \mu_u(r) &\le \frac{1}{\ln k}\{M_u(kr) - u(0)\},\ \text{if } n = 2.\end{aligned} \tag{2.1.3}$$

In the absence of the assumption that the function u is harmonic in a neighbourhood of the origin, we have the equality

$$\mathfrak{M}_u(0,r) - \mathfrak{M}_u(0,\varepsilon) = (n-2)\int_\varepsilon^r \frac{\mu_u(t)}{t^{n-1}}\,dt,\ 0 < \varepsilon < r, \tag{2.1.4}$$

for $n > 2$ and

$$\mathfrak{M}_u(0,r) - \mathfrak{M}_u(0,\varepsilon) = \int_\varepsilon^r \frac{\mu_u(t)}{t}\,dt, \qquad 0 < \varepsilon < r,$$

for $n = 2$. Correspondingly:

---

[6] We used this fact in **Chapter 1** for $G = \mathbb{R}^2$.

$$\mu_u(r) \le C(k)r^{n-2}\{M_u(kr) - \mathfrak{M}_u(0,\varepsilon)\} \quad \text{if } n > 2,$$
$$\mu_u(r) \le \frac{1}{\ln k}\{M_u(kr) - \mathfrak{M}_u(0,\varepsilon)\} \qquad \text{if } n = 2. \tag{2.1.5}$$

If the positive measure μ has compact support, then the function $v_\mu(x)$ defined by

$$v_\mu(x) = -\int_{\mathbb{R}^n} \frac{1}{|x-y|^{n-2}}\, d\mu(y)$$

for $n > 2$ and

$$v_\mu(x) = \int_{\mathbb{R}^2} \ln|x-y|\, d\mu(y)$$

for $n = 2$ is called the potential of μ. The potential of a positive measure μ is a subharmonic function, and its Riesz associated measure coincides with μ. Thus, if $u \in SH(G)$, $G' \subset\subset G$, $\mu' = \mu_u|_{G'}$ is the restriction[7)] of μ_u to G', then the function $H(x) = u(x) - v_{\mu'}(x)$ is harmonic in G'.

If the domain is bounded and its boundary is (in some sense) good while either the function $u \in SH(G)$ is continuous in $\bar{G}$ or $u \in SH(G^0)$, where $G^0 \supset\supset G$, then side by side with the representation $u = H + v_{\mu'}$ (the Riesz representation) the function $u(x)$ can be represented as

$$u(x) = -\int_G g(x,y)\,d\mu_u(y) - \int_{\partial G} \frac{\partial g(x,y)}{\partial n_y} u(y)\,dS_y, \quad \forall x \in G, \tag{2.1.6}$$

where $g(x,y)$ is the Green's function of the domain G, $\frac{\partial g}{\partial n}$ is the derivative along the interior normal, and dS is the $(n-1)$-dimensional "area" element of the surface ∂G. In particular, if $n > 2$, $G = B_R$ and $U \in SH(B_{R'})$, $R' > R$, then

$$u(x) = -\int_{B_R} G_R(x,y)\,d\mu_u(y) + \int_{S_R} u(y)P_R(x,y)\,d\sigma(y), \quad \forall x \in B_r, \tag{2.1.7}$$

---

7) The measure μ' equal to μ on E and vanishing on $\mathbb{R}^n \setminus E$ is called the restriction $\mu|_E$ of μ to the Borel set E.

where

$$P_R(x,y) = \frac{R^2-|x|^2}{\sigma_{n-1}R} \frac{1}{|x-y|^n},$$

$$G_R(x,y) = \frac{1}{\sigma_{n-1}|x-y|^{n-2}} - \frac{(R|y|)^{n-2}}{\sigma_{n-1}|x|y|^2-R^2y|^{n-2}}.$$

2. $\mathcal{D}'$-convergence. Now we give adduce some facts concerning the convergence of subharmonic functions in $\mathcal{D}'(G)$.

Lemma 2.1.1. Let the positive measures $\mu_t \in \mathcal{D}'(G)$, $t > 1$, be such that $\exists \mathcal{D}'\text{-}\lim\limits_{t\to\infty} \mu_t \left(\overset{def}{=} \mu\right)$. Then for any domain $G' \subset\subset G$ such that $\mu(\partial G') = 0$ the equality

$$\lim_{t\to\infty} \mu_t(G') = \mu(G')$$

holds and, moreover, as $t \to \infty$ the measures μ_t converge in G' to the measure μ as functionals on the space $C(\bar{G}')$.

A proof of this lemma can be found, for example, in M.Landkof [1], p.21 and L.Ronkin [1], p.30.

Lemma 2.1.2. If the positive measures μ_t, $t > 1$, are such that $supp\mu_t \subset B_R$, $\forall t$, and $\exists \mathcal{D}'\text{-}\lim\limits_{t\to\infty} \mu_t$ $(= \mu)$ then

$$v_\mu = \mathcal{D}'\text{-}\lim_{t\to\infty} v_{\mu_t}.$$

Proof. Let $\varphi \in \mathcal{D}(\mathbb{R}^n)$, $supp\varphi \subset B_r$, and let a function $\alpha \in \mathcal{D}(\mathbb{R}^n)$ be such that $\alpha(x) = 1$, $\forall x \in B_R$. Then

$$\int_{\mathbb{R}^n} \varphi(x) v_{\mu_t}(x)dx = -\int_{B_r} \varphi(x)\left(\int_{B_R} \frac{1}{|x-y|^{n-2}} d\mu_t(y)\right)dx =$$

$$= -\int_{B_R} d\mu_t(y) \int_{B_r} \frac{\varphi(x)}{|x-y|^{n-2}} dx = -\int_{\mathbb{R}^n} \Phi(y)d\mu_t(y),$$

where

$$\Phi(y) = \alpha(y)\int_{\mathbb{R}^n} \frac{\varphi(x)}{|x-y|^{n-2}} dx.$$

It is obvious that $\Phi\in\mathcal{D}(\mathbb{R}^n)$, and since $\mu_t \xrightarrow{\mathcal{D}'} \mu$, we have

$$\lim_{t\to\infty}\int_{\mathbb{R}^n}\varphi(x)v_{\mu_t}(x)dx = -\lim_{t\to\infty}\int_{\mathbb{R}^n}\Phi(y)d\mu_t(y) =$$

$$= -\int_{\mathbb{R}^n}\Phi(y)d\mu(y) = \int_{\mathbb{R}^n}\varphi(x)v_\mu(x)dx.$$

Thus the proof of the lemma is finished.

Lemma 2.1.3. Let $u_j\in SH(G)$, $j = 1, 2, \ldots$ be a sequence of functions which is uniformly bounded above on every compact subset of the domain G. Suppose also that there is a compact subset of G such that $\{u_j\}$ does not uniformly converge on it to $-\infty$ as $j\to\infty$. Then we can choose a subsequence $\{u_{j_k}\}_{k=1}^{\infty}$ from the sequence $\{u_j\}_{j=1}^{\infty}$ such that the u_{j_k} converge in $\mathcal{D}'(G)$ to some function $u\in SH(G)$ as $k\to\infty$.

Proof. Since a family of measures of bounded variation is weakly compact and (as mentioned above) the weak limit of subharmonic functions is a subharmonic function, to prove the lemma it suffices to show that

$$\sup_j\int_{G'}|u_j(x)|dx < \infty,\ \forall G' \subset\subset G,$$

or, equivalently,

$$\sup_j\int_{B_r(a)}|u_j(x)|dx < \infty,\ \forall a,r:\ B_r(a) \subset\subset G.$$

Let us denote by A the set of points $x\in G$ such that $u_j \not\rightrightarrows -\infty$ in a neighbourhood of x. This set is open. Now note that if $B_{r'}(x') \subset B_r(x)$, $v\in SH(B_r(x))$, and $\sup\{v(y): y\in B_r(x)\} = c < \infty$ then

$$V_n r^n(v(x)-c) \le \int_{B_r(x)}(v(y)-c)dy \le \int_{B_{r'}(x')}(v(y)-c)dy.$$

Thus the following inequality is valid in this situation:

$$v(x) \le \left(\frac{r'}{r}\right)^n\cdot\mathfrak{M}_u(x',r') + \left(1-\left(\frac{r'}{r}\right)^n\right)c. \tag{2.1.8}$$

From this it follows that $u_j \not\rightrightarrows -\infty$ in every ball $B_r(x^0) \subset\subset G$ which contains at least one point $x'\in A$. Therefore the set A is closed in G. Hence either $A = G$ or $A = \emptyset$. Since by the condition of the lemma there is a compact subset of G such that $\{u_j\}$ does not converge to $-\infty$ on it, $A = \emptyset$. Thus, arbitrarily close to any point $a\in G$ there is such a point x' such that $\inf_j u_j(x') > -\infty$. In this case, if $r' \geq r + |a-x'|$ and $B_{r'}(x') \subset\subset G$, then

$$\sup_j \int_{B_r(a)} |u_j(x)|dx \leq \sup_j \int_{B_{r'}(x')} |u_j(x)|dx \leq$$

$$\leq V_n r'^n \cdot \sup_j \mathfrak{M}_{|u_j|}(x',r') \leq V_n r'^n \sup_j\{2\mathfrak{M}_{u_j^+}(x',r') - u_j(x')\} < \infty.$$

This finished the proof.

Lemma 2.1.4. If a uniformly bounded above family of functions $u_t\in SH(G)$, $t > 1$, is such that $\exists\mathcal{D}'-\lim_{t\to\infty} u_t = u$[8], then

i) $u = reg \overline{\lim_{t\to\infty}} u_t$;

ii) $u_t(x) \to u(x)$ in $L^1(G')$, $\forall G'\subset\subset G$, as $t \to \infty$;

iii) if we assume in addition that the $u_t(x)$ are harmonic, then they converge uniformly to $u(x)$ on every compact subset of the domain G.

Proof. First of all we note that without loss of generality we may assume that $G = B_R$, $G' = B_r$, $r < R$. Now let $\alpha(x)$ be such a function such that $\alpha\in C^\infty(\mathbb{R}^n)$, $\alpha(x) \geq 0$, $supp\alpha \subset B_1$, $\alpha(x) = \alpha(|x|,0,\dots,0)$, $\int\alpha(x)dx = 1$, and set $\alpha_\varepsilon(x) = \varepsilon^{-n}\alpha\left(\frac{x}{\varepsilon}\right)$.

Put

$$\int_{B_R} u_t(y)\alpha_\varepsilon(x-y)dy = u_t*\alpha_\varepsilon,\ y\in B_r,\ \varepsilon < R-r,$$

and

---

[8] Note that the subharmonic function $u(x) \equiv -\infty$ does not belong to the space of distributions.

$$v(x) = reg\ \overline{lim}_{t \to \infty} u_t(x).$$

Since $u_t \xrightarrow{\mathcal{D}'} u$ and $\alpha_\varepsilon(x-y) \in \mathcal{D}(B_R)$, $\forall y \in B_r$, we have

$$\lim_{t \to \infty} u_t * \alpha_\varepsilon = u * \alpha_\varepsilon.$$

Using **Fatou's Lemma** we conclude that

$$u*\alpha_\varepsilon \le \left(\overline{lim}_{t \to \infty} u_t\right) * \alpha_\varepsilon \le v*\alpha_\varepsilon. \tag{2.1.9}$$

Since u and v are subharmonic functions, $v*\alpha_\varepsilon \downarrow v$, $u*\alpha_\varepsilon \downarrow u$ as $\varepsilon \downarrow 0$. Hence it follows from (2.1.9) that $u(x) \le v(x)$, $\forall x \in B_r$. Now note that the functions $u_t * \alpha_\varepsilon$ are subharmonic and from a family which is uniformly bounded above. Therefore we can apply **Hartogs' Lemma** to show that the inequality

$$\delta + \alpha_\varepsilon * u - u_t \ge 0 \tag{2.1.10}$$

holds in B_r for all $\delta > 0$ and all $t > t_0 = t_0(\delta, r)$. In turn, from (2.1.10) it follows that

$$v(x) \le \delta + (\alpha_\varepsilon * u)(x),\ \forall x \in B_r,\ \varepsilon < R - r.$$

If we let $\delta \to 0$, $\varepsilon \to 0$ in the last inequality, then we find that $v(x) \le u(x)$, $\forall x \in B_R$. Hence $u(x) = v(x)$, $\forall x \in B_R$, and thus (i) is proved.

To prove (ii) we estimate $|u - u_t|$ using (2.1.10). We have

$$|u - u_t| \le |u - u*\alpha_\varepsilon - \delta| + |u*\alpha_\varepsilon + \delta - u_t| \le$$

$$\le 2(u*\alpha_\varepsilon + \delta - u_t),\ \forall x \in B_r, t > t_0(\delta, r).$$

Taking into account this estimate we conclude that $\forall \chi \in D(B_r)$, $\chi \ge 0$,

$$\overline{lim}_{t \to \infty} \int_{B_r} |u-u_t| \chi dx \le 2 \int_{B_r} (u*\alpha_\varepsilon + \delta - u)\chi dx.$$

Since the numbers r, δ are arbitrary and $(\alpha_\varepsilon * u) \downarrow u$ as $\varepsilon \to 0$, we have

$$\lim_{t \to \infty} \int_{B_r} |u-u_t| dx = 0. \tag{2.1.11}$$

Thus (ii) is proved.

In the case when the u_t are harmonic we have $v_t(x) = \mathfrak{N}(x,\varepsilon)$, where $v_t = u_t - u$. Hence, for $x \in B_r$, $\varepsilon + r = r' < R$,

$$|v_t(x)| \le \mathfrak{N}_{|v_t|}(x,\varepsilon) \le \frac{1}{V_n \varepsilon^n} \int_{B_{r'}} |v_t| dx.$$

To complete the proof of the uniform convergence of u_t to u it suffices to refer to (2.1.11).

Thus the proof of the lemma is finished.

Lemma 2.1.5. Let a family of functions $u_t \in SH(B_R)$ which is uniformly bounded above satisfy the condition $\exists \mathcal{D}'\text{-}\lim_{t\to\infty} u_t = u$. Then $\forall r \in (0,R)$,

$$\lim_{t\to\infty} \int_{S_r} \varphi u_t d\sigma = \int_{S_r} \varphi u d\sigma, \qquad \forall \varphi \in C(S_r).$$

Proof. Take r' in (r,R) such that $\mu_u(S_{r'}) = 0$ and consider the Riesz representation in the disc $B_{r'}$ for the functions u_t and u. Namely,

$$u = v_{\mu'} + H,$$
$$u_t = v_{\mu'_t} + H_t, \tag{2.1.12}$$

where μ' and μ'_t are the restrictions of the measures μ and μ_t to $B_{r'}$, and H and H_t are the corresponding harmonic functions. Then

$$\int_{S_r} \varphi u_t d\sigma = \int_{S_r} v_{\mu'_t} \varphi d\sigma + \int_{S_r} H_t \varphi d\sigma = \int_{B_{r'}} F(y) d\mu'_t(y) + \int_{S_r} H_t \varphi d\sigma, \tag{2.1.13}$$

where

$$F(y) = \int_{S_r} \frac{\varphi(x)}{|x-y|^{n-2}} d\sigma(x) \quad \text{when } n > 2,$$

and

$$F(y) = \int_{S_r} \varphi(x) \ln|x-y| d\sigma \quad \text{when } n = 2.$$

Since $\mu_u(S_{r'}) = 0$ and $\mu_{u_t} \xrightarrow{\mathcal{D}'} \mu$, then the measures μ'_t converge as functionals on the space $C(\bar{B}_{r'})$ to the measure μ' as $t \to \infty$ (see **Lemma 2.1.1**). From here, taking into account the readily proved continuity of the function, F(y) we conclude that

$$\lim_{t \to \infty} \int_{B_{r'}} F(y)d\mu'_t(y) = \int_{B_{r'}} F(y)d\mu'(y) = \int_{S_r} \varphi v_{\mu'} d\sigma. \qquad (2.1.14)$$

Further, since $u = \mathcal{D}'\text{-}\lim_{t \to \infty} u_t$ and in accordance with **Lemma 2.1.2** $v_{\mu'} = \mathcal{D}'\text{-}\lim_{t \to \infty} v_{\mu'_t}$, it follows from **Lemma 2.1.4** that

$$\lim_{t \to \infty} \int_{B_{r'}} |u_t - u| dx = 0,$$

$$\lim_{t \to \infty} \int_{B_{r'}} |v_{\mu'_t} - v_\mu| dx = 0.$$

Hence

$$\lim_{t \to \infty} \int_{B_{r'}} |H - H_t| dx = 0.$$

Therefore (see **Lemma 2.1.4**, (iii)) the functions H_t converge uniformly to the function H in the ball $\bar{B}_r$ as $t \to \infty$. Thus

$$\lim_{t \to \infty} \int_{S_r} H_t \varphi d\sigma = \int_{S_r} H \varphi d\sigma.$$

From here and (2.1.12)-(2.1.14) it follows that

$$\lim_{t \to \infty} \int_{S_r} u_t \varphi d\sigma = \int_{S_r} u \varphi d\sigma.$$

This finishes the proof.

To formulate and prove **Lemma 2.1.6** we need the concepts of Carleson measure and $(\varepsilon,\alpha,\alpha')$-normal point.

Let $E \subset \mathbb{R}^n$ and let the balls $B_{r_j}(x_j)$ be a cover of E. The Carleson α-measure of E is

$$mes_C^\alpha(E) = \inf \sum_j r_j^\alpha ,$$

where *inf* is taken over all covers $\{B_{r_j}(x_j)\}$ of E.

We say that a family of functions u_t, $t > 1$, defined in a domain $G \subset \mathbb{R}^n$, converges to the function u with respect to the Carleson α-measure as $t \to \infty$ if for any $\varepsilon > 0$ the sets

$$E_{t,\varepsilon} = \{x \in G: |u_t(x)-u(x)| > \varepsilon\}$$

satisfy the condition

$$\lim_{t \to \infty} mes_C^{\alpha}(E_{t,\varepsilon}) = 0.$$

Let μ be a positive measure in $\mathbb{R}^n$ with compact support and let $\alpha > 0$, $\alpha' > 0$. Then we set $\mu(B_s(x)) = \mu(s;x)$. A point $x \in \mathbb{R}^n$ is called a $(\varepsilon,\alpha,\alpha')$-normal point of μ if the inequality $\mu(s;x) < \varepsilon^{-\alpha'} s^{\alpha}$ is valid for all $s \le \varepsilon$. We denote by $\tilde{E}(\varepsilon,\alpha',\alpha,\mu)$ the set of all points x which are not $(\varepsilon,\alpha',\alpha)$-normal points of μ.

Lemma 2.1.6. Let a uniformly bounded above family of functions u_t, $t > 1$, be such that $\exists \mathcal{D}'\text{-}lim u_t = u$. Then the functions $u_t(x)$ converge to $u(x)$ with respect to the Carleson α-measure for any α in the interval $(n-2,n]$ as $t \to \infty$.

Proof. First we note that without loss of generality we may assume $G = B_R$, $G' = B_r$. Further, under the condition of the lemma the functions H_t in the Riesz representation $u_t = v_{\mu'_t} + H_t$ in the ball $B_{r'}$, $r < r' < R$, converge uniformly to a function H (see the proof of **Lemma 2.1.5**). Thus it is sufficient to consider only the case $u_t = v_{\mu_t}$, $u = v_{\mu}$, where the measures μ_t are positive, $supp\mu_t \subset \bar{B}_{r'}$, $\mu(S_{r'}) = 0$, $\mathcal{D}'\text{-}lim\, \mu_t = \mu$. We estimate $|u_t(x) - u(x)|$ in some point $x \in B_r \setminus \hat{E}_t$, where $\hat{E}_t = \tilde{E}(\varepsilon,\alpha',\alpha,\mu) \cup \tilde{E}(\varepsilon,\alpha',\alpha,\mu_t)$, $\varepsilon + r < r'$, $0 < \alpha' < \alpha-n+2$. Here we assume that $n > 2$. The case $n = 2$ is similar and we omit it. Taking into account the above said we have

$$|u_t(x)-u(x)| = |v_{\mu_t}(x)-v_{\mu}(x)| \le$$

$$\le \left| \int_{B_{r'}} min\left\{\frac{1}{\varepsilon^{n-2}}, \frac{1}{|x-y|^{n-2}}\right\} d\mu_t(y) - \int_{B_{r'}} min\left\{\frac{1}{\varepsilon^{n-2}}, \frac{1}{|x-y|^{n-2}}\right\} d\mu(y) \right| +$$

$$+ \left| \int_{B_\varepsilon(x)} \frac{1}{|x-y|^{n-2}} d\mu_t(y) - \int_{B_\varepsilon(x)} \frac{1}{|x-y|^{n-2}} d\mu(y) \right|$$

$$+ \frac{1}{\varepsilon^{n-2}}\left| \int_{B_\varepsilon(x)} d\mu_t(y) - \int_{B_\varepsilon(x)} d\mu(y) \right| =$$

$$= J_{1,t} + J_{2,t} + J_{3,t}.$$

Since $\mu_t \xrightarrow{\mathcal{D}'} \mu$ as $t \to \infty$ and $\mu(S_{r'}) = 0$, we have $\mu_t \to \mu$ in the sense of weak convergence of functionals on the space $C(\bar{B}_{r'})$. Therefore $J_{1,t} \to 0$ as $t \to \infty$. Moreover, using the **Banach-Steinhaus Theorem** it is easy to show that this convergence is uniform on $\bar{B}_r$. Thus, for any $\delta > 0$ and $\varepsilon < r'-r$, $\varepsilon > 0$, there is a number $t_0 = t_0(\delta,\varepsilon)$ such that $J_{1,t}(x) < \delta$, $\forall t > t_0$, $x \in B_r$.

Now we estimate $J_{2,t}$. Taking into account that the point x is assumed to be $(\varepsilon,\alpha',\alpha)$-normal for the measures μ_t and μ, we have

$$J_{2,t}(x) = \left| \int_0^\varepsilon s^{-n+2} d\mu_t(s;x) - \int_0^\varepsilon s^{-n+2} d\mu(s;x) \right| \le$$

$$\le \varepsilon^{-n+2}\mu_t(\varepsilon;x) + (n-2)\int_0^\varepsilon s^{-n+1}\mu_t(s;x)ds +$$

$$+ \varepsilon^{-n+2}\mu(\varepsilon;x) + (n-2)\int_0^\varepsilon s^{-n+1}\mu(s;x)ds \le$$

$$\le 2\left(1 + \frac{n-2}{\alpha-n+2}\right)\varepsilon^{\alpha-\alpha'+n-2}.$$

It is also obvious that

$$J_{3,t}(x) \le 2\varepsilon^{\alpha-\alpha'+n-2}, \quad \forall x \in B_r \setminus \hat{E}_t.$$

Combining these estimates we conclude that for $t > t_0 = t_0(\delta,\varepsilon)$ and $x \in B_r \setminus E_t$

$$|u_t(x) - u(x)| \le \varepsilon^{\alpha-\alpha'-n+2}\left(4 + \frac{2(n-2)}{\alpha-n+2}\right) + \delta = \varepsilon'.$$

Thus

$$(B_r \cap E_{t,\varepsilon'}) \subset (B_r \cap \hat{E}_t), \quad \forall t > t_0.$$

Note that a given $\delta > 0$ it is possible to find $\varepsilon_0 > 0$ such that $\varepsilon' < 2\delta$, $\forall\varepsilon < \varepsilon_0$. Hence the following inclusion is valid for the sets $E_{t,\delta} = \{x\in G: |u_t(x) - u(x)| > \delta\}$, whose behaviour determines the convergence with respect to the measure:

$$(B_r \cap E_{t,2\delta}) \subset (E_t \cap B_r), \ \forall t > t_0, \ \varepsilon < \varepsilon_0. \tag{2.1.15}$$

Now we estimate the Carleson measure of $\hat{E}_t \cap B_r$.

For brevity we set $\tilde{E}(\varepsilon,\alpha',\alpha,\mu_t) = \tilde{E}_{\mu_t}$. It is clear that

$$mes_C^{\alpha}(\hat{E}_t \cap B_r) \le mes_C^{\alpha}(\tilde{E}_{\mu_t} \cap B_r) + mes_C^{\alpha}(\tilde{E}_{\mu} \cap B_r).$$

Further, according to the definition of the sets $\tilde{E}$, for every $x\in(\tilde{E}_{\mu_t} \cap B_r)$ there is a $\delta(x) \le \varepsilon$ such that $\mu_t(B_{\delta(x)}(x)) > \varepsilon^{-\alpha}(\delta(x))^{\alpha}$. Let us choose a countable system of balls $B^{(j)} = B_{s_j}(x_j)$, $x_j\in(\tilde{E}_{\mu_t} \cap B_r)$, $s_j = s_j(x)$ covering $\tilde{E}_{\mu_t} \cap B_r$ with finite multiplicity. This is possible by **Besicovitch's Theorem** on covers of finite multiplicity, and the multiplicity of the cover is bounded above by some constant $d(n)$ which depends only on the dimension of the space. Thus we have

$$mes_C^{\alpha}(\tilde{E}_{\mu_t} \cap B_r) \le \sum_{j=1}^{\infty} s_j^{\alpha} \le \varepsilon^{\alpha} \sum_{j=1}^{\infty} \mu_t(B^{(j)}) \le d(n)\varepsilon^{\alpha'} \mu_t(\bar{B}_{r'}).$$

From the assumptions on the measures μ_t it follows that $\lim_{t\to\infty} \mu_t(\bar{B}_{r'}) = \mu(\bar{B}_{r'})$ and therefore for all t larger than some t_1 the inequality $\mu_t(\bar{B}_{r'}) \le 2\mu(\bar{B}_{r'})$ is valid. Hence

$$E \subset \bigcup_j B_{r_j}(x_j) \subset B_{R'},$$

and

$$\sum r_j^n \le c_1 mes_C^n E,$$

where c_1 is a constant which depends only on R' and n. We set

$$D = \bigcup_j B_{r_j}(x_j).$$

It is obvious that

$$mes_n D \le \sum_j mes_n B_{r_j}(x_j) \le V_n c_1^n mes_C E \qquad (2.1.16)$$

(mes_n is Lebesgue measure in $\mathbb{R}^n$).

We $\int_D |v|dx$. According to (2.1.7) we can represent the function v in the ball $B_{R''}$ as follows:

$$v(x) = \frac{1}{\sigma_{n-1}} \int_{S_{R''}} v(y) \frac{R''-|x|^2}{R''|x-y|^n} d\sigma(y) +$$

$$-\frac{1}{\sigma_{n-1}} \int_{B_{R''}} \left\{ \frac{1}{|x-y|^{n-2}} - \left(\frac{R''|y|}{|x|y|^2-R''y|} \right)^{n-2} \right\} d\mu_v(y) = J_1 + J_2.$$

If $|x| < R'$, then

$$|J_1(x)| \le \mathfrak{M}_{|v|}(0,R'')\cdot(R'')^{n-1}\cdot \max_{\substack{|x|\le R' \\ |y|=R''}} \frac{R''^2-|x|^2}{R''|x-y|^n} \le$$

$$\le \{2\mathfrak{M}_{v^+}(0,R'') - v(0)\}\left(\frac{R''}{R''-R'}\right)^n \le (2|M| + |K|)\left(\frac{R''}{R''-R'}\right)^n = c_2.$$

Hence

$$\int_D |J_1|dx \le c_2 mes_n D. \qquad (2.1.17)$$

To estimate J_2 we first of all note that if $|x| \le R'$, $|y| \le R''$, then the inequality

$$\frac{R''|y|}{|x|y|^2-R''y|} \le \frac{R''}{R''-R'} = c_3$$

is valid. Therefore

$$|J_2(x)| \le \int_{B_{R''}} \frac{1}{|x-y|^{n-2}} d\mu_v(y) + c_3\mu_v(B_{R''}), \; x\in B_{R'}$$

and thus

$$\int_D |J_2(x)|dx \le \int_{B_{R''}} d\mu_v(y)\int_D \frac{dx}{|x-y|^{n-2}} + c_3\mu_v(B_{R''})mes_n D. \tag{2.1.18}$$

Now we estimate $\mu_v(B_{R'})$. By (2.1.3),

$$\mu_v(B_{R'}) \le ((R')^{2-n} - (R'')^{2-n})^{-1}(M_v(R'') - v(0)) \le$$

$$\le ((R')^{2-n} - (R'')^{2-n})^{-1}(|M| + |K|) = c_4. \tag{2.1.19}$$

To estimate $\int_D |x-y|^{2-n}dx$ we use the $\mathbb{R}^n$-variant, $n > 2$, of **Lemma 7.2**, in A.Gol'dberg, I.Ostrovskiĭ [1], Chapter 1. This lemma states that for any function $\lambda(t)$ monotone nonincreasing on $(0,R)$ and any measurable set $A \subset B_r$ the following inequality is valid:

$$\int_A \lambda(|x|)dx \le \int_{B_r} \lambda(|x|)dx,$$

where r is defined by the condition $mes_n A = mes_n B_r$ [9]. Thus we have

---

[9] This fact may be proved as follows. Put $\sigma_A(t) = \int_{S_t \cap A} d\sigma$ and $\sigma(t) = \sigma_{n-1}t^{n-1}$. Then $mes A = \int_0^\infty \sigma_A(t)dt = \int_0^r \sigma(t)dt$, and therefore

$$\int_0^r(\sigma_A(t)-\sigma(t))dt + \int_r^\infty \sigma_A(t)dt = 0.$$

This and the fact that $\sigma_A(t) \le \sigma(t)$, $\forall t$, give

$$\int_A\lambda(|x|)dx = \int_0^\infty\lambda(t)\sigma_A(t)dt =$$

$$= \int_0^r\lambda(t)\sigma(t)dt + \int_0^r\lambda(t)(\sigma_A(t)-\sigma(t))dt + \int_r^\infty\lambda(t)\sigma_A(t)dt \le$$

$$\le \int_A\lambda(|x|)dx + \lambda(r)\int_0^r(\sigma_A(t)-\sigma(t))dt + \lambda(r)\int_r^\infty\sigma_A(t)dt = \int_{B_r} \lambda(|x|)dx.$$

$$\int_D |x-y|^{2-n}dx \le \int_{B_r} |x|^{2-n}dx =$$

$$= \sigma_{n-1}\left(\frac{1}{V_n} mes_n D\right)^{2/n} = c_5(mes_n D)^{2/n}.$$

This and (2.1.18), (2.1.19) imply that

$$mes_C^{\alpha}(\hat{E}_{\mu_t} \cap B_r) \le 2d(n)\varepsilon^{\alpha'}\mu(\bar{B}_{r'}), \ \forall t > t_1.$$

Quite similarly,

$$mes_C^{\alpha}(\tilde{E}_{\mu} \cap B_r) \le d(n)\varepsilon^{\alpha'}\mu(\bar{B}_{r'}),$$

and therefore

$$mes_C^{\alpha}(\hat{E}_t \cap B_r) \le 3d(n)\varepsilon^{\alpha'}\mu(\bar{B}_{r'}) = a\varepsilon^{\alpha'}.$$

This taking into account (2.1.15), gives

$$mes_C^{\alpha}(E_{t,2\delta} \cap B_r) < a\varepsilon^{\alpha'},$$

$\forall t > max(t_1,t_0)$, $0 < \varepsilon < \varepsilon'$, $0 < \alpha' < \alpha-n+2$.

Hence

$$\overline{lim}_{t\to\infty} \ mes_C^{\alpha}(E_{t,\delta} \cap B_r) = 0, \ \forall\delta > 0, \ n-2 < \alpha < n.$$

This finishes the proof of the lemma.

To prove the statement converse to **Lemma 2.1.6** we need the following lemma.

Lemma 2.1.7. Let $v(x) \in SH(B_R)$, $R > 1$; $v(x) \le M < \infty$, $\forall x \in B_R$; $v(0) \ge K > -\infty$. Then there is constant c depending only on n, R, M, and K such that for any measurable set $E \subset B_1$ the inequality

$$\int_E |v(x)|dx \le c(mes_C^n E)^{2/n}$$

holds.

Proof. Let E be a measurable set in B_1 and let $1 < R' < R'' < R$. The definition of Carleson α-measure easily implies that we can choose sequences of points x_j and numbers r_j such that

$$\int_D |J_2|dx \le c_4c_5(mes_n D)^{2/n} + c_4c_3 mes_n D.$$

Combining the estimates obtained we conclude

$$\int_E |v|dx \le \int_D |v|dx \le \int_D (|J_1| + |J_2|)dx \le$$

$$\le c_2 mes_n D + c_4c_3 mes_n D + c_4c_5(mes_n D)^{2/n} \le$$

$$\le c_6 mes_C^n E + c_7(mes_n E)^{2/n} \le (c_6 + c_7)(mes_C^n)^{2/n}.$$

The proof is finished.

The following lemma is the converse of **Lemma 2.1.6.**

Lemma 2.1.8. Let $\{u_t\}_{t>1}$ be a uniformly bounded above family of subharmonic functions in a domain G. Suppose the functions $u_t(x)$ converge with respect to the Carleson α-measure to $u \in L^1_{loc}(G)$ for some $\alpha \in (n-2, n]$. Then the equality $u(x) = \lim_{t\to\infty} u_t(x)$ holds in $L^1_{loc}(G)$. In particular, this implies that $u = \mathcal{D}'\text{-}\lim_{t\to\infty} u_t(x)$, and $u(x)$ is a subharmonic function up to a summand which is equivalent to zero in $L^1_{loc}(G)$.

Proof. Without loss of generality we may assume that $G = B_{1+2\delta}$, $\delta > 0$. Taking this into account, we put

$$E'_{t,\varepsilon} = \{x \in B_1 : |u_t(x) - u(x)| > \varepsilon\}$$

Since by the assumption of the lemma, for any $\varepsilon > 0$,

$$\lim_{t\to\infty} mes_C^{\alpha}(E'_{t,\varepsilon}) = 0,$$

we find that for all t larger than some t_0 the set $B_\delta \setminus E'_{t,\varepsilon} \ne \emptyset$. Since $u \in L^1_{loc}(B_{1+2\delta})$, there is $K > -\infty$ such that

$$sup\{u_t(x) : x \in B_\delta \setminus E'_{t,\varepsilon}\} > K, \ \forall t > t_0.$$

Thus, every function $u_t(x)$ satisfies the conditions of **Lemma 2.1.7** with K independent of t in some ball $B_{1+\delta}(x_t)$, where $x_t \in B_\delta \setminus E'_{t,\varepsilon}$. Applying this lemma we find that

$$\int_{E'_{t,\varepsilon}} |u_t(x)|dx \le c(mes_C^{\alpha} E'_{t,\varepsilon})^{2/n}, \ \forall t > t_0. \tag{2.1.20}$$

We estimate $\|u_t - u\|_{L^1(B_1)}$, using the last inequality. We have

$$\|u_t - u\|_{L^1(B_1)} = \int_{B_1} |u_t(x) - u(x)|dx =$$

$$= \int_{B_1 \setminus E'_{t,\varepsilon}} |u_t(x) - u(x)|dx + \int_{E'_{t,\varepsilon}} |u_t(x) - u(x)|dx \le$$

$$\le \varepsilon V_n + \int_{E'_{t,\varepsilon}} |u_t(x)|dx + \int_{E'_{t,\varepsilon}} |u(x)|dx \le$$

$$\le \varepsilon V_n + c(mes_C^n E'_{t,\varepsilon})^{2/n} + \int_{E'_{t,\varepsilon}} |u(x)|dx.$$

It is clear that if $\lim_{t\to\infty} mes_C^{\alpha} E'_{t,\varepsilon} = 0$ for some α, then $\lim_{t\to\infty} mes_C^{\beta} E'_{t,\varepsilon} = 0$ for every $\beta \ge \alpha$. Moreover, it is obvious that if $\lim_{t\to\infty} mes_n E'_{t,\varepsilon} = 0$, then $\lim_{t\to\infty} mes_C^n E'_{t,\varepsilon} = 0$. Taking all this into account as well as inequality (2.1.20), we conclude that $\|u_t - u\|_{L^1(B_1)} \longrightarrow 0$ as $t \to 0$. It is clear that the same is true if we replace the ball B_1 by an arbitrary ball B_r, $r < 1 + \delta$.

The proof of the lemma is finished.

3. Derivatives of subharmonic functions. Generally speaking. a subharmonic function is not differentiable. But as a locally summable function it has derivatives in the distributional sense. Let us prove some properties of such derivatives.

Lemma 2.1.9. Let $u \in SH(G)$. Then the distributions $\frac{\partial u}{\partial x_i}$, $i = 1, \ldots, n$, as elements of $\mathcal{D}'(G)$, are locally summable functions.

Proof. We assume that $n > 2$. The case $n = 2$ is similar and we omit it.

It follows from the Riesz representation of a subharmonic function that the proof may be restricted to the case $u = v_\mu$, where μ is a measure with compact support. Without loss of generality we may assume that $supp\mu \subset B_r$. In this case we us set

$$\Phi_1(x) = (n-2)\int_{\mathbb{R}^n} \frac{x_1-y_1}{|x-y|^n}\, d\mu(y).$$

Since

$$\int_{B_R} dx \int_{\mathbb{R}^n} \frac{|x_1-y_1|}{|x-y|^n}\, d\mu(y) = \int_{\mathbb{R}^n} d\mu(y) \int_{B_R} \frac{|x_1-y_1|}{|x-y|^n}\, dx \le$$

$$\le \int_{\mathbb{R}^n} d\mu(y) \int_{B_{r+R}} \frac{|x_1-y_1|}{|x-y|^n}\, dx \le \mu(\mathbb{R}^n)(R+r)\sigma_{n-1} < \infty,\ \forall R, \qquad (2.1.21)$$

the integral defining the function $\Phi_1(x)$ converges almost everywhere and $\Phi_1(x)$ is a locally summable function. We show that, as an element of $\mathcal{D}'(\mathbb{R}^n)$, $\Phi_1(x)$ satisfies the condition $\frac{\partial u}{\partial x_1} = \Phi_1$. Note that if $'x = (x_2,\dots,x_n) \ne (y_2,\dots,y_n)$, then

$$(n-2)\int_{-\infty}^{\infty} \varphi(x) \frac{x_1-y_1}{|x-y|^n}\, dx_1 = \int_{\infty}^{\infty} \frac{\partial \varphi}{\partial x_1} \frac{dx_1}{|x-y|^{n-2}},\ \forall \varphi \in D(\mathbb{R}^n),$$

and therefore

$$(n-2)\int_{\mathbb{R}^n} \varphi(x) \frac{x_1-y_1}{|x-y|^n}\, dx = \int_{\mathbb{R}^n} \frac{\partial \varphi}{\partial x_1} \frac{dx}{|x-y|^{n-2}},\ \forall \varphi \in D(\mathbb{R}^n).$$

From this taking into account (2.1.21), we conclude that

$$\int_{\mathbb{R}^n} v_\mu \frac{\partial \varphi}{\partial x_1}\, dx = -\int_{\mathbb{R}^n} d\mu(y) \int_{\mathbb{R}^n} \frac{1}{|x-y|^{n-2}} \frac{\partial \varphi}{\partial x_1}\, dx =$$

$$= -(n-2)\int_{\mathbb{R}^n} d\mu(y) \int_{\mathbb{R}^n} \varphi(x) \frac{x_1-y_1}{|x-y|^n}\, dx = -\int_{\mathbb{R}^n} \varphi(x)\Phi_1(x)dx.$$

According to the definition of distributional derivative this equality means that $\frac{\partial v_\mu}{\partial x_1} = \Phi_1$.

Thus the proof is finished.

Remark. We could take functions $\varphi\in C^1(\mathbb{R}^n)$, *supp*$\varphi \subset\subset \mathbb{R}^n$, instead of functions $\varphi\in\mathcal{D}(\mathbb{R}^n)$ in the proof. Then we would get obtain equality

$$\int_{\mathbb{R}^n} u\,\frac{\partial\varphi}{\partial x_1}\,dx = \int_{\mathbb{R}^n} \Phi_1\varphi dx,\ \forall\varphi\in C^1(\mathbb{R}^n),\ supp\varphi \subset\subset \mathbb{R}^n.$$

Lemma 2.1.10. Let $u = \lim_{j\to\infty} u_j$, where $u_j\in(SH(B_R)\cap C^\infty(B_R))$, $u_{j+1} \le u_j$, $j = 1, 2, \dots$. Then for any $r\in(0,R)$ satisfying the condition $\mu_u(S_r) = 0$ and any function $\varphi\in C^2(S_r)$, the limit

$$\lim_{j\to\infty} \int_{S_r} \frac{\partial u_j}{\partial n}\varphi d\sigma \stackrel{\text{def}}{=} \int_{S_r} \frac{\partial u}{\partial n}\varphi d\sigma$$

exists and does not depend on the sequence u_j. Here $\frac{\partial u}{\partial n}$ is the interior normal derivative.

Proof. Let $\varphi\in C^2(S_r)$ and $\alpha(t)\in C^\infty([0.r])$, *supp*$\alpha \subset [r/2,r]$, $\alpha'(r) = 0$, $\alpha(r) = 1$. Setting $\psi(x) = \alpha(|x|)\varphi\left(\frac{x}{|x|}r\right)$, we have

$$\int_{B_r} \psi\Delta u_j dx - \int_{B_r} u_j\Delta\psi dx = - \int_{S_r} \varphi\,\frac{\partial u_j}{\partial n}d\sigma. \tag{2.1.22}$$

Since the u_j converge monotonically to u, we have

$$\lim_{j\to\infty}\int_{B_r} u_j\Delta\psi dx = \int_{B_r} u\Delta\psi dx,$$

and since the Δu_j converge to the measure $\Delta u = \theta_n\mu_u$ in $\mathcal{D}'(B_R)$ the equality

$$\lim_{j\to\infty}\int_{B_r} \psi\Delta u_j dx = \theta_n\int_{B_r} \psi d\mu_u$$

holds for any r such that $\mu_u(S_r) = 0$. From this and (2.1.22) follows the existence of the require limit follows. Moreover, the equality is true:

$$\lim_{j\to\infty}\int_{S_r} \frac{\partial u_j}{\partial n}\varphi d\sigma = \int_{B_r} u\Delta\psi dx - \theta_n \int_{B_r} \psi d\mu_u, \tag{2.1.23}$$

where the right-hand side does not depend on the sequence u_j.

Thus the proof of the lemma is finished.

Lemma 2.1.11. Let the functions u_j, u be as in **Lemma 2.1.8**. Let us set

$$\Phi(x) = -\sum_{j=1}^{n} \frac{x_j}{|x|} \Phi_j(x),$$

where Φ_j, $j = 1, \dots, n$, are locally summable functions defining $\frac{\partial u}{\partial x_j}$ as elements of $\mathcal{D}'(B_r)$. Then for any function $\varphi \in C^2(B_R)$ and almost all $r \in (0,R)$ the equality

$$\lim_{j\to\infty} \int_{S_r} \varphi \frac{\partial u_j}{\partial n} d\sigma = \int_{S_r} \varphi \Phi d\sigma$$

holds.

Proof. From the definitions of the functions Φ and u_j it follows that

$$\mathcal{D}' - \lim_{j\to\infty} \frac{\partial u_j}{\partial |x|} = -\Phi.$$

Moreover, according to the **Remark** following **Lemma 2.1.9** the functions $\frac{\partial u_j}{\partial |x|}$ converge to $-\Phi(x)$ also as functionals on the space $C_0^1(\mathbb{R}^n)$. Thus, if $\alpha(t) \in \mathcal{D}((0,r))$, $\varphi \in C^1(B_R)$, then

$$\lim_{j\to\infty} \int_{B_r} \alpha(|x|) \frac{\partial u_j}{\partial |x|} \varphi(x) dx = -\int_{B_r} \Phi(x)\varphi(x)\alpha(|x|)dx$$

or, equivalently,

$$\lim_{j\to\infty} \int_0^R \alpha(r)dr \int_{S_r} \frac{\partial u_j}{\partial n} \varphi d\sigma = \int_0^R \alpha(r)dr \int_{S_r} \Phi\varphi d\sigma. \tag{2.1.24}$$

Note that by the previous lemma, for all $r \in (0,\infty)$, outside an at most countable set the limit

$$\lim_{j\to\infty} \int_{S_r} \frac{\partial u_j}{\partial n} \varphi d\sigma$$

exists. By (2.1.24) it is easy to show that

$$\sup_{j,r}\left|\int_{S_r}\frac{\partial u_j}{\partial n}\varphi d\sigma\right| < \infty.$$

Therefore limit transition under the integral sign in (2.1.24) is possible. Hence

$$\int_0^R \alpha(r)\left\{\lim_{j\to\infty}\int_{S_r}\frac{\partial u_j}{\partial n}\varphi d\sigma\right\}dr = \int_0^R \alpha(r)dr\int_{S_r}\Phi\varphi d\sigma.$$

Since $\alpha(t)$ is arbitrary, we obtain that for almost all $r\in(0,R)$,

$$\lim_{j\to\infty}\int_{S_r}\frac{\partial u_j}{\partial n}\varphi d\sigma = \int_{S_r}\Phi\varphi d\sigma.$$

This finishes the proof.

Lemma 2.1.12. Let $u\in SH(B_R)$ and let r, r' be numbers such that $0 < r' < r < R$ and $\mu_u(S_{r'}) = \mu_u(S_r) = 0$. Then

$$\theta_n\int_{B_r}\varphi d\mu_u - \int_{B_r}u\Delta\varphi dx = \int_{S_r}u\,\frac{\partial\varphi}{\partial n}d\sigma - \int_{S_r}\varphi\,\frac{\partial u}{\partial n}d\sigma,\ \forall\varphi\in C^2(\bar{B}_r), \tag{2.1.25}$$

and

$$\theta_n\int_{B_r\setminus B_{r'}}\varphi d\mu_u - \int_{B_r\setminus B_{r'}}u\Delta\varphi dx =$$

$$= \int_{\partial(B_r\setminus B_{r'})}u\,\frac{\partial\varphi}{\partial n}d\sigma - \int_{\partial(B_r\setminus B_{r'})}\varphi\,\frac{\partial u}{\partial n}d\sigma,\ \forall\varphi\in C^2(\bar{B}_r\setminus B_{r'}). \tag{2.1.26}$$

Proof. It suffices to write the corresponding Green's formula for the functions u_j and φ, where u_j are as in **Lemma 2.1.10**, and to take the limit as $j\to\infty$.

Lemma 2.1.13. Let a uniformly bounded family $u_t\in SH(B_R)$ satisfy the condition : $\exists\mathcal{D}'-\lim_{t\to\infty}u_t = u$. If $\mu_u(S_r) = \mu_{u_t}(S_r) = 0$, $0 < r < R$, $\forall t > t_0$; then

$$\lim_{t\to\infty}\int_{S_r}\varphi\,\frac{\partial u_t}{\partial n}d\sigma = \int_{S_r}\varphi\,\frac{\partial u}{\partial n}d\sigma,\ \forall\varphi\in C^2(S_r).$$

Proof. Let us denote by $\Phi(x)$ the solution of the **Dirichlet Problem** in the ball B_r with the boundary condition $\Phi(x) = \varphi(x)$, $x \in S_r$. Then by (2.1.25) we conclude that

$$\int_{S_r} \varphi \frac{\partial u_t}{\partial n} d\sigma = \int_{S_r} u \cdot \frac{\partial \Phi}{\partial n} d\sigma - \theta_n \int_{B_r} \Phi d\mu_u .$$

In this equality we take the limit as $t \to \infty$. This is possible by **Lemma 2.1.10** and **Lemma 2.1.1**. Then we obtain

$$\lim_{t \to \infty} \int_{S_r} \varphi \frac{\partial u_t}{\partial n} d\sigma = \int_{S_r} u_t \frac{\partial \varphi}{\partial n} d\sigma - \theta_n \int_{B_r} \varphi d\mu_{u_t} .$$

From the using (2.1.25) once more, we conclude that

$$\lim_{t \to \infty} \int_{S_r} \varphi \frac{\partial u_t}{\partial n} d\sigma = \int_{S_r} \varphi \frac{\partial u}{\partial n} d\sigma .$$

This finishes the proof.

4. Functions of finite order. Let $u \in SH(\mathbb{R}^n)$. Recall that in §2.1 we introduced the notation

$$M_u(r) = \sup_{x \in B_r} u(x).$$

Note that by the maximum principle

$$M_u(r) = \sup_{x \in S_r} u(x).$$

The order of growth, or simply the order, of the function $u(x)$ is the value ρ_u equal to the order of growth of $M_u(r)$. Thus

$$\rho_u = \rho(M_u(r)) = \overline{\lim_{r \to \infty}} \frac{\ln^+ M_u(r)}{\ln r} .$$

The type σ_u of the function $u(x)$ with respect to the order $\rho > 0$ is defined similarly. Namely,

$$\sigma_u = \sigma(M_u(r), \rho) = \overline{\lim_{r \to \infty}} \frac{M_u(r)}{r^\rho} .$$

It is also obvious in what way the concepts of functions of

minimal, normal and maximal type must be defined. We denote by $SH(\mathbb{R}^n,\rho]$ the set of all functions $u\in SH(\mathbb{R}^n)$ that have at most normal type with respect to the order $\rho > 0$. Thus

$$u\in SH(\mathbb{R}^n,\rho] \iff (\exists\, a < \infty,\ b < \infty:\ u(x) \le a|x|^\rho + b,\ \forall x\in\mathbb{R}^n).$$

In the sequel we shall often use the family of functions $u^{[t]}(x)$, $t > 1$, constructed from a function $u\in SH(\mathbb{R}^n)$ as follows:

$$u^{[t]}(x) = \frac{u(tx)}{t^\rho}.$$

It is obvious that

$$u\in SH(\mathbb{R}^n,\rho] \iff \sup_{x\in B_1,\, t>1} u^{[t]}(x) < \infty.$$

As in the case of entire functions of one variable it is also possible to give a more refined characterization of the growth of a function u by means of its indicator (with respect to the order ρ)

$$\mathcal{L}_u(x) = \mathcal{L}(x;u,\rho) \overset{\text{def}}{=} \lim_{t\to\infty} u^{[t]}(x).$$

For $n > 2$ this indicator is neither continuous nor upper semicontinuous. Therefore, instead of $\mathcal{L}_u(x)$ its regularization

$$\mathcal{L}_u^*(x) = \lim_{\varepsilon\to 0}\ \sup_{|x'-x|<\varepsilon} \mathcal{L}_u(x')$$

is often used. $\mathcal{L}_u^*(x)$ is also called an indicator (or regularized indicator). It is easy to see that $\mathcal{L}_u^*(x)$ is a subharmonic function in $\mathbb{R}^n$. It is also obvious that $\mathcal{L}_u(x)$ and $\mathcal{L}_u^*(x)$ are positively homogeneous of degree ρ, i.e. $\mathcal{L}_u(tx) = t^\rho\mathcal{L}_u(x)$, $\mathcal{L}_u^*(tx) = t^\rho\mathcal{L}_u^*(x)$, $\forall t > 0$.

In addition, note that $\mathcal{L}_u^*(x) = \mathcal{L}_u(x)$ for almost all $x\in\mathbb{R}^n$, and, moreover, **Cartan's Theorem** on regularization (see, for example, Ronkin [1]) implies that the set $\{x\in\mathbb{R}^n:\ \mathcal{L}_u(x) < \mathcal{L}_u^*(x)\}$ has capacity zero.

For $n > 2$ the regularized indicator is need not be continuous. This fact is an essential difference between the spatial and planer cases, and some other differences are based upon it. In particular, there are functions $u\in SH(\mathbb{R}^n,\rho]$ for which the inequality

$$u(x) \le \mathfrak{L}^*_u(x) + o(|x|^\rho),\ |x| \to \infty, \tag{2.1.27}$$

is not valid. Note that (2.1.27) is valid for $n = 2$ and plays an important role in that case. However, (2.1.27) is valid under the additional assumption of continuity of $\mathfrak{L}^*_u(x)$. This follows easily from **Hartogs' Lemma**.

Now we consider a positive measure μ in $\mathbb{R}^n$. The order of the function $r^{2-n}\mu(r)$, where $\mu(r) = \mu(\bar{B}_r)$ will be referred to as the order ρ_μ of the measure μ. It is easy to see that the order of a measure coincides with its convergence exponent

$$\gamma_\mu = \inf\left\{\alpha: \int^\infty r^{2-n+\alpha} d\mu(r) < \infty\right\} = \inf\left\{\alpha: \int^\infty r^{1-n+\alpha}\mu(r)dr < \infty\right\}.$$

It follows from Jensen's formula for subharmonic functions that $\gamma_{\mu_u} \le \rho_u$, $\forall u \in SH(\mathbb{R}^n)$.

It is possible to associate with a measure μ that has finite convergence exponent some special potential. This potential plays the same role as the canonical Weierstrass product does in the theory of entire functions. When $n = 2$, the kernel of this potential, which we shall call canonical, coincides with $ln|G_q|$, where G_q is the primary Weierstrass factor. When $n > 2$ the kernel is constructed from the fundamental solution of the **Laplace Equation**, i.e. from the function $-|x-y|^{2-n}$, similarly to the construction of $G_q(z)$, $z \in \mathbb{C}$, from $(1-z)$. Namely, we denote by $D_m(x,y)$ the homogeneous polynomial of degree m in the Taylor expansion of $-|x-y|^{2-n}$ in the powers of $x_1, \ldots, x_n$. It is obvious that $D_m(x,y)$ is a harmonic function with respect to x. We set

$$h_q(x,y) = -\frac{1}{|x-y|^{n-2}} + \sum_{m=0}^{q} D_m(x,y),$$

and call this function a canonical kernel. It is clear that $h_q(x,y)$ is a subharmonic function with respect to x in $\mathbb{R}^n$ and that its Riesz associated measure is the unit mass at the point y. In other words,

$$\Delta_x h_q(x.y) = \theta_n \delta(x-y).$$

We give an explicit expression for the functions $D_m(x,y)$ and $h_q(x,y)$

in terms of by $|x|$, $|y|$ and the angle ψ (or, equivalently, $(x\widehat{\ ,}y)$) between the vectors x and y.

We put:

$$v = \frac{|x|}{|y|},\quad A_m^\alpha = \frac{\alpha(\alpha+1)*\ldots*(\alpha+m-1)}{m!}$$

and

$$g_m^n(\psi) = \sum_{k+l=m} A_k^{n/2-1} A_l^{n/2-1} \cos(k-l)\psi. \tag{2.1.28}$$

The following equalities hold:

$$D_m(x,y) = -\frac{g_m^n(\psi)v^m}{|y|^{n-2}},$$

$$h_q(x,y) = -\frac{1}{|y|^{n-2}}(1 - 2v\cos\psi + v^2)^{1-n/2} + \frac{1}{|y|^{n-2}} + \sum_{m=1}^{q}\frac{g_m^n(\psi)v^m}{|y|^{n-2}},\ y \neq 0,$$

and

$$h_q(x,y) = -\sum_{m=q+1}^{\infty}\frac{g_m^n(\psi)v^m}{|y|^{n-2}},\ |x| < |y|. \tag{2.1.29}$$

The canonical kernel is estimated by

$$h_q(x,y) \le \text{const}\cdot\frac{|y|^{2-n-q}|x|^{q+1}}{|x| + |y|},\ \forall x,y,\ n \ge 2, \tag{2.1.30}$$

and

$$|h_q(x,y)| \le c(\tau)|y|^{1-n-q}|x|^{q+1},\ |x| < \tau|y|,\ \tau > 1,\ n \ge 2. \tag{2.1.31}$$

Theorem 2.1.1. Let μ be a positive measure in $\mathbb{R}^n$, $n \ge 2$, with a finite convergence exponent and let it vanish in some neighbourhood of the origin. Moreover, we set $\mu(r) = \mu(\overline{B}_r)$ and denote by q the least nonnegative integer α satisfying the condition

$$\int \frac{\mu(r)}{r^{q+n}}dr < \infty.$$

Then

1) the integral

$$J_\mu(x) = \int_{\mathbb{R}^n} h_q(x,y)d\mu(y)$$

converges at every point $x \in \mathbb{R}^n$, except at the points where it diverges to $-\infty$;

2) $J_q(x) \in SH(\mathbb{R}^n)$;

3) the Riesz associated measure of the function $J_m(x)$ coincides with μ;

4) the estimate

$$J_\mu(x) \le \text{const}\cdot\left\{r^q \int_0^r \frac{\mu(t)}{t^{q+n-1}}\, dt + r^{q+1} \int_r^\infty \frac{\mu(t)}{t^{q+n}}\, dt\right\}$$

holds.

The integral $J_\mu(x)$ figuring in this theorem is called the canonical potential of μ. From the estimate of the canonical potential contained the this theorem it follows that $\rho_{J_\mu} \le \rho_\mu$, and hence $\rho_{J_\mu} = \rho_\mu$. In the case $\rho_\mu \notin \mathbb{Z}$ we can also state that $J_\mu(x)$ is of minimal, normal or maximal type if the function $r^{2-n}\mu(r)$ is of the corresponding type. As for entire functions the case $\rho_\mu \in \mathbb{N}$ is more complicated.

If $\rho_\mu = q+1$, then the function J_μ is of minimal type. If $\rho_\mu = q$, then J_μ is of minimal, normal or maximal type according whether the value

$$\max\left\{ \overline{\lim_{r\to\infty}} \frac{\mu(r)}{r^{n+q-2}},\ \overline{\lim_{r\to\infty}} \sup_{|y|=1} \int_{|x|<r} |x|^{2-n-q} g_q^n(\psi) d\mu(x)\right\}$$

or, equivalently,

$$\max\left\{ \overline{\lim_{r\to\infty}} \frac{\mu(r)}{r^{n+q-2}},\ \overline{\lim_{r\to\infty}} \sup_{|y|=1} \left|\int_{|x|<r} \frac{g_q^n(\psi)}{|x|^{n+q-2}}\, d\mu(x)\right|\right\} \tag{2.1.32}$$

is equal 0, differs from 0 and ∞, or is equal to ∞ respectively.

The following theorem is the analogue of **Hadamard's Theorem** on the factorization of entire functions.

Theorem 2.1.2. Let $u(x) \in SH(\mathbb{R}^n)$ be a function of finite order ρ. Then it can be represented as

$$u(x) = v_{\mu'}(x) + J_{\mu''}(x) + P(x),$$

where μ' is the restriction of the measure μ to some ball B_ε, μ'' is the restriction of μ_u to $\mathbb{R}^n \setminus B_\varepsilon$, and $P(x)$ is a harmonic polynomial in $x_1, \ldots, x_n$ of degree at most ρ.

§2 Criteria for regularity of growth in $\mathbb{R}^n$

1. Regularity of growth and convergence with respect to a Carleson measure. Similarly to the definition of entire function of c.r.g. in $\mathbb{C}$ given in **Chapter 1**, it is natural to define a subharmonic function of c.r.g. in $\mathbb{R}^n$ as a subharmonic function that is close to its indicator outside some exceptional set. However, this analogy does not extend to the construction of the exceptional set, i.e. it is not correct to consider exceptional sets in the form $\{x \in \mathbb{R}^n: |x| \in E\}$ as in the case of functions in $\mathbb{C}$. At the same time a modification of the concept of C_0-set, which plays a role in one of the criteria for c.r.g. in $\mathbb{C}$, may be used in the multidimensional case.

A set $E \subset \mathbb{R}^n$ is called a C_0^α-set if there are balls $B_{r_j}(x_j)$, $j = 1, 2, \ldots$, such that

$$E \subset \bigcup_j B_{r_j}(x_j)$$

and

$$\lim_{R \to \infty} \frac{1}{R^\alpha} \sum_{j: |x_j| < R} r_j^\alpha = 0.$$

A function $u \in SH(\mathbb{R}^n, \rho]$ is called a function of c.r.g. (in $\mathbb{R}^n$ with respect to the order ρ) if there a exists C_0^{n-1}-set E such that

$$\lim_{R \to \infty} \sup_{|x|=R,\, x \notin E} \left| \frac{u(x)}{|x|^\rho} - \mathfrak{L}_u^*\left(\frac{x}{|x|}\right) \right| = 0. \tag{2.2.1}$$

As will be shown below, this definition is equivalent to the following one.

A function $u \in SH(\mathbb{R}^n, \rho]$ is called a function of c.r.g. (in $\mathbb{R}^n$, with respect to the order ρ) if the functions[10] $u^{[t]}(x)$ converge to $\mathcal{L}_u^*(x)$ in the ball B_1 with respect to the Carleson (n-1)-measure as $t \to \infty$.

Moreover, the equivalence of (2.2.1) to the convergence of the functions $u^{[t]}(x)$ with respect to the Carleson measure is true in the case when E is a C_0^α-set and mes_C^{n-1} is replaced by mes_C^α. Indeed, let

$$E_\varepsilon = \{x \in \mathbb{R}^n:\ |u(x) - \mathcal{L}_u^*(x)| > \varepsilon |x|^\rho\}$$

and

$$E_{t,\varepsilon} = \{x \in \mathbb{R}^n:\ |u^{[t]}(x) - \mathcal{L}_u^*(x)| > \varepsilon\}.$$

If $x_0 \in E_\varepsilon$ and $2^{m-1} \le |x^0| \le 2^m$, then

$$\varepsilon < |x^0|^{-\rho}|u(x^0) - \mathcal{L}_u^*(x^0)| =$$

$$= 2^{m\rho}|x^0|^{-\rho}\left|2^{-m\rho}u\left(\frac{x^0 2^m}{2^m}\right) - \mathcal{L}_u^*\left(\frac{x^0}{2^m}\right)\right| \le$$

$$\le 2^{-\rho}\left|u^{[2^m]}\left(\frac{x^0}{2^m}\right) - \mathcal{L}_u^*\left(\frac{x^0}{2^m}\right)\right|,$$

and hence $(2^{-m}x^0) \in E_{2^m, 2^\rho\varepsilon}$. Therefore, if $2^{k-1} \le R \le 2^k$, then the inclusion

$$E_\varepsilon \cap B_R \subset B_1 \cup \left(\bigcup_{m=1}^{k}\left\{x \in \mathbb{R}^n:\ \frac{x}{2^m} \in E_{2^m, 2^\rho\varepsilon},\ 2^{m-1} \le |x| < 2^m\right\}\right). \tag{2.2.2}$$

takes place. It is also obvious that

$$E_\varepsilon \backslash B_1 \subset \bigcup_{m=1}^{\infty}\left\{x \in \mathbb{R}^n:\ \frac{x}{2^m} \in E_{2^m, 2^\rho\varepsilon},\ 2^{m-1} \le |x| < 2^m\right\}. \tag{2.2.3}$$

On the other hand, if $x^0 \in E_{t,\varepsilon} \cap B_1$, then

$$\varepsilon < |t^{-\rho}u(tx^0) - \mathcal{L}_u^*(x^0)| = \left|\frac{u(tx^0)}{|tx^0|^\rho} - \mathcal{L}_u^*\left(\frac{tx^0}{|tx^0|^\rho}\right)\right||x^0|^\rho \le$$

---

[10] Remind that $u^{[t]}(x) = \dfrac{u(tx)}{t^\rho}$

$$\le \left| \frac{u(tx^0)}{|tx^0|^\rho} - \frac{\mathcal{E}_u^*(tx^0)}{|tx^0|^\rho} \right|,$$

and thus $tx^0 \in E_\varepsilon$. Hence

$$E_{t,\varepsilon} \cap B_1 \subset \frac{1}{t} (E_\varepsilon \cap B_1). \tag{2.2.4}$$

According to the definition of mes_C^α, there are balls $\tilde{B}^{(j,t)}$, $j = 1, 2, \ldots$, covering $E_{t,2^\rho\varepsilon} \cap B_1$ and with radii $r_{j,t}$ satisfying the condition

$$\sum_j r_{j,t}^\alpha \le 2 mes_C^\alpha \{E_{t,2^\rho\varepsilon}\}.$$

From among the balls $2^m \tilde{B}^{(j,t)}$ we pick out balls $\hat{B}^{(j,t)}$ with centres in the spherical layer $B_{2^{m+1}} \backslash B_{2^{m-2}}$. Without loss of generality we may assume that $r_{j,t} < \frac{1}{4}$ and thus these balls form a cover of the set

$$\left\{ x \in \mathbb{R}^n : \frac{x}{2^m} \in E_{2^m, 2^\rho\varepsilon}, \ 2^{m-1} \le |x| < 2^m \right\}.$$

Hence the inclusion

$$E_\varepsilon \backslash B_R \subset \bigcup_m \bigcup_j \hat{B}^{(j,m)}$$

is valid and if $2^{k-1} \le |x| \le 2^k$, then the inclusion

$$E_\varepsilon \backslash B_R \subset B_1 \cup \{ \bigcup_{m=1}^{k+1} \bigcup_j \hat{B}^{(j,m)} \}$$

is also valid. Taking into account the way of choosing the balls $\tilde{B}^{(j,m)}$ and $\hat{B}^{(j,m)}$ we obtain the following estimate for the sums of the radii $\hat{r}_{j,m}$ of the balls $\hat{B}^{(j,m)}$:

$$\frac{1}{R^\alpha} \sum_{m=1}^k \sum_j (\hat{r}_{j,m})^\alpha \le \frac{1}{2^{(k-1)\alpha}} \sum_{m=1}^k 2^{m\alpha} \sum_j \left(\frac{1}{2^m} \hat{r}_{j,2^m} \right)^\alpha \le$$

$$\le \frac{2}{2^{(k-1)\alpha}} \sum_{m=1}^k 2^{m\alpha} mes_C^\alpha \left\{ E_{2^m, 2^\rho\varepsilon} \right\}.$$

Therefore, if

$$\lim_{m \to \infty} mes_C^\alpha\left\{E_{2^m, 2^\rho\varepsilon}\right\} = 0, \ \forall\varepsilon > 0,$$

then

$$\lim_{R \to \infty} \frac{1}{R^\alpha} \sum_{m=1}^{k} \sum_j (\hat{r}_{j,m})^\alpha = 0;$$

and hence for any $\varepsilon > 0$ E_ε is C_0^α-set. From this follows the existence of a C_0^α-set E for which condition (2.2.1) is valid. Now we assume that condition (2.2.1) is valid for some C_0^α-set E. Then it is obvious that for any $\varepsilon > 0$, E_ε is also a C_0^α-set. Let $B_{r_j}(x_j)$ be the balls corresponding to E_ε in the definition of C_0^α-set. Then the balls $B^{j,R} = B_{R_j}(X_j)$, where $X_j = \frac{1}{R}\cdot x_j$, $R_j = \frac{r_j}{R}$ for $|x| < 2R$, form a cover of the set $\frac{1}{R}\cdot(E_\varepsilon \cap B_R)$. Their radii satisfy the condition

$$\lim_{R \to \infty} \sum_{j:|x_j|<2R} R_j^\alpha = 0.$$

This and imply (2.2.4) it follows that

$$\lim_{t \to \infty} mes_C^\alpha E_{t,\varepsilon} = 0.$$

Thus the above-mentioned equivalence of convergence with respect to the Carleson α-measure and convergence outside a C_0^α-set is proved. In particular, the equivalence of the definitions of f.c.r.g. given in this section is proved as well. Note also the obvious fact that everywhere above we could take a ball B_d with arbitrary $d > 0$ instead of the ball B_1.

2. Regularity of growth and convergence of the functions $u^{[t]}$ in $\mathcal{D}'$. Similarly to the case of functions in $\mathbb{C}$ considered in **Chapter 1**, regularity of growth of functions in $\mathbb{R}^n$ may be characterized in terms of the weak convergence.

Theorem 2.2.1. In order that $u \in SH(\mathbb{R}^n)$ be a function of c.r.g. it is necessary and sufficient that

$$\mathcal{D}' - \lim_{t \to \infty} u^{[t]}(x) = \mathfrak{L}_u^*(x).$$

By i) of **Lemma 2.1.4**, this theorem can be formulated in another way.

Namely:

Theorem 2.2.1'. In order that $u \in SH(\mathbb{R}^n)$ be a function of c.r.g. it is necessary and sufficient that $\mathcal{D}'\text{-}\lim_{t\to\infty} u^{[t]}(x)$ exist.

If we recall what was said in §2.1 about the definition of f.c.r.g. then these theorems may be considered as statements on the relationship between the convergence of functions $u^{[t]}$ in the $\mathcal{D}'$-topology and with respect to a Carleson measure. These statements clearly follow from the following theorem:

Theorem 2.2.2. Let $u \in SH(\mathbb{R}^n)$. Then:

1) if $\mathcal{D}'\text{-}\lim_{t\to\infty} u^{[t]}(x)$ exists, then the functions $u^{[t]}(x)$, $t \to \infty$, in the ball B_1 converge to the indicator $\mathcal{L}^*_u(x)$ with respect to the Carleson α-measure for any $\alpha \in (n-2,n]$;

2) if for some $\alpha \in (n-2,n]$ the functions $u^{[t]}(x)$, $t \to \infty$, converge in the ball B_1 to $v \in L^1_{loc}(B_1)$ with respect to Carleson α-measure, then $reg\, v(x) = \mathcal{L}^*_u(x)$ and

$$\mathcal{D}'\text{-}\lim_{t\to\infty} u^{[t]}(x) = \mathcal{L}^*_u(x).$$

Before passing to the proof of this theorem, let us mention two obvious corollaries.

Corollary 1. Convergence of $u^{[t]}(x)$, $t \to \infty$, with respect to some Carleson α_1-measure, $\alpha_1 \in (n-2,n]$, is equivalent to convergence of these functions with respect to any other Carleson α-measure, $\alpha \in (n-2,n]$.

Corollary 2. The class of f.c.r.g. will not change if in the definitions of f.c.r.g. given in §2.1, C_0^{n-1}-set would be replaced by C_0^{α}-set with $\alpha \in (n-2,n]$, or convergence with respect to Carleson $(n-1)$-measure would be replaced by convergence with respect to a Carleson α-measure, $\alpha \in (n-2,n]$.

Note also that in the formulation of **Theorem 2.2.2** any ball B_R, $R > 0$, may be taken instead of B_1. The proof of this theorem, as well as the others in this chapter, will be given for $n > 2$. The case $n = 2$ differs from the case $n > 2$ in certain details related only with the form of the kernel $h_q(x,y)$. So we omit this case.

Proof of Theorem 2.2.2. The function $u(x)$ figuring in the theorem has at most normal type with respect to the order ρ. Hence the family of functions $u^{[t]}(x)$ is bounded above. Therefore the lemmas concerning the relationships between the different types of convergence, proved above, may be applied. Thus, statement 1) is obtained as an immediate consequence of **Lemma 2.1.6** and **Lemma 2.1.4** (point i)), and statement 2) follows from 1) and **Lemma 2.1.8**.

The proof is finished.

In relation to **Theorem 2.2.2** and, especially, with its **Corollary 2**, the problem arises on the most refined characteristic of a set E outside which a function $u \in SH(\mathbb{R}^n, \rho]$ of c.r.g. converges uniformly to its indicator. We give without proof two theorems related to this problem. They deal with functions of minimal type. Note that if $u \in SH(\mathbb{R}^n, \rho]$ is of minimal type, then the average $\mathfrak{M}_{|u|}(0,r)$ is of minimal type too (see (2.1.9)). In turn, this implies that $\mathcal{D}'\text{-}lim\ u^{[t]} = 0$, and thus u(x) is a function of c.r.g. with indicator $\mathfrak{L}^*_u(x) \equiv 0$.

Let CapE denote the capacity of a set $E \subset \mathbb{R}^n$.

Theorem 2.2.3. If a function $u \in SH(\mathbb{R}^n, \rho)$ is of minimal type with respect to the order ρ, then $|x|^{-\rho}u(x) \to 0$ as $x \to 0$ outside some set E satisfying the condition

$$\lim_{R \to \infty} \frac{\mathrm{Cap}(E \cap B_R)}{\mathrm{Cap}(B_R)} = 0. \qquad (2.2.5)$$

Theorem 2.2.4. Let A be a closed set satisfying condition (2.2.5). Then for any $\rho \neq 2k$, $k = 0, 1, 2, \ldots$ (and in the case $n = 2$ for arbitrary $\rho > 0$), there is function $u \in SH(\mathbb{R}^n, \rho]$ of minimal type with respect to its order ρ and

$$\lim_{x \to \infty,\, x \in A} \frac{u(x)}{|x|^{\rho}} = -\infty .$$

Note that **Theorem 2.2.3** is not valid for functions of normal type.

3. Regularity of growth and regular distribution of the associated measure. We denote by $\mathfrak{C}(\rho)$ the set of all positive measures μ in $\mathbb{R}^n$ such that the functions $t^{2-n}\mu(t)$ have at most normal type with respect to the order ρ, $\rho > 0$. We associate to every measure $\mu \in \mathfrak{C}(\rho)$ the family of measures $\mu^{[t]}$, $t > 1$, where the $\mu^{[t]}$ are defined on the Borel sets $E \subset \mathbb{R}^n$ by

$$\mu^{[t]}(E) = \frac{\mu(tE)}{t^{\rho+n-2}} .$$

Let $\mu\in\mathfrak{C}(\rho)$ be such that

$$\exists\mathcal{D}'-\lim_{t\to\infty}\mu^{[t]} \overset{\text{def}}{=} \hat{\mu}. \tag{2.2.6}$$

It is clear that $\hat{\mu}$ is positively homogeneous of degree $\rho+n-2$, i.e.

$$\hat{\mu}(tE) = t^{\rho+n-2}\hat{\mu}(E),\ \forall t > 0.$$

Therefore $d\hat{\mu} = d(t^{\rho+n-2})\otimes d\hat{\nu}$, where $\hat{\nu} = \hat{\nu}_{\mu}$ is the measure on S_1 defined on the Borel sets $E \subset S_1$ by the equality

$$\hat{\nu}(E) = \hat{\mu}\left(\left\{x: \frac{x}{|x|}\in E\right\} \cap B_1\right).$$

We set $K^E = \left\{x\in\mathbb{R}^n: \frac{x}{|x|}\in E\right\}$. If Γ is a domain on S_1 such that $\hat{\nu}(\partial\Gamma) = 0$, then $\hat{\mu}(\partial(K^\Gamma\cap B_1)) = 0$. Therefore it follows from **Lemma 2.1.1** that

$$\exists \lim_{t\to\infty} \mu^{[t]}(K^\Gamma\cap B_1) = \hat{\mu}(K^\Gamma\cap B_1) = \hat{\nu}(\Gamma),$$

or, equivalently,

$$\exists \lim_{t\to\infty} \frac{1}{t^{\rho+n-2}}\mu(K^\Gamma\cap B_1) = \hat{\nu}(\Gamma). \tag{2.2.7}$$

It is easy to see that the converse statement is also valid. Namely, if $\lim_{t\to\infty} t^{-\rho-n+2}\mu(K^\Gamma\cap B_1)$ exists for every domain $\Gamma \subset S_1$ such that $\nu(\partial\Gamma) = 0$, where ν is some positive measure on S_1, then $\exists\mathcal{D}'-\lim_{t\to\infty}\mu^{[t]}$.

Taking all this into account, it is natural to call the measure $\hat{\nu}$ the cone density of the measure μ, and to interpret the existence of $\mathcal{D}'-\lim\mu^{[t]}$ as the existence of the cone density of the measure μ.

Note that if $n = 2$ and $\mu = \tilde{n}$, i.e. if μ is defined by the zero distribution of an entire function, the concept of cone density coincides with the concept of angular density of the zero set of the function, as defined in **Chapter 1**.

Theorem 2.2.5. Let $u\in SH(\mathbb{R}^n,\rho]$, $\rho\notin\mathbb{Z}$. In order that $\mathcal{D}'-\lim_{t\to\infty} u^{[t]}$ exist, it is necessary and sufficient that the measure μ_u have a cone density. If it exists, the equality $\hat{\mu} = \mu_{\mathfrak{L}^*_u}$ is valid.

Proof. Since $\mu_u = \frac{1}{\theta_u}\cdot\Delta u$, the necessity part of is an obvious consequence of the continuity of the differentiation operator in the space of distributions, this being true for any $\rho > 0$. Note also that, as mentioned above, the existence of $\mathcal{D}'\text{-}lim\ u^{[t]}$ implies the equality $\mathcal{D}'\text{-}lim\ u^{[t]} = \mathfrak{L}^*_u$, and hence $\hat{\mu}_u = \mu_{\mathfrak{L}^*_u}$.

To prove the sufficiency we use **Theorem 2.1.2**. It follows from this theorem that

$$u^{[t]}(x) = -\frac{1}{t^\rho}\int_{B_1}\frac{1}{|tx-y|^{n-2}}\,d\mu_u(y) + \frac{1}{t^\rho}\int_{\mathbb{R}^n\setminus B_1} h_q(tx,y)d\mu_u(y) + \frac{1}{t^\rho}P(tx),$$

where $q < \rho < q + 1$, $q\in\mathbb{Z}_+$, and $P(tx)$ is a polynomial of degree less than ρ. It is obvious that the first and third summands on the right converge uniformly to zero as $|x| \to \infty$. Therefore, proving sufficiency is reduced to proving the implication

$$\exists\mathcal{D}'\text{-}\lim_{t\to\infty}\mu^{[t]} \Rightarrow \exists\mathcal{D}'\text{-}\lim_{t\to\infty}\frac{1}{t^\rho}\int_{\mathbb{R}^n\setminus B_1} h_q(tx,y)d\mu_u(y).$$

Taking into account that $h_q(tx,ty) = t^{2-n}h_q(x,y)$, we represent the potential figuring in this implication as

$$\frac{1}{t^\rho}\int_{\mathbb{R}^n\setminus B_1} h_q(tx,y)d\mu_u(y) = \frac{1}{t^\rho}\int_{1<|y|<\varepsilon t} h_q(tx,y)d\mu_u(y) +$$

$$+ \int_{\varepsilon\le|\xi|\le R} h_q(x,\xi)d\mu_u^{[t]}(\xi) + \frac{1}{t^\rho}\int_{Rt<|y|} h_q(tx,y)d\mu_u(y) = J_1 + J_2 + J_3.$$

We estimate $\int_{B_1}|J_1|dx$. By (2.1.29) and since $\mu_u\in\mathfrak{C}(\rho)$, we have

$$\int_{B_1}|J_1|dx \le t^{-\rho}\int_{B_1}dx\int_{1<|y|<\varepsilon t}|h_q(tx,y)|d\mu_u(y) \le$$

$$\le t^{-\rho}\int_{B_1} dx \int_{1<|y|<\varepsilon} |tx-y|^{2-n}d\mu_u(y) + t^{-\rho}\sum_{m=1}^{q}\int_{B_1} dx \int_{1<|y|<\varepsilon t} \frac{|g_m^n(\psi)|v^m t^m}{|y|^{n-2}}d\mu_u(y) \le$$

$$\le \frac{\mu_u(\varepsilon t)}{t^{\rho+n-2}}\int_{B_1}\frac{dx}{|x|^{n-2}} + c_1\sum_{m=1}^{q} t^{m-\rho}\left\{\frac{\mu_u(s)}{s^{m+n-2}}\Bigg|_1^{\varepsilon t} + (m+n-1)\int_1^{\varepsilon t}\frac{\mu_u(s)}{s^{m+n-1}}ds\right\} \le$$

$$\le c_2\varepsilon^{\rho+n-2} + c_1\sum_{m=1}^{q} t^{m-\rho}\left\{(\varepsilon t)^{\rho-m} + \frac{m+n-1}{\rho-m}(\varepsilon t)^{\rho-m}\right\} =$$

$$= c_2\varepsilon^{\rho+n-2} + c_3\varepsilon^{\rho-q}. \tag{2.2.8}$$

Here c_1, c_2, and c_3 are constants depending on the function u only.

Quite similarly, but using inequality (2.1.31) instead of (2.1.29), the integral $\int_{B_1}|J_3|dx$ can be estimated. Then we obtain

$$\int_{B_1}|J_3|dx \le c_4R^{\rho-q-1}. \tag{2.2.9}$$

Now, let $\varphi\in C(\bar{B}_1)$ be an arbitrary function. We consider the integral

$$\int_{B_1}\varphi(x)J_2(x)dx = \int_{\varepsilon\le|\xi|\le R} d\mu_u^{[t]}(\xi)\int_{B_1} h_q(x,\xi)\varphi(x)dx.$$

It can be verified directly that the function $\Phi(x) = \int_{B_1} h_q(x,\xi)\varphi(x)dx$ is continuous in $\mathbb{R}^n$. Therefore if

$$\exists\mathcal{D}' - \lim_{t\to\infty}\mu^{[t]} \overset{\text{def}}{=} \hat{\mu}_u \tag{2.2.10}$$

then

$$\lim_{t\to\infty}\int_{\varepsilon\le|\xi|\le R}\Phi(\xi)d\mu_u^{[t]}(\xi) = \int_{\varepsilon\le|\xi|\le R}\Phi(\xi)d\hat{\mu}_u(\xi),$$

and hence

$$\lim_{t\to\infty} \int_{B_1} \varphi(x) J_2(x)dx = \int_{B_1} \varphi(x)dx \int_{\varepsilon\le|\xi|\le R} h_q(x,\xi)d\hat{\mu}_u(\xi).$$

This and the estimates (2.2.8) and (2.2.9) imply, since ε and R are arbitrary, that for any function $\varphi\in C(\overline{B}_1)$,

$$\exists \lim_{t\to\infty} \int_{B_1} \varphi(x)u^{[t]}(x)dx = \int_{B_1} \varphi(x)dx \int_{\mathbb{R}^n} h_q(x,y)d\hat{\mu}_u(y).$$

Thus

$$\exists \mathcal{D}'\text{-}\lim_{t\to\infty} u^{[t]} = \int_{\mathbb{R}^n} h_q(x,y)d\hat{\mu}_u(y). \tag{2.2.11}$$

The proof is finished.

Remark. Equality (2.2.11) means also that the indicator of a function u(x) satisfying the conditions of **Theorem 2.2.4** can be represented as

$$\mathfrak{L}^*_u(x) = \int_{\mathbb{R}^n} h_q(x,y)d\hat{\mu}_u(y). \tag{2.2.12}$$

Since $d\hat{\mu}_u(x) = d(|x|^{n+\rho-2})\otimes d\hat{\nu}_u\left(\frac{x}{|x|}\right)$, this representation implies

$$\mathfrak{L}^*_u(x) = |x|^\rho \int_{S_1} \tilde{h}_q\left(\frac{x}{|x|},y\right)d\hat{\nu}_u(y),$$

where

$$\tilde{h}_q(x,y) = -(\rho+n-2)\int_0^1 \frac{s^{\rho+n-3}}{|sx-y|^{n-2}}ds + \sum_{m=1}^{q} \frac{\rho+n-2}{\rho-m} g^n_m(\psi) +$$

$$+ \sum_{m=q+1}^{\infty} \frac{\rho+n-2}{m-\rho} g^n_m(\psi), \ \forall x\in S_1,\ y\in S_1. \tag{2.2.13}$$

When $\rho\in\mathbb{Z}$, as in the case of entire functions, the existence of the cone density of the measure μ_u is not sufficient for $\mathcal{D}'$-convergence of the functions $u^{[t]}(x)$. To obtain the corresponding result some special symmetry of the measure μ_u is needed. This symmetry can be described using the trigonometrical polynomial $g^n_m(\psi)$, $\psi = (x\widehat{\ ,}y)$ or, equivalently, the harmonic polynomial

$$D_m(x,y) = -\frac{g_m^n(\psi)|x|^m}{|y|^{n-2+m}} = \frac{1}{m!}\left(\frac{d^m}{dt^m}\frac{1}{|tx-y|^{n-2}}\right)\Bigg|_{t=0} .$$

Theorem 2.2.6. Let $u\in SH(\mathbb{R}^n,p]$, $p\in\mathbb{N}$. Then for the limit

$$\mathcal{D}'-\lim_{t\to\infty} u^{[t]} \tag{2.2.14}$$

to exist it is necessary and sufficient that the measure μ_u has a cone density and

$$\exists \lim_{t\to\infty}\int_{|y|<t} |y|^{2-n-p} g_p^n(\psi)d\mu_u(y),\ \forall x\in S_1. \tag{2.2.15}$$

If these conditions are satisfied, then the following equalities holds:

$$\int_{S_1} g_p^n(\psi)d\hat{\nu}_u(y) = 0,\ \forall x\in S_1, \tag{2.2.16}$$

where $\hat{\nu}_u$ is the cone density of the measure μ_u, and

$$\lim_{t\to\infty}\int_{|y|<t} |y|^{2-n-p} g_p^n(\psi)d\mu_u(y) =$$

$$= \theta_n^{-1}(n-2+2p)\int_{S_1} g_p^n(\psi)\mathcal{E}_u^*(y)d\sigma(y) +$$

$$-\ \theta_n^{-1}\lambda^{2-n-p}\int_{S_\lambda} g_p^n(\psi)\frac{\partial u}{\partial n}\,d\sigma(y) +$$

$$-\ \theta_n^{-1}(n-2+p)\lambda^{1-n-p}\int_{S_\lambda} g_p^n(\psi)u(y)d\sigma(y), \tag{2.2.17}$$

where λ, $0<\lambda<1$, is an arbitrary number satisfying the condition $\mu_u(S_\lambda) = 0$.

Proof. The existence of the cone density of the measure μ_u if the limit (2.2.14) exists was noted in the proof of **Theorem 2.2.4**. We shall use **Lemma 2.1.12** to prove that condition (2.2.15) holds in this situation. Put $E = \{t>0\colon \mu_u(S_t) > 0\}$. Taking into account that $D_p(x,y)$ is harmonic in $\mathbb{R}^n\setminus\{0\}$ (with respect to y), by **Lemma 2.1.12** we have for $\lambda \notin E$ and $t \notin E$,

$$\theta_n \int_{B_t} |y|^{2-n-p} g_p^n(\psi) d\mu_u(y) =$$

$$= -t^{2-n-p} \int_{S_t} g_p^n(\psi) \frac{\partial u}{\partial n} d\sigma(y) - (n-2+p)t^{1-n-p} \int_{S_t} g_p^n(\psi) u(y) d\sigma(y) +$$

$$+ \lambda^{2-n-p} \int_{S_\lambda} g_p^n(\psi) \frac{\partial u}{\partial n} d\sigma(y) + (n-2+p)\lambda^{1-n-p} \int_{S_\lambda} g_p^n(\psi) u(y) d\sigma(y) =$$

$$= - \int_{S_1} g_p^n(\psi) \frac{\partial u^{[t]}}{\partial n} d\sigma(y) - (n-2+p) \int_{S_1} g_p^n(\psi) u^{[t]}(y) d\sigma(y) +$$

$$+ \lambda^{2-n-p} \int_{S_\lambda} g_p^n(\psi) \frac{\partial u}{\partial n} d\sigma(y) + (n-2+p)\lambda^{1-n-p} \int_{S_\lambda} g_p^n(\psi) u(y) d\sigma(y).$$

Using Lemma 2.1.5 and Lemma 2.1.13, this implies that

$$\lim_{t\to\infty,\, t\notin E} \int_{B_t} |y|^{2-n-p} g_p^n(\psi) d\mu_u(y) =$$

$$= - \theta_n^{-1} \int_{S_1} g_p^n(\psi) \frac{\partial \mathcal{L}_u^*}{\partial n} d\sigma(y) - (n-2+p)\theta_n^{-1} \int_{S_1} g_p^n(\psi) \mathcal{L}_u^*(y) d\sigma(y) +$$

$$- \theta_n^{-1} \lambda^{2-n-p} \int_{S_\lambda} g_p^n(\psi) \frac{\partial u}{\partial n} d\sigma(y) + \theta_n^{-1}(n-2+p)\lambda^{1-n-p} \int_{S_\lambda} g_p^n(\psi) u(y) d\sigma(y).$$

To get rid of the restriction $t \notin E$ in the left-hand limit, it is sufficient to note that the set E is at most countable and

$$\lim_{t\to\infty} \int_{S_t} |y|^{2-n-p} g_p^n(\psi) d\mu_u(y) =$$

$$= \lim_{t\to\infty} \left\{ \lim_{s\to t+0,\, s\notin E} \int_{B_s} |y|^{2-n-p} g_p^n(\psi) d\mu_u(y) \right.$$

$$- \lim_{s\to t-0,\, s\notin E} \int_{B_s} |y|^{2-n-p} g_p^n(\psi)d\mu_u(y) \Big\} = 0.$$

Thus, it has been proved that condition (2.2.15) is satisfied and equality (2.2.1) is true.

From the existence of the limit (2.2.15) it follows that

$$\lim_{t\to\infty} \int_{\frac{t}{2}<|y|<t} |y|^{2-n-p} g_p^n(\psi)d\mu_u(y) = 0.$$

On the other hand,

$$\lim_{t\to\infty} \int_{\frac{1}{2}<|y|<1} |y|^{2-n-p} g_p^n(\psi)d\mu_u(y) =$$

$$= \lim_{t\to\infty} \int_{\frac{t}{2}<|y|<t} |y|^{2-n-p} g_p^n(\psi)d\mu_u^{[t]}(y) =$$

$$= \int_{\frac{1}{2}<|y|<1} |y|^{2-n-p} g_p^n(\psi)d\hat{\mu}_u(y) = \int_{/2}^{1} \frac{d(t^p)}{t^{n-2+p}} \int_{S_1} g_p^n(\psi)d\hat{\nu}_u(y).$$

Hence

$$\int_{S_1} g_p^n(\psi)d\hat{\nu}_u(y) = 0,\ \forall x\in S_1,$$

i.e. equality (2.2.16) is true.

Now we show that the conditions (2.2.6) and (2.2.15) are sufficient for the existence of the limit (2.2.14). As in the case $\rho\notin\mathbb{Z}$, we arrive at the equality

$$\int_{\mathbb{R}^n} u^{[t]}(x)\varphi(x)dx = -\frac{1}{t^p}\int_{\mathbb{R}^n} dx \int_{B_1} \frac{\varphi(x)}{|tx-y|^{n-2}} d\mu_u(y) + \frac{1}{t^p}\int_{\mathbb{R}^n} P(tx)\varphi(x)dx +$$

$$+ \int_{\varepsilon<|y|<R} d\mu_u^{[t]}(y) \int_{\mathbb{R}^n} h_q(x,y)\varphi(x)dx + \frac{1}{t^p} \int_{1<|y|<\varepsilon t} d\mu_u(y) \int_{\mathbb{R}^n} h_q(tx,y)\varphi(x)dx +$$

$$+ \frac{1}{t^p} \int_{|y|\geq Rt} d\mu_u(y) \int_{\mathbb{R}^n} h_q(tx,y)\varphi(x)dx.$$

Here, as in the case $\rho\notin\mathbb{Z}$, the first three right-hand summands tend to limits as $t \to \infty$, but there are no estimates similar to (2.2.8) and (2.2.9) for the last two summands. Thus we must additionally investigate that the use of condition (2.2.15) is needed. In the case $\rho = p \in \mathbb{Z}$ two situations are possible: $p = q$ and $p = q + 1$. We first consider the case $p = q$. Then

$$\frac{1}{t^p} \int_{1\leq|y|<\varepsilon t} d\mu_u(y) \int_{\mathbb{R}^n} h_q(tx,y)\varphi(x)dx =$$

$$= \frac{1}{t^p} \int_{1\leq|y|<\varepsilon t} d\mu_u(y) \int_{\mathbb{R}^n} h_{p-1}(tx,y)\varphi(x)dx +$$

$$+ \frac{1}{t^p} \int_{1\leq|y|<\varepsilon t} d\mu_u(y) \int_{\mathbb{R}^n} D_p(tx,y)\varphi(x)dx = J_1 + J_2.$$

By (2.2.15) and the fact that $\mu_u \in \mathfrak{C}(\rho)$, it follows that $\lim\limits_{t\to\infty} J_2$ exists. At the same time it is obvious that an estimate similar to (2.2.8) is valid for J_1; namely,

$$|J_1| \leq c_\varphi^1 \, \varepsilon.$$

Quite similarly,

$$\frac{1}{t^p} \int_{\mathbb{R}^n} \varphi(x)dx \int_{|y|\geq Rt} h_q(tx,y)d\mu_u(y) \leq c_\varphi^2 \cdot \frac{1}{R}.$$

Combining all this we conclude that *lim* $\int\varphi(x)u^{[t]}(x)dx$ exists, i.e. condition (2.2.14) is satisfied.

Now we consider the case $p = q + 1$. Here

$$t^{-p}\int_{\mathbb{R}^n} \varphi(x)dx \int_{|y|\geq Rt} h_q(tx,y)d\mu_u(y) = t^{-p}\int_{\mathbb{R}^n} \varphi(x)dx \int_{|y|\geq Rt} h_p(tx,y)d\mu_u(y) +$$

$$- t^{-p}\int_{\mathbb{R}^n}\varphi(x)dx \int_{|y|\ge Rt} D_p(tx,y)d\hat{\mu}_u(y) = I_1 + I_2.$$

Estimating in a standard way (see (2.2.8)) the value I_1, we obtain

$$|I_1| \le c^3_\varphi \frac{1}{R}.$$

At the same time, since $\lim_{t\to\infty} J_2$ exists, we have $\lim_{t\to\infty} I_2 = 0$. This and the above-said imply that (2.2.14) is satisfied in the case $p = q + 1$ as well.

This finishes the proof.

Remark. The reasoning carried out in the process of proving **Theorem** 2.2.5 shows also that the indicator of the function $u(x)$ can be represented as

$$\mathfrak{L}^*_u(x) = \int_{|y|<\varepsilon} h_{p-1}(x,y)d\hat{\mu}_u(y) + \int_{|y|\ge\varepsilon} h_p(x,y)d\hat{\mu}_u(y) + \Phi(x), \qquad (2.2.18)$$

where $\Phi(x)$ is a homogeneous harmonic polynomial of degree p.

In the turn, from this representation it follows that

$$\mathfrak{L}^*_u(x) = |x|^p\int_{S_1} h^{(1)}_p\left(\frac{x}{|x|},y\right)d\hat{\nu}_u(y) + |x|^p\int_{S_1} h^{(2)}_p\left(\frac{x}{|x|},y\right)d\hat{\nu}_u(y), \qquad (2.2.19)$$

where

$$h^{(1)}_p(x,y) = -(p+n-2)\int_0^1 \frac{s^{p+n-3}ds}{|sx-y|^{n-2}} + \sum_{m=1}^{p-1}\frac{p+n-2}{p-m}\, g^n_m(\psi),\ \forall x\in S_1, y\in S_1,$$

$$h^{(2)}_p(x,y) = \sum_{m=p+1}^{\infty}\frac{p+n-2}{m-p}\, g^n_m(\psi),\ \forall x\in S_1, y\in S_1 .$$

The statement on the relationship between the regularity of growth and the character of its associated measure follows from **Theorems** 2.2.1, 2.2.5, 2.2.6 as an obvious consequence. To formulate this statement more conveniently, we introduce the concept of regular distribution of a measure.

A measure $\mu\in\mathfrak{E}(\rho)$ is said to be regularly distributed if its cone density exists and, in addition, in the case $\rho = p \in \mathbb{N}$ the limit

$$\lim_{t \to \infty} \int_{B_t} |y|^{2-n-p} g_p^n(\psi) d\mu(y) \tag{2.2.20}$$

exists for any $x \in S_1$.

Theorem 2.2.7. In order that a function $u \in SH(\mathbb{R}^n, \rho]$ be of c.r.g. it is necessary and sufficient that the measure μ_u be regularly distributed.

To finish this section we note that since the function

$$\int_{B_t} |x|^p |y|^{2-n-p} g_p^n(\psi) d\mu(y) = \int_{B_t} D_p(x,y) d\mu(y)$$

is a polynomial in x for any t, the condition of existence of the limit (2.2.20), i.e. the condition of existence of the limit of the family of functions family is equivalent to the condition of existence of the limit of a finite number of number sequences; namely,

$$\lim_{t \to \infty} \left\{ \int_{B_t} \frac{\partial^{k_1 + \ldots + k_n}}{\partial x_1^{k_1} \ldots \partial x_n^{k_n}} \left(\frac{1}{|x-y|^{n-2}} \right) d\mu(y) \right\} \Bigg|_{x=0}, \quad k_1 + \ldots + k_n \le p.$$

§3 Rays of completely regular growth and limit sets

1. Definition of ray of c.r.g. As has been shown by S.Favorov [1], there are functions $u \in SH(\mathbb{R}^n, \rho]$ of c.r.g. in $\mathbb{R}^n$ such that for any $x \in S_1$ and any $\varepsilon > 0$ the set $\{t > 0\colon |u(tx) - \mathfrak{L}_u^*(x)| > \varepsilon t^\rho\}$ is not a set of relative measure zero. Without going into the details of the rather delicate construction of such a function, we note that by **Theorem 2.2.4** the problem reduces to constructing a set A satisfying the condition (2.2.5) and such that its intersection with any ray $\{tx\colon t > 0\}$ does not have relative measure zero.

The existence of such functions shows that the direct generalization of the definition of ray of c.r.g., given in **Chapter 1** for entire function of one variable, to the case of functions $u \in SH(\mathbb{R}^n, \rho]$ is not fruitful. To arrive at a definition such that the described situation is impossible and which at the same time is equivalent to the one for the case n = 2, we prove the following lemmas.

Lemma 2.3.1. Let $u \in SH(\mathbb{R}^n, \rho]$. Then

$$\lim_{\delta \to 0} \overline{\lim_{r \to \infty}} r^{-\rho} \mathfrak{M}_u(rx_0, \delta r) = \mathfrak{L}_u^*(x_0), \ \forall x_0. \tag{2.3.1}$$

Proof. From **Hartogs' Lemma** it follows that if $A > \mathfrak{L}_u^*(x_0)$ then there exist $\varepsilon_0 > 0$ and $t_0 > 0$ such that for all $x \in B_{\varepsilon_0}(x_0)$ and all $t > t_0$ the inequality $t^{-\rho} u(tx) \le A$ is valid. This immediately implies that

$$\lim_{\delta \to 0} \overline{\lim_{r \to \infty}} r^{-\rho} \mathfrak{M}_u(rx_0, \delta r) \le \mathfrak{L}_u^*(x_0).$$

To prove the opposite inequality we use the estimate (2.1.8) for the function $u^{[r]}(x)$, and conclude that for every $x \in B_{\delta'}(x_0)$, $\delta' < \delta$, and for every $r > 1$,

$$u^{[r]}(x) \le \left(\frac{\delta}{\delta' + \delta}\right)^n \mathfrak{M}_{u^{[r]}}(x_0, \delta) + \left(1 - \left(\frac{\delta}{\delta' + \delta}\right)^n\right) c,$$

where

$$c = \sup \{ u^{[r]}(x) : r > 1, x \in B_{|x_0| + 2\delta} \} < \infty.$$

Note that

$$\mathfrak{M}_{u^{[r]}}(x_0, \delta) = r^{-\rho} \mathfrak{M}_u(rx_0, \delta r)$$

and

$$\overline{\lim_{r \to \infty}} u^{[r]}(x) = \mathfrak{L}_u(x).$$

Thus

$$\mathfrak{L}_u(x) \le \left(\frac{\delta}{\delta' + \delta}\right)^n \overline{\lim_{r \to \infty}} r^{-\rho} \mathfrak{M}_u(rx_0, \delta r) + \left(1 - \left(\frac{\delta}{\delta' + \delta}\right)^n\right) c, \ \forall x \in B_{\delta'}(x_0).$$

Hence

$$\mathfrak{L}_u^*(x) \le \overline{\lim_{r \to \infty}} r^{-\rho} \mathfrak{M}_u(rx_0, \delta r), \ \forall \delta > 0$$

and therefore

$$\mathfrak{L}_u^*(x) \le \lim_{\delta \to 0} \overline{\lim_{r \to \infty}} r^{-\rho} \mathfrak{M}_u(rx_0, \delta r).$$

The proof is finished.

Lemma 2.3.2. Let $u(z) \in SH(\mathbb{C}, \rho]$. Then in order that there is a set E of relative measure zero and such that

$$\lim_{r\to\infty, r\notin E} r^{-\rho}u(re^{i\theta_0}) = h_u(\theta_0) \qquad (2.3.2)$$

for any given $\theta_0\in[0,2\pi)$, it is necessary and sufficient that

$$\lim_{\delta\to 0}\ \underline{\lim}_{r\to\infty} r^{-\rho}\mathfrak{N}_u(re^{i\theta_0},\delta r) = h_u(\theta_0). \qquad (2.3.3)$$

Proof. Assume that condition (2.3.2) is satisfied. Then for every $r > 0$, we can choose $r'=r'(r)$ such that $r'\notin E$ and $r-r' = o(r)$ as $r\to\infty$. For brevity, we put

$$\delta r-|r'-r| = r'',$$

$$\delta r+|r'-r| = r''',$$

$$U(r'e^{i\theta_0},r'') = U'\ ^{11)},$$

$$U(re^{i\theta_0},\delta r)\backslash U' = U'',$$

$$U(r'e^{i\theta_0},r''') = U'''.$$

Taking into account (2.3.2) and (2.1.2) we conclude[12] that

$$\lim_{\delta\to 0}\ \underline{\lim}_{r\to\infty} r^{-\rho}\mathfrak{N}_u(re^{i\theta_0},\delta r) =$$

$$= \lim_{\delta\to 0}\ \underline{\lim}_{r\to\infty} \frac{1}{\pi(\delta r)^2}\left\{\int_{U'} u d\omega_z + \int_{U''} u d\omega_z\right\} \ge$$

$$\ge \lim_{\delta\to 0}\ \underline{\lim}_{r\to\infty}\left[\left(\frac{r'}{r}\right)^{\rho}\left(\frac{r''}{r}\right)^2\frac{1}{r'^{\rho}}\mathfrak{N}_u(r'e^{i\theta_0},r'')\right] -$$

$$-\ \overline{\lim}_{\delta\to 0}\ \overline{\lim}_{r\to\infty}\frac{1}{\pi r^{\rho+2}\delta^2}\int_{U''}|u|d\omega_z \ge$$

---

11) Recall that $U(z^0,r) = \{z\in\mathbb{C}: |z-z^0|<r\}$.

12) Recall that $d\omega_z$ is the area element in the complex z-plane.

$$\geq h_u(\theta_0) - \overline{\lim_{\delta \to 0}} \ \overline{\lim_{r \to \infty}} \frac{1}{\pi r^{\rho+2}\delta^2} \int_{U''\setminus U'} |u| d\omega_z \geq$$

$$\geq h_u(\theta_0) - \overline{\lim_{\delta \to 0}} \ \overline{\lim_{r \to \infty}} \frac{2}{r^{\rho+2}\delta^2} \int_{r''}^{r'''} (2M_{u^+}(r'e^{i\theta_0},t) - u(r'e^{i\theta_0}))tdt \geq$$

$$\geq h_u(\theta_0) - \overline{\lim_{\delta \to 0}} \ \overline{\lim_{r \to \infty}} \frac{2}{r^{\rho+2}\delta^2} \int_{r''}^{r'''} (2M_{u^+}(r',t) - u(r'e^{i\theta_0}))tdt =$$

$$= h_u(\theta_0).$$

This and (2.3.1) imply that equality (2.3.3) is valid. Thus the necessary part of the lemma is proved.

To prove sufficiency we assume that condition (2.3.2) is not satisfied for any set E of relative measure zero. Then it is easy to see that for some $\varepsilon > 0$ and $\kappa > 0$ the set $E_\varepsilon(\theta_0) = \{r > 0\colon u(re^{i\theta_0}) < (h_f(\theta_0)-\varepsilon)r^\rho\}$ satisfies the condition

$$\overline{\lim_{r \to \infty}} \frac{1}{r} \, mes\{E_\varepsilon(\theta_0) \cap (0,r)\} = \kappa.$$

In turn, this implies that for any $\delta \in (0,1)$ there is a sequence $r_j \uparrow \infty$ such that

$$\lim_{j \to \infty} \frac{1}{2\delta r_j} \, mes\{E_\varepsilon(\theta_0) \cap (r_j - \delta r_j, r_j + \delta r_j)\} > \frac{\kappa}{2}.$$

Clearly, the set $E_\varepsilon(\theta_0)$ is open. Therefore, for any $j = 1, 2, \ldots$, it is possible to choose a set E_j in such a way that E_j is the union of finitely many intervals, $E_j \subset E_\varepsilon(\theta_0) \cap (r_j - \delta r_j, r_j + \delta r_j)$ and $mesE_j \geq \delta r_j \kappa$. Note that such a choice of numbers r_j and sets E_j dependent on $\delta \in (0,1)$ is possible for any δ. In the sequel we take $\delta > 0$ small such that for given $\varepsilon_1 > 0$

$$u(z) \leq (h_u(\theta_0)+\varepsilon_1)r^\rho, \ \forall z\colon |argz - \theta_0| < \delta. \qquad (2.3.4)$$

This possible because $h_u(\theta)$ is continuous and the estimate (2.1.22) holds. For this case we now estimate $\mathfrak{N}_u(r_j e^{i\theta_0}, \delta r_j)$, using the harmonic measure $W_{E'_j}(z)$ of the set $E'_j = \{r: (\delta r_j r + r_j) \in E_j\}$ with respect to the semidisc $U^+ = \{z \in \mathbb{C}: |z| < 1, \ Im z > 0\}$. According to the definition of harmonic measure, the function $W_{E'_j}(z)$ is harmonic and bounded in U^+ and has boundary value 1 at the points from E'_j and boundary value 0 on $\partial U^+ \setminus \bar{E}'_j$. We show that the integral

$$\int_{U^+} W_{E'_j}(z) d\omega_z$$

is bounded away from zero by some positive constant κ_1 which depends on κ only. In fact, according to the standard representation of a harmonic function by Green's function we have

$$W_{E'_j}(z) = \int_{E'_j} \frac{\partial G(z,\zeta)}{\partial n_\zeta} d\zeta ,$$

where $G(z,\zeta)$, $z = x+iy$, $\zeta = \xi+i\eta$, is the Green's function of U^+. It is obvious that

$$min\left\{\frac{\partial G(z,\xi)}{\partial n_\zeta}: \ \frac{1}{4} \le |z| \le \frac{3}{4}, \ \frac{1}{4}\cdot\pi \le argz \le \frac{3}{4}\cdot\pi, \ -1 \le \xi \le 1\right\} = \kappa_2 \ge 0.$$

Hence

$$min\left\{W_{E'_j}(z): \ \frac{1}{4} \le |z| \le \frac{3}{4}, \ \frac{1}{4}\cdot\pi \le argz \le \frac{3}{4}\cdot\pi\right\} \ge \kappa_2 mesE'_j \ge \frac{\kappa_2\kappa}{2},$$

and therefore

$$\int_{U^+} W_{E'_j}(z) d\omega_z \ge \kappa_1 = \frac{\kappa_2\kappa}{4*2} \cdot \frac{\pi}{2} > 0. \qquad (2.3.5)$$

Let us assume, for the sake of being specific, that $h_u(\theta_0) - \varepsilon > 0$. Then

$$u(re^{i\theta_0}) \le (h_u(\theta_0)-\varepsilon)r^\rho \le (h_u(\theta_0)-\varepsilon)(1+\delta)^\rho r_j^\rho , \ \forall r \in E_j.$$

This and (2.3.4), taking into account the properties of $W_{E'_j}(z)$, imply

$$u(r_j e^{i\theta_0} + \delta r_j e^{i\theta_0}\xi) \le (1-W_{E'_j}(\xi))(h_f(\theta_0)+\varepsilon_1)r_j^\rho +$$

$$+ (h_u(\theta_0)-\varepsilon)(1+\delta)^\rho r_j^\rho W_{E'_j}(\xi), \ \forall \xi \in U^+.$$

The analogous inequality is valid in the semidisc $U^- = -U^+$. Using these inequalities and the estimate (2.3.5) we conclude that

$$r_j^{-\rho}\mathfrak{N}_u(r_j e^{i\theta_0}, \delta r_j) = \frac{1}{\pi r_j^\rho}\int_{B_1} u(r_j e^{i\theta_0}(1+\delta\xi))d\omega_\xi \le$$

$$\le h_u(\theta_0) + \frac{2\varepsilon_1}{\pi}\int_{U^+}(1-W_{E'_j}(\xi))d\omega_\xi + \frac{2}{\pi}h_u(\theta_0)[(1+\delta)^\rho - 1]\int_{U^+} W_{E'_j}(\xi)d\omega_\xi +$$

$$-\varepsilon(1+\delta)^\rho\int_{U^+} W_{E'_j}(\xi)d\omega_\xi \le h_u(\theta_0) + h_u(\theta_0)[(1+\delta)^\rho - 1] - \varepsilon(1+\delta)^\rho\kappa_1.$$

Hence

$$\lim_{\delta\to 0}\ \underline{\lim}_{r\to\infty} r^{-\rho}\mathfrak{N}_u(re^{i\theta_0}, \delta r) \le h_u(\theta_0) + \varepsilon_1 - \varepsilon\kappa_1.$$

Choosing $\varepsilon_1 < \varepsilon\kappa_1$, we now obtain that

$$\lim_{\delta\to 0}\ \underline{\lim}_{r\to\infty} r^{-\rho}\mathfrak{N}_u(re^{i\theta_0}, \delta r) < h_u(\theta_0).$$

Thus the condition (2.3.2) is sufficient for the validity of (2.3.1) and some set E of relative measure zero.

The proof is completed.

Now it is quite naturally to introduce the following definition, due to L.Gruman.

A ray $l_{x_0} = \{ x\in\mathbb{R}^n: x = tx_0, t > 0 \}$, $x_0 \ne 0$, is called a ray of c.r.g. of a function $u\in SH(\mathbb{R}^n,\rho]$, $n \ge 2$, if the equality

$$\lim_{\delta\to 0}\ \underline{\lim}_{r\to\infty} \frac{1}{r^\rho}\mathfrak{N}_u(rx_0,\delta r) = \mathfrak{L}_u^*(x_0)$$

is valid at the point x_0.

Note that, as in the case n = 2 the inequality

$$\lim_{\delta\to 0}\ \overline{\lim}_{r\to\infty} \frac{1}{r^\rho}\mathfrak{N}_u(rx_0,\delta r) \le \mathfrak{L}_u^*(x_0)$$

is true for any function $u\in SH(\mathbb{R}^n,\rho]$ (this follows from **Hartogs' Lemma** in view of the upper semicontinuity of the function $\mathfrak{L}_u^*(x)$).

It follows from **Lemma 2.3.2** that for n = 2 the rays of c.r.g. of a subharmonic function $u = \ln|f(z)|$, $f\in H(\mathbb{C},\rho]$, coincides with the rays of c.r.g. of the entire function f(z). Besides, analysis of the proof of **Lemma 2.3.2** shows that the part concerning the necessity of condition (2.3.2) can be generalized to the multidimensional case without essential changes. Thus, the following sufficient condition for c.r.g. is obtained.

If a function $u\in SH(\mathbb{R}^n,\rho]$ satisfies on a ray l_{x_0} the condition

$$\lim_{t\to\infty,\ t\notin E} \frac{1}{t^\rho} u(tx_0) = \mathfrak{L}_u^*(x_0),$$

where E is a set in $\mathbb{R}_+$ of relative measure zero, then the ray l_{x_0} is a ray of c.r.g. for the function u.

It is convenient (but not necessary) to state the properties of rays of c.r.g. using so-called limit sets of subharmonic functions, introduced by V.Azarin [2], [3]. This will be done in §3.3.

2. Limit sets. Let $u\in SH(\mathbb{R}^n,\rho]$. Then the set of subharmonic functions v that can be represented in the form

$$v(x) = \mathcal{D}'\text{-}\lim_{j\to\infty} u^{[t_j]}(x),$$

where $t_j \to \infty$ as $j\to\infty$, is called the limit set of u and is denoted by *Fru*.

Let us note some properties of the sets *Fru*.

An obvious consequence of **Lemma 2.1.3** is the following property

a) $Fru \neq \emptyset$, $\forall u\in SH(\mathbb{R}^n,\rho]$, $u \not\equiv -\infty$.

It follows immediately from the definition of *Fru* that

b) $\sup\{v(x)\colon v\in Fru,\ x\in B_R\} \le \sigma_u R^\rho$, $\forall R > 0$,

where σ_u is the type of u with respect to the order ρ.

c) $\sup\{\mathfrak{M}_{|v|}(0,R)\colon v\in Fru\} \le 2\sigma_u R^\rho$, $\forall R > 0$.

d) $v^{[t]}\in Fru$, $\forall v\in Fru$.

e) $Fr(u_1+u_2) \subset Fru_1+Fru_2 \overset{\text{def}}{=} \{v\in SH(\mathbb{R}^n,\rho]\colon v=v_1+v_2,\ v_i\in Fru_i,\ i=1,2\}$.

f) $Fru(x) = Fru(x+x_0)$, $\forall x_0 \in \mathbb{R}^n$.

From c) it follows that

g) $v(0) = 0$, $\forall v \in Fru$.

From point ii) of **Lemma 2.1.4** it follows that

h) The set *Fru* will not change if in its definition $\mathcal{D}'$-convergence is replaced by convergence in the topology of the space $L^1_{loc}(\mathbb{R}^n)$. Further, the following property holds:

i) The set *Fru* is compact in $L^1_{loc}(\mathbb{R}^n)$.

Indeed, by g) and h) and **Lemma 2.1.3**, from any sequence of functions $v_j \in Fru$, $j = 1, 2, \dots$, we can extract out a subsequence converging in $\mathcal{D}'(\mathbb{R}^n)$ or, equivalently (see **Lemma 2.1.4**), in $L^1_{loc}(\mathbb{R}^n)$. Therefore, to prove i) it is sufficient to show that *Fru* is closed in $L^1_{loc}(\mathbb{R}^n)$ or $\mathcal{D}'(\mathbb{R}^n)$. Let $v = \mathcal{D}' - \lim_{j \to \infty} v_j$, $v_j \in Fru$. Then according to the definition of *Fru* and by ii) of **Lemma 2.1.4**, for any $j = 1, 2, \dots$, there exists $t_j > j$ such that

$$\int_{|x|<j} |u^{[t_j]} - v_j| dx < \frac{1}{j} .$$

Now it is clear that $v = \lim_{j \to \infty} u^{[t_j]}$ and therefore $v \in Fru$. From point i) of **Lemma 2.1.4** it clearly follows that

j) $v(x) \le \mathfrak{L}^*_u(x)$, $\forall v \in Fru$, $\forall x \in \mathbb{R}^n$.

Let us show that

k) $reg\{sup v(x):\ v \in Fru\} = \mathfrak{L}^*_u(x)$.

For an arbitrary prescribed $x_0 \in \mathbb{R}^n \setminus \{0\}$ we choose $t_j \uparrow \infty$ so that $\lim_{j \to \infty} u^{[t_j]}(x_0) = \mathfrak{L}_u(x_0)$. Then we extract pick out from the sequence $\{t_j\}$ a subsequence $\{t_{j_k}\}_{k=1}^{\infty}$ such that $\exists \mathcal{D}'\text{-}\lim u^{[t_{j_k}]}$. This possible by **Lemma 2.1.3**. Then from i) of **Lemma 2.1.4** it follows that the subharmonic function $v = \mathcal{D}'\text{-}\lim u^{[t_{j_k}]}$ satisfies the inequality $v(x) \ge$

$\overline{\lim_{k\to\infty}} u^{[t_{j_k}]}(x)$, $\forall x \in \mathbb{R}^n$. Thus, we have shown that for any $x_0 \in \mathbb{R}^n\setminus\{0\}$ there exists $v \in Fru$ such that $v(x_0) \ge \mathfrak{L}_u(x_0)$. This and j) imply k).

Before passing to the next section we note that we have listed only the most elementary properties of *Fru* which will be used in the sequel.

3. Limit sets and functions of completely regular growth. First of all we reformulate **Theorem 2.2.1** in terms of limit sets.

Theorem 2.3.1. In order that $u \in SH(\mathbb{R}^n, \rho]$ be a function of c.r.g. in $\mathbb{R}^n$ it is necessary and sufficient that the set *Fru* consist of one element. This unique element is $\mathfrak{L}_u^*(x)$.

Moreover, rays of c.r.g. can be transparently described in terms of limit sets.

Theorem 2.3.2. In order that a ray $l_{x_0} = \{x \in \mathbb{R}^n: x = tx_0,\ t > 0\}$, $x_0 \ne 0$, be a ray of c.r.g. it is necessary and sufficient that on this ray the equality $v(x) = \mathfrak{L}_u^*(x)$ be true for every $v \in Fru$.

Proof. We assume that there is a function $v \in Fru$ such that $v \ne \mathfrak{L}_u^*$ in some point $x' \in l_{x_0}$. According to property j) in the previous section, this means that $v(x') < \mathfrak{L}_u^*(x')$. Hence, for some ε and a, $\varepsilon > 0$, $v(x') < a < \mathfrak{L}_u^*(x')$, the inequality $v(x') < a$ is valid for all $x \in B_\varepsilon(x')$. Since $v \in Fru$, $v = \mathcal{D}'\text{-}\lim u^{[t_j]}$. Then, according to i) of **Lemma 2.1.4** it follows that $v(x) \ge \overline{\lim_{j\to\infty}} u^{[t_j]}(x)$, $\forall x \in B_\varepsilon(x')$. Therefore $\overline{\lim_{j\to\infty}} u^{[t_j]}(x) \le a$, $\forall x \in B_\varepsilon(x')$. From this we conclude that for some j_0, ε_1 and a_1, $a < a_1 < \mathfrak{L}_u^*(x')$, $0 < \varepsilon < \varepsilon_1$ the inequality

$$u^{[t_j]}(x) \le a_1,\ \forall x \in B_{\varepsilon_1}(x'),\ j \ge j_0$$

is valid. Hence

$$\lim_{\delta\to 0} \underline{\lim_{t\to\infty}} \mathfrak{N}_{u^{[t]}}(x',\delta) < \mathfrak{L}_u^*(x'), \tag{2.3.6}$$

and since

$$\mathfrak{N}_{u^{[t]}}(x',\delta) = \frac{1}{t^\rho} \mathfrak{N}_u(tx',\delta t), \tag{2.3.7}$$

(2.3.6) means that the ray $l_{x'} = l_{x_0}$ is not of c.r.g.

The necessity part is proved.

To prove sufficiency, we assume that $\mathit{lim}_{\delta\to 0}\ \overline{\mathit{lim}}_{t\to\infty}\ t^{-\rho}\mathfrak{N}_u(tx_0,\delta t) < \mathfrak{L}^*_u(x_0)$. Then there are $\delta > 0$, $\varepsilon > 0$ and a sequence $t_j\uparrow\infty$ such that

$$\mathfrak{N}_{u^{[t_j]}}(x_0,\delta) = t_j^{-\rho}\mathfrak{N}_u(t_jx_0,\delta t_j) < \mathfrak{L}^*_u(x_0) - \varepsilon,\ j = 1,\ 2,\ \dots\ .$$

Using inequality (2.1.8) we conclude that for $|x-x_0| < \eta$, $\delta' = \delta+\eta$ the inequality

$$u^{[t_j]}(x) \le \left(\frac{\delta}{\delta'}\right)^n \mathfrak{N}_{u^{[t_j]}}(x_0,\delta) + \left(1-\left(\frac{\delta}{\delta'}\right)^n\right)c$$

holds, where $c = \mathit{sup}\{u^{[t]}(x):\ t > 1,\ x\in B_{\delta+2\eta}(x_0)\} < \infty$.

We choose $\eta > 0$ small enough such that the inequality

$$\left(\frac{\delta}{\delta'}\right)^n \mathfrak{N}_{u^{[t_j]}}(x_0,\delta) + \left(1-\left(\frac{\delta}{\delta'}\right)^n\right)c < \mathfrak{L}^*_u(x_0) - \frac{\varepsilon}{2}$$

becomes true. Then

$$u^{[t_j]}(x) \le \mathfrak{L}^*_u(x_0) - \frac{\varepsilon}{2},\ \forall x\in B_\eta(x_0). \qquad (2.3.8)$$

Now we extract from the sequence $\{t_j\}$ a subsequence $\{t_{j_k}\}$ such that

$$\exists\mathcal{D}'-\mathit{lim}_{k\to\infty}\ u^{[t_{j_k}]} = v.$$

Then

$$v = \mathit{reg}\left\{\overline{\mathit{lim}}_{k\to\infty}\ u^{[t_{j_k}]}(x)\right\},$$

and by (2.3.8)

$$v(x_0) < \mathfrak{L}^*_u(x_0) - \frac{\varepsilon}{2}\ .$$

Hence coincidence of all functions $v\in \mathit{Fr}u$ on the ray l_{x_0} is sufficient for this ray to be of c.r.g. for the function u.

The proof is finished.

The following statements are obvious corollaries of **Theorems 2.3.1** and 2.3.2.

Theorem 2.3.3. In order that the function $u \in SH(\mathbb{R}^n,\rho]$ be of c.r.g. in $\mathbb{R}^n$ it is necessary and sufficient that u be of c.r.g. on every ray l_{x_0}, $x_0 \in S_1$.

Theorem 2.3.4. Let the indicator $\mathfrak{L}_u^*(x)$ of a function $u \in SH(\mathbb{R}^n,\rho]$ be a harmonic function in the open connected cone $K^\Gamma = \{ x: x = tx_0,\ x_0 \in \Gamma,\ t > 0 \}$, $\Gamma \subset S_1$. Let the function u be of c.r.g. on some ray $l_{x_0} \subset K^\Gamma$. Then u is of c.r.g. on every ray $l_{x'} \subset K^\Gamma$.

Theorem 2.3.5. If the function $u \in SH(\mathbb{R}^n,\rho]$ is of c.r.g. on almost all rays from an open connected cone K^Γ, then

$$\mathcal{D}'-\lim_{t\to\infty} u^{[t]} = \mathfrak{L}_u^*$$

in this cone.

Theorem 2.3.6. If two sets $E \subset \mathbb{R}^n\backslash\{0\}$ and $E_1 \subset \mathbb{R}^n\backslash\{0\}$ are such that $\bar{E} \subset E_1$ and if the restriction $\mathfrak{L}_u^*|_{E_1}$ of the indicator $\mathfrak{L}_u^*(x)$ of a function $u \in SH(\mathbb{R}^n,\rho]$ to the set E_1 is continuous (on E_1), then the following implication is true:

$$l_x \text{ is a ray of c.r.g., } \forall x \in E \Rightarrow l_x \text{ is a ray of c.r.g., } \forall x \in \bar{E}.$$

In particular, if the indicator $\mathfrak{L}_u^*(x)$ of the function $u \in SH(\mathbb{R}^n,\rho]$ is continuous in $\mathbb{R}^n$, then the set of all rays l_{x_0} on which the function is of c.r.g. is closed.

Note that for n = 2 the indicator $\mathfrak{L}_u$ (or, equivalently, $h_u(\theta)$) is continuous for any function $u \in SH(\mathbb{C},\rho]$. In particular, the proof of the necessity part in **Theorem 1.3.4** is contained in that of **Theorem 2.3.6**. Recall that **Theorem 1.3.4** was given without proof.

§4 Addition of indicators

Directly from the definition of indicator of a subharmonic function it follows that

$$\mathfrak{L}_{u_1+u_2}^*(x) \le \mathfrak{L}_{u_1}^*(x) + \mathfrak{L}_{u_2}^*(x),\ \forall u_i \in SH(\mathbb{R}^n,\rho],\ i=1,2,\ x \in \mathbb{R}^n. \tag{2.4.1}$$

Similar to the case of entire functions, the assumption on c.r.g. of one of the functions u_i allows us to strengthen this statement.

Theorem 2.4.1. If the function $u_1 \in SH(\mathbb{R}^n, \rho]$ is of c.r.g. in $\mathbb{R}^n$, then

$$\mathfrak{L}^*_{u_1+u_2}(x) = \mathfrak{L}^*_{u_1}(x) + \mathfrak{L}^*_{u_2}(x), \ \forall u_2 \in SH(\mathbb{R}^n, \rho], \quad x \in \mathbb{R}^n. \tag{2.4.2}$$

Proof. Using the properties of limit sets and **Theorem 2.3.1** we find that

$$\mathfrak{L}^*_{u_1+u_2}(x) = reg\ sup\ \{\ v(x):\ v \in Fr(u_1+u_2)\} \ge$$

$$\ge reg\{\ \mathfrak{L}^*_{u_1} + sup\{\ v_2(x):\ v_2 \in Fru_2\}\ \} = \mathfrak{L}^*_{u_1}(x) + \mathfrak{L}^*_{u_2}(x).$$

This and (2.4.1) imply the equality (2.4.2).

This finishes the proof.

Theorem 2.4.2. If the function $u_1 \in SH(\mathbb{R}^n, \rho]$ is of c.r.g. on the ray l_{x_0}, $x_0 \neq 0$, then

$$\mathfrak{L}^*_{u_1+u_2}(x_0) = \mathfrak{L}^*_{u_1}(x_0) + \mathfrak{L}^*_{u_2}(x_0), \ \forall u_2 \in SH(\mathbb{R}^n, \rho].$$

Proof. From **Lemma 2.3.1** and the definition of ray of c.r.g. it follows that

$$\mathfrak{L}^*_{u_1+u_2}(x) = \lim_{\delta \to 0} \overline{\lim_{r \to \infty}}\ r^{-\rho} \mathfrak{N}_{u_1+u_2}(rx_0, \delta r) =$$

$$= \lim_{\delta \to 0} \overline{\lim_{r \to \infty}}\ r^{-\rho}(\mathfrak{N}_{u_1}(rx_0, \delta r) + \mathfrak{N}_{u_2}(rx_0, \delta r)) \ge$$

$$\ge \lim_{\delta \to 0} \left\{ \underline{\lim_{r \to \infty}}\ r^{-\rho} \mathfrak{N}_{u_1}(rx_0, \delta r) + \overline{\lim_{r \to \infty}}\ r^{-\rho} \mathfrak{N}_{u_2}(rx_0, \delta r) \right\} =$$

$$= \lim_{\delta \to 0} \underline{\lim_{r \to \infty}}\ r^{-\rho} \mathfrak{N}_{u_1}(rx_0, \delta r) + \lim_{\delta \to 0} \overline{\lim_{r \to \infty}}\ r^{-\rho} \mathfrak{N}_{u_2}(rx_0, \delta r) =$$

$$= \mathfrak{L}^*_{u_1}(x_0) + \mathfrak{L}^*_{u_2}(x_0)\ .$$

This and (2.4.1) imply that

$$\mathfrak{L}^*_{u_1}(x_0) + \mathfrak{L}^*_{u_2}(x_0) = \mathfrak{L}^*_{u_1+u_2}(x_0).$$

This finishes the proof.

Note that **Theorem 2.4.1** is a consequence of **Theorems 2.4.2** and **2.3.3**.

The property of addition of indicators comprised in the theorems given above is a characteristic property of the functions of c.r.g. under consideration. Namely, the following theorems holds:

Theorem 2.4.3. If a function $u_1 \in SH(\mathbb{R}^n, \rho]$ is such that

$$\mathfrak{L}^*_{u_1+u_2}(x) = \mathfrak{L}^*_{u_1}(x) + \mathfrak{L}^*_{u_2}(x), \ \forall u_2 \in SH(\mathbb{R}^n, \rho], \quad x \in \mathbb{R}^n,$$

then it is of c.r.g. in $\mathbb{R}^n$.

Theorem 2.4.4. If a function $u_1 \in SH(\mathbb{R}^n, \rho]$ is such one that

$$\mathfrak{L}^*_{u_1+u_2}(x_0) = \mathfrak{L}^*_{u_1}(x_0) + \mathfrak{L}^*_{u_2}(x_0), \ \forall u_2 \in SH(\mathbb{R}^n, \rho],$$

at some point $x_0 \in \mathbb{R}^n \setminus \{0\}$, then it is of c.r.g. on the ray l_{x_0}.

Note that **Theorem 2.4.3** is an obvious consequence of **Theorems 2.4.4** and **2.3.3**.

To prove **Theorem 2.4.4** we assume the contrary, i.e. that the ray l_{x_0} is not a ray of c.r.g. of the function u_1. Then there are a sequence $t_j \uparrow \infty$, $t_{j+1} > 2t_j$, and numbers $\delta > 0$, $\varepsilon > 0$ such that

$$\mathfrak{N}_{u_1}(t_j x_0, \delta t_j) < (\mathfrak{L}^*_{u_1}(x_0) - \varepsilon) t_j^\rho$$

or, equivalently,

$$\mathfrak{N}_{u_1^{[t_j]}}(x_0, \delta) < \mathfrak{L}^*_{u_1}(x_0) - \varepsilon, \ \forall j. \tag{2.4.3}$$

From this inequality it follows by (2.1.8) that for $\delta' > 0$ small enough,

$$u_1^{[t_j]}(x) \le \mathfrak{L}^*_{u_1}(x_0) - \frac{\varepsilon}{2}, \ \forall x \in B_{\delta'}(x_0), \ \forall j.$$

To finish the proof of the theorem it is now sufficient to construct a function $u_2 \in SH(\mathbb{R}^n, \rho]$ that for some $j_0 \in \mathbb{N}$, $\gamma > 0$, and for arbitrary η, $0 < \eta < \delta'$, satisfies the condition

$$u_2(x) \le \mathfrak{L}^*_{u_2}(x) - \gamma |x|^\rho, \ \forall x \in K_\eta(x_0) \setminus A,$$

where

$$K_\eta(x_0) = \{ x: x=tx', x'\in B_\eta(x_0), t>0 \},$$

$$A = \bigcup_{j>j_0} B_{\eta t_j}(t_j x_0),$$

Indeed since inequalities

$$1 - \frac{\eta}{|x^0|} \le t \le 1 + \frac{\eta}{|x^0|}$$

are valid for $tx\in B_{\eta t_j}(t_j x_0)$, in this situation we have

$$\varlimsup_{\substack{t\to\infty \\ t:tx\in A}} \{ u_1^{[t]}(x) + u_2^{[t]}(x) \} \le$$

$$\le \left(1 \pm \frac{\eta}{|x^0|}\right)\left(\mathcal{L}^*_{u_1}(x_0) - \frac{\varepsilon}{2}\right) + \mathcal{L}^*_{u_2}(x),\ \forall x\in B_\eta(x_0), \qquad (2.4.4)$$

where the sign in $\left(1 \pm \frac{\eta}{|x^0|}\right)$ must be chosen opposite to the sign of the number $\mathcal{L}^*_{u_1}(x_0) - \frac{\varepsilon}{2}$. Moreover,

$$\lim_{\substack{t\to\infty \\ t:tx\notin A}} \{ u_1^{[t]}(x) + u_2^{[t]}(x) \} \le$$

$$\le \mathcal{L}^*_{u_1}(x) + \mathcal{L}^*_{u_2}(x\ ) - \gamma|x|^\rho,\ \forall x\in B_\eta(x_0). \qquad (2.4.5)$$

From (2.2.4) and (2.4.5) it follows that

$$\mathcal{L}^*_{u_1+u_2}(x_0) \le$$

$$\le \max\left\{\left(1 \pm \frac{\eta}{|x_0|}\right)^{-\rho}\left(\mathcal{L}^*_{u_1}(x_0) - \frac{\varepsilon}{2}\right)+\mathcal{L}^*_{u_2}(x_0),\ \mathcal{L}^*_{u_1}(x_0) + \mathcal{L}^*_{u_2}(x_0) - \gamma|x_0|^\rho\right\}.$$

Hence, since $\eta > 0$ is arbitrary, we obtain the inequality

$$\mathcal{L}^*_{u_1+u_2}(x_0) < \mathcal{L}^*_{u_1}(x_0) + \mathcal{L}^*_{u_2}(x_0),$$

contradicting the condition of the theorem.

Thus, it remains to construct a function u_2 with the properties mentioned above. We look for it in the form

$$u_2(x) = |x|^{\rho} + \gamma \sum_{j=1}^{\infty} |x|^{\rho}\chi\left(\frac{x-t_jx_0}{2t_j\eta}\right),$$

where $\chi(x)$ is a function in $C^{\infty}(\mathbb{R}^n)$ satisfying the following conditions: $\chi(x) = 1$, $\forall x\in B_{1/2}$; $\chi(x) = 0$, $\forall x\notin B_1$; $0 < \chi(x) \le 1$, $\forall x\in\mathbb{R}^n$.

We show that if $\gamma > 0$ is small enough, then $u_2(x)$ is subharmonic function. To proof this we estimate its Laplacian. We use the following notation: $a = max\{|\Delta\chi(x)|: x\in\mathbb{R}^n\}$ and $b = \{|grad\chi(x)|: x\in\mathbb{R}^n\}$. It can be shown by direct calculation that

$$\Delta u_2(x) \ge$$

$$\ge \Delta|x|^{\rho} - a\gamma \sum_{j=j_0}^{\infty} \frac{1}{4\eta^2 t_j^2}\, \chi\left(\frac{x-t_jx_0}{4t_j\eta}\right) - \gamma b|grad|x|^{\rho}| \sum_{j=1}^{\infty} \chi\left(\frac{x-t_jx_0}{4t_j\eta}\right) \ge$$

$$\ge |x|^{\rho-2}\left(\rho(\rho+n-2) - \gamma\left[\frac{a}{4\eta^2}(|x_0|+\eta)^2 + \frac{bn\rho}{2\eta}\right]\sum_{j=1}^{\infty}\chi\left(\frac{x-t_jx_0}{4t_j\eta}\right)\right) \ge$$

$$\ge |x|^{\rho-2}\left(\rho(\rho+n-2) - \gamma\left[\frac{a}{4\eta^2}(|x_0|+\eta)^2 + \frac{bn\rho}{2\eta}\right]\right) =$$

$$= |x|^{\rho-2}(\rho(\rho+n-2) - \gamma c) .$$

Therefore, if $\gamma < \frac{1}{c}\cdot\rho(\rho+n-2)$, then the function $u_2(x)$ is subharmonic in $\mathbb{R}^n$. Then it is obvious that $u_2\in SH(\mathbb{R}^n,\rho]$. Moreover, it follows directly from the construction of the function u_2 that if $x\in B_{\eta}(x_0)$ then $\mathfrak{L}^*_{u_2}(x) = |x|^{\rho} + \gamma|x|^{\rho}$ and $u_2(x) = |x|^{\rho} \le \mathfrak{L}^*_{u_2}(x) - \gamma|x|^{\rho}$ if $x\in K_{\eta}(x_0)\setminus A$.

The proof is finished.

Notes

The term "Carleson α-measure" was introduced by V.Azarin [1]. **Lemma 2.1.6** concerning this notion is due V.Azarin too (see V.Azarin [2]). A statement which is close to **Lemma 2.1.8** in the case n = 2, is contained in the paper by P.Agranovič and L.Ronkin [2]. A similar statement for the case n > 2 can found in the paper P.Agranovič [1], which contains in fact **Lemma 2.1.7**. The form of **Lemma 2.1.8** given here was were obtained by V.Azarin [2]. Sources to which readers may turn to in connection with §1.3 are unknown to the author. **Theorems 2.1.1** and **2.1.2** were first obtained by M.Brelot [1], and were repeated by P.Lelong [1] and Hayman-Kennedy [1].

The theory of functions of completely regular growth was extended to the case of subharmonic functions in $\mathbb{R}^n$, $n \geq 3$, by V.Azarin [5]-[8] without using $\mathcal{D}'$-convergence. As already mentioned, the relationship between the regularity of growth of functions $u \in SH(\mathbb{R}^n,\rho]$ and convergence in $\mathcal{D}'$ of the functions $u^{[t]}$ was ascertained by P.Agranovič and L.Ronkin [1]-[3], V.Azarin [1]-[2], and P.Agranovič [1], almost simultaneously. Moreover, at the same time the functions $u \in SH(\mathbb{R}^n,\rho]$ for which $\exists \mathcal{D}'\text{-}lim\ u^{[t]}$ were considered by C.Kiselman in his unpublished papers, but a relationship between such functions and functions of c.r.g. was not established. **Theorem 2.2.1** was obtained by P.Agranovič [1], and in part earlier by P.Agranovič and L.Ronkin [1] (see also [2], [3]). **Theorem 2.2.2** follows from more general results concerning limit sets due to V.Azarin [2]. **Theorems 2.2.3** and **2.2.4** were obtained by S.Favorov [1],[2]. **Theorem 2.2.7** was proved by V.Azarin [5], [6]. **Theorems 2.2.5** and **2.2.6**, which preceded **Theorem 2.2.7**, can be regarded to as consequences of **Theorems 2.2.1 and 2.2.7**. The definition of ray of c.r.g. for functions $u \in SH(\mathbb{R}^n,\rho]$ given in §3 is due to L.Gruman [1]. The limit sets *Fru* for functions $u \in SH(\mathbb{R}^n,\rho]$ were introduced by V.Azarin in [1] for the case n = 2 and in [2] for the case n > 2. However the relationship between the sets *Fru* and rays of c.r.g. in $\mathbb{R}^n$ (in Gruman's sense) was not considered. **Theorems 2.3.4** and **2.3.6** (for n > 2) were obtained by L.Gruman [1] (see also Lelong-L.Gruman [1]). **Theorem 2.3.5** is contained in the book of Lelong-Gruman [1], in which the authors define a function of c.r.g. on a set of rays as a function in $SH(\mathbb{R}^n,\rho]$

that it is of c.r.g. on every of the rays under consideration. **Theorem 2.4.1** was obtained simultaneously by S.Favorov [3] and C.Kiselman (unpublished). **Theorem 2.4.3** is due to S.Favorov [3]. **Theorems 2.4.2** and **2.4.4** are contained in the book of Lelong-Gruman [1] mentioned above.

CHAPTER 3

ENTIRE FUNCTIONS OF COMPLETELY REGULAR GROWTH IN $\mathbb{C}^n$

Here and in the sequel we use the standard notation of the multidimensional complex analysis. In particular: $d = \partial+\bar{\partial}$, $d^c = \frac{i}{4}(\partial-\bar{\partial})$, $\mathbb{P}^{n-1}$ is projective space, i.e. the space of complex rays $l_\lambda = \{z\in\mathbb{C}^n: z = \lambda w,\ w\in\mathbb{C}\}$, $\lambda\in\mathbb{C}^n\setminus\{0\}$, $\omega = dd^c \ln|z|^2$ is the Fubini-Study form, and $dw_{2n-2} = \frac{1}{(n-1)!}\omega^{n-1}$ is the volume element (form) of the space $\mathbb{P}^{n-1}$ in the Fubini-Study metric[13]. For $k = (k_1,\dots,k_n)\in\mathbb{Z}^n_+$, $z = (z_1,\dots,z_n)\in\mathbb{C}^n$ we set $\|k\| = k_1+\dots+k_n$, $k! = k_1!\cdot\ \dots\ \cdot k_n!$, $z^k = z_1^{k_1}\cdot\dots\cdot z_n^{k_n}$, $\langle z,\xi\rangle = z_1\xi_1+\dots+z_n\xi_n$, $\frac{\partial^{\|k\|}}{\partial z^k} = \frac{\partial^{k_1+\dots+k_n}}{\partial z_1^{k_1}\dots\partial z_n^{k_n}}$.

The space of all holomorphic functions in a domain $G\subset\mathbb{C}^n$ is denoted by $H(G)$ and the set of plurisubharmonic functions[14] in $G\subset\mathbb{C}^n$ is denoted by $PSH(G)$. The volume element (form) in $\mathbb{C}^n$ is denoted by $d\omega_z$. In other words, $d\omega_z = d\omega_{z_1}\wedge\dots\wedge d\omega_{z_n} = \left(\frac{i}{2}\right)^n dz_1\wedge d\bar{z}_1\wedge\dots\wedge dz_n\wedge d\bar{z}_n$.

---

13) For a more detailed description of projective space and integration in it see, for example, E.Čirka [1], L.Ronkin [1].

14) Recall that a function $u(z)$, $z\in\mathbb{C}^n$, is called plurisubharmonic in a domain $G\subset\mathbb{C}^n$ if it is upper semicontinuous and if for any $a\in\mathbb{C}^n$, $b\in\mathbb{C}^n$ the function $\Phi_{a,b}(w) = u(aw+b)$, where $w\in\mathbb{C}$, is subharmonic on the set $\{w\in\mathbb{C}: aw+b\in G\}$. For more detailed information on plurisubharmonic functions see, for example, V.Vladimirov [2], P.Lelong [2], L.Ronkin[1], Lelong-Gruman [1].

§1 Functions of c completely regular growth on complex rays

1. Definition of entire function of c. r.g. and general remarks. Let $f(z)$ be an entire function in $\mathbb{C}^n$, i.e. $f\in H(\mathbb{C}^n)$. Then the function $ln|f(z)|$ is plurisubharmonic, and is therefore subharmonic in $\mathbb{C}^n = \mathbb{R}^{2n}$. We agree to relate all characteristics of the subharmonic function $ln|f(z)|$ directly to the function f. For example, the respective order and type of the function $u(z) = ln|f(z)|$ will be called the order ρ_f and type σ_f of the entire function f(z). The indicators $\mathfrak{L}_f(z)$ and $\mathfrak{L}^*_f(z)$ are taken to be the corresponding indicators of the function $ln|f(z)|$. An entire function f(z) is called a function of at most normal type with respect to the order $\rho > 0$ if $ln|f|\in SH(\mathbb{R}^{2n},\rho] = SH(\mathbb{C}^n,\rho]$. The set of all such functions will be denoted by $H(\mathbb{C}^n,\rho]$.

Let us call $f\in H(\mathbb{C}^n,\rho]$ a function of completely regular growth in $\mathbb{C}^n$ or, on a ray $l_{z^0} = \{z\in\mathbb{C}^n: z = tz^0, t > 0\}$, if $ln|f(z)|$ as a function from $SH(\mathbb{C}^n,\rho]$ is a function of c.r.g. in $\mathbb{C}^n$ (on the ray l_{z^0}, respectively).

Note that $ln|f(z)|$ is not only subharmonic but is also plurisubharmonic. Therefore its indicators $\mathfrak{L}^*_f(z)$ and $\mathfrak{L}_f(z)$ (which are obviously also plurisubharmonic) have all the properties mentioned above of indicators of subharmonic functions, as well as some additional properties. So the set $\{y\in\mathbb{R}^n: \mathfrak{L}_f(iy) < \mathfrak{L}^*_f(iy)\}$ has Lebesgue measure zero in $\mathbb{R}^n$ (see, for example, L.Ronkin [1]), and, moreover, as follows from the positive solution of the Second Lelong Problem by E.Bedford and B.Taylor [1], the set $\{z\in\mathbb{C}^n: \mathfrak{L}_f(z) < \mathfrak{L}^*_f(z)\}$ is pluripolar[15].

---

[15] A set $E \subset \mathbb{C}^n$ is called pluripolar ($\mathbb{C}^n$-polar) if there exists a function $u\in PSH(\mathbb{C}^n)$, $u \not\equiv -\infty$, such that $E \subset \{z\in\mathbb{C}^n: u(z) = -\infty\}$.

The Second Lelong Problem is customarily the question on the pluripolarity of the set $\{z\in\mathbb{C}^n: v(z) < v^*(z)\}$, where $v(z) = \sup_\alpha u_\alpha(z)$, and $\{u_\alpha(z)\}$ is some uniformly bounded set of functions which are plurisubharmonic in $G \subset \mathbb{C}^n$.

As in the case n = 1, the Riesz-associated measure $\mu_{ln|f|}$ of the function $ln|f(z)|$ is closely (and geometrically simply) connected with the zero distribution of f(z). More exactly, it is connected with the divisor Z_f. Recall that the divisor Z_f of an entire (or holomorphic in a domain) function f(z) is the pair$(|Z_f|,\gamma_f(z))$ consisting of the support $|Z_f| \overset{def}{=} \{z\in\mathbb{C}^n: f(z) = 0\}$ and the multiplicities (at the point z) $\gamma_f(z) \overset{def}{=} max\left\{ p: \frac{\partial^{\|k\|}f(z)}{\partial z^k} = 0, \forall k: \|k\|<p \right\}$. We denote by $V_f(G)$ the (2n-2)-dimensional Euclidean volume of the divisor Z_f in the domain G, i.e.

$$V_f(G) = \int_{G\cap Z_f} \gamma_f(z)dV_{2n-2},$$

where dV_{2n-2} is the (2n-2)-dimensional Euclidean volume element of Z_f or, equivalently,

$$dV_{2n-2} = \frac{1}{(n-1)!}\left(\frac{i}{2}(dz_1\wedge d\bar{z}_1+\ldots+dz_n\wedge d\bar{z}_n)\right)^{n-1}\Big|_{Z_f}.$$

The above-mentioned relationship between the measure $\mu_{ln|f|}$ and the divisor Z_f lies in the existence of the equality[16)]

$$\mu_{ln|f|}(G) = \frac{2\pi}{\theta_{2n}}\cdot V_f(G), \forall G\subset\mathbb{C}^n,$$

which can be naturally seen as the coincidence of $\mu_{ln|f|}$ with the volume of the divisor Z_f (for a more detailed account of this subject see, for example, Lelong-Gruman [1], E.Čirka [1], L.Ronkin [1]). So it is natural to interpret cone density and regular distribution of the measure $\mu_{ln|f|}$ as cone density and regular distribution of the divisor Z_f, and to write $\hat{\nu}_f$ instead of $\hat{\nu}_{ln|f|}$. Thus, if the cone density of the divisor Z_f exists (i.e. the weak limit of the corresponding measures exists), then for any domain Γ on the sphere S_1 such that $\hat{\nu}_f(\partial\Gamma) = 0$ the following equality is true:

---

16) Recall that $\theta_{2n} = (2n-2)\int_{S_1} dS_1$, $n > 1$, $\theta_2 = 2\pi$.

$$\hat{\nu}_f(\Gamma) = \lim_{r \to \infty} V_f(K^{\Gamma} \cap B_r).$$

Taking all this into account, each theorem in **Chapter 2** concerning subharmonic functions of c.r.g. can be reformulated to a theorem concerning entire functions. Statements obtained in such a manner, with the natural exception of the converse theorems concerning addition of indicators, are consequences of the corresponding statements for subharmonic functions.The converse theorems on the addition of indicators need special consideration, which we will give in §3. As an example we state one of the mentioned consequences.

Theorem 3.1.1. In order that a function $f \in H(\mathbb{C}^n, \rho]$, $\rho > 0$, be a function of c.r.g. in $\mathbb{C}^n$ it is necessary and sufficient that the divisor Z_f be regularly distributed.

2. Functions of c.r.g. on complex rays. A traditional problem in the theory of entire functions of several variables is the problem of describing properties of entire function under the condition that its restrictions to some family of complex lines have prescribed properties. For example, the following statement is a result in this direction: A restriction on the growth of the restrictions $f|_{L_\lambda}$ of the function $f(z)$ to the rays $L_\lambda = \{z \in \mathbb{C}^n: z = \lambda w, w \in \mathbb{C}\}$, $\lambda \in \mathbb{C}^n$, can in some sense be generalized to the growth of the function $f(z)$ itself, more exactly, to the growth of the function $M_f(r) = \max_{|z|=r} |f(z)|$ (see O.Sire [1], Lelong-Gruman [1], L.Ronkin [1],S.Favorov [1], etc.). The similar problem with respect to c.r.g. is also natural.

Theorem 3.1.2. Let $f \in H(\mathbb{C}^n, \rho]$, $\rho > 0$, and let for almost all rays $L_\lambda \in \mathbb{P}^{n-1}$ the function $f|_{L_\lambda}$ be a function of c.r.g. in $\mathbb{C}$. Then $f(z)$ is of c.r.g. in $\mathbb{C}^n$.

Proof. Put $u_t(z) = t^{-\rho} \ln|f(tz)|$. In accordance with**Theorem 2.2.1'** it is sufficient to prove the existence of

$$\lim_{t \to \infty} \int_{B_R} u_t(z)\varphi(z)d\omega_z, \ \forall \varphi \in D(B_R), \ B_R = \{z \in \mathbb{C}^n: |z| < R\}. \qquad (3.1.1)$$

It is well-known (see, for example, E.Čirka [1], p.124) that the equality

$$\int_{S_R} v(z)d\sigma(z) = \int_{\mathbb{P}^{n-1}} dW_{2n-2}(L_\lambda) \int_0^{2\pi} v\left(\frac{\lambda}{|\lambda|} Re^{i\theta}\right)d\theta \qquad (3.1.2)$$

holds for any function $v(z)$ which is integrable on the sphere $S_R = \{z\in\mathbb{C}^n\colon |z| = R\}$, as does

$$\int_{B_R} v(z)d\omega_z = \int_{\mathbb{P}^{n-1}} dW_{2n-2}(L_\lambda) \int_{|w|<R} v\left(\frac{\lambda}{|\lambda|} w\right)|w|^{2n-2}d\omega_w \quad (z\in\mathbb{C}^n) \qquad (3.1.2')$$

for any integrable function $v(z)$ in B_R. Therefore

$$\int_{B_R} u_t(z)\varphi(z)d\omega_z = \int_{B_{1/t}} u_t(z)\varphi(z)d\omega_z + \int_{\frac{1}{t}\le|z|\le R} u_t(z)\varphi(z)d\omega_z =$$

$$= \frac{1}{t^{\rho+2n}} \int_{B_1} ln|f(z)|\varphi\left(\frac{z}{t}\right)d\omega_z +$$

$$+ \int_{\mathbb{P}^{n-1}} dW_{2n-2}(L_\lambda) \int_{\frac{1}{t}\le|w|\le R} u_t\left(\frac{\lambda}{|\lambda|} w\right)\varphi\left(\frac{\lambda}{|\lambda|} w\right)|w|^{2n-2}d\omega_w. \qquad (3.1.3)$$

From the estimate $u_t(\lambda w) < c_1|\lambda|^\rho|w|^\rho+c_2$, which holds since $f\in H(\mathbb{C}^n,\rho]$, and inequality (2.1.2) it follows that for $t > 1$,

$$\left|\int_{\frac{1}{t}\le|w|\le R} u_t\left(\frac{\lambda}{|\lambda|} w\right)\varphi\left(\frac{\lambda}{|\lambda|} w\right)|w|^{2n-2}d\omega_w\right| =$$

$$=\left|\int_{1/t}^{R} r^{2n-1}dr \int_0^{2\pi} u_t\left(\frac{\lambda}{|\lambda|} re^{i\theta}\right)\varphi\left(\frac{\lambda}{|\lambda|} re^{i\theta}\right)d\theta\right| \le$$

$$\le \|\varphi\|_\infty \int_{1/t}^{R} r^{2n-1}dr \int_0^{2\pi} \left|u_t\left(\frac{\lambda}{|\lambda|} re^{i\theta}\right)\right|d\theta \le$$

$$\le 4\pi\|\varphi\|_\infty \int^{R} (c_1r^{\rho+2n-1}+c_2r^{2n-1})dr +$$

$$-\|\varphi\|_\infty \int_{1/t}^{R} r^{2n-1}dr \int_0^{2\pi} u_t\left(\frac{\lambda}{|\lambda|}\, e^{i\theta}\right)d\theta \le$$

$$\le c(R) + \int_0^{R} r^{2n-1}dr \int_0^{2\pi} \left|u\left(\frac{\lambda}{|\lambda|}\, e^{i\theta}\right)\right|d\theta =$$

$$= c(R) + \frac{R^n}{n} \int_0^{2\pi} \left|u\left(\frac{\lambda}{|\lambda|}\, e^{i\theta}\right)\right|)|d\theta = \Phi(L_\lambda,R). \tag{3.1.4}$$

Since

$$\int_{\mathbb{P}^{n-1}} dW_{2n-2}(L_\lambda)\int_0^{2\pi} \left|u\left(\frac{\lambda}{|\lambda|}e^{i\theta}\right)\right|d\theta = \int_{S_1} |u(z)|d\sigma(z) < \infty,$$

the function $\Phi(L_\lambda,R)$, bounding the function

$$\left|\int_{\frac{1}{t}\le|w|\le R} u_t\left(\frac{\lambda}{|\lambda|}\, w\right)\varphi\left(\frac{\lambda}{|\lambda|}\, w\right)|w|^{2n-2}d\omega_w\right|$$

for any $t > 1$, satisfies the condition: $\int_{\mathbb{P}^{n-1}} \Phi(L_\lambda,R)dW_{2n-2}(L_\lambda) < \infty$. At the same time as it follows from **Theorem 2.2.1** and that for almost all $L_\lambda \in \mathbb{P}^{n-1}$,

$$\lim_{t\to\infty} \int_{\frac{1}{t}\le|w|\le R} u_t\left(\frac{\lambda}{|\lambda|}\, w\right)\varphi\left(\frac{\lambda}{|\lambda|}\, w\right)d\omega_w$$

exists. Thus we may apply to the right-hand part of (3.1.3) the theorem on the dominated convergence. Hence the limit (3.1.1) exists.

The proof is finished.

As in the case of subharmonic functions, the converse theorem is not true. This follows from a theorem of S.Favorov [5], cited here without proof.

Theorem 3.1.3. For any given $\rho > 0$ there exists a function $f(z)\in H(\mathbb{C}^2,\rho]$ of c.r.g. in $\mathbb{C}^2$ which for almost all L_λ is not of c.r.g. in $\mathbb{C}$ as a function $f|_{L_\lambda}$.

3. Functions of c.r.g. in distinguished variable. By considering some parameterization of the space $\mathbb{P}^{n-1}$ (more exactly, of its charts) we actually reduce the problem on the relationship between the functions f and $f|_{L_\lambda}$ to the problem on the relationship between the properties of the function $\Phi(\lambda,w) = f(\lambda w)$ in the large and its properties as a function of w for fixed λ. A similar problem can be formulated in the general case as well, i.e. for functions $\Phi(\lambda,w)$, $\lambda\in\mathbb{C}^n$, $w\in\mathbb{C}$, which are not necessary generated as above by some entire function of n variables. Here we give two statements concerning functions f(z,w) of completely regular growth in w. The proofs will be omitted, since they are quite evident, given **Chapter 2** and §§ 1.1, 1.2 of this chapter.

Let f(z,w), $z\in\mathbb{C}^n$, $w\in\mathbb{C}$, be an entire function. We set

$$M_f(R,r) = \max_{|z|=R,\,|w|=r} |f(z,w)|.$$

We denote by $H(\mathbb{C}^{n+1},w,\rho]$ the set of entire functions f(z,w) satisfying the condition

$$\overline{\lim_{r\to\infty}} \frac{\ln M_f(R,r)}{r^\rho} < \infty, \ \forall R>0.$$

We introduce indicators with respect to w of a function $f\in H(\mathbb{C}^{n+1},w,\rho]$ similarly to the case of the radial indicators $\mathcal{L}_f$ and $\mathcal{L}_f^*$, as follows:

$$\mathcal{L}_{f,1}(z,w) = \overline{\lim_{t\to\infty}} \frac{\ln|f(z,tw)|}{t^\rho},$$

$$\mathcal{L}^*_{f,1}(z,w) = \lim_{\varepsilon\to 0} \sup\{\mathcal{L}_{f,1}(z',w'):\ |z-z'|<\varepsilon,\ |w-w'|<\varepsilon\}.$$

It is clear that $\mathcal{L}^*_{f,1}(z,w)$ is a plurisubharmonic function. Hence the distribution $\frac{1}{2\pi}\frac{\partial^2}{\partial w\partial\bar{w}}\mathcal{L}^*_{f,1}(z,w)$ is positive, and is therefore a positive measure. We denote this measure by $\mu_{f,1}$.

In the situation under consideration, i.e. when the variable w is distinguished, it is natural to describe the zero distribution of f(z,w) using the function $n_f(t,\theta_1,\theta_2,z)$ equal to the number of zeros

$$n_f(t;\theta_1,\theta_2;G) = \int_G n_f(t,\theta_1,\theta_2,z)d\omega_z,$$

$$A(G,\theta_1,\theta_2) = \{(z,w): z\in G,\ \theta_1 < \arg w < \theta_2\}.$$

Theorem 3.1.4. Let $f(z,w)\in H(\mathbb{C}^{n+1},w,\rho]$, $\rho > 0$, and let for almost all $z\in\mathbb{C}^n$ the function $f(z,w)$, as function of w, be of c.r.g. in $\mathbb{C}$. Then

1) $\exists\mathcal{D}'-\lim\limits_{t\to\infty} \dfrac{\ln|f(z,tw)|}{t^\rho} = \mathfrak{L}^*_{f,1}(z,w)$;

2) if the domain $A = A(G,\theta_1,\theta_2)$ is such that $\mu_{f,1}(\partial A) = 0$, then

$$\exists \lim_{t\to\infty} \frac{n_f(t;\theta_1,\theta_2;G)}{t^\rho} = \mu_{f,1}(G\times Y(1;\theta_1,\theta_2)).$$

§2 Addition of indicators

The following theorems are obvious consequences of **Theorems 2.4.1** and **2.4.2** on the addition of indicators of subharmonic functions.

Theorem 3.2.1. If $f_1\in H(\mathbb{C}^n,\rho]$, $\rho > 0$, is a function of c.r.g. then

$$\mathfrak{L}^*_{f_1f_2} = \mathfrak{L}^*_{f_1} + \mathfrak{L}^*_{f_2},\quad \forall f_2\in H(\mathbb{C}^n,\rho].$$

Theorem 3.2.2. If $f_1\in H(\mathbb{C}^n,\rho]$ is a function of c.r.g. on the ray $l_{z^0} = \{z:\ z=tz^0,\ t > 0\}$, $z^0 \neq 0$, then

$$\mathfrak{L}^*_{f_1f_2}(z^0) = \mathfrak{L}^*_{f_1}(z^0) + \mathfrak{L}^*_{f_2}(z^0),\quad \forall f_2\in H(\mathbb{C}^n,\rho].$$

Since the set of functions of the form $\ln|f|$, where $f\in H(\mathbb{C}^n,\rho]$, is a proper subset of $SH(\mathbb{C}^n,\rho]$, the converse theorems are not included in the corresponding theorems for subharmonic functions. However, as will be shown, they are valid, i.e. the following theorems hold.

Theorem 3.2.3. If $f_1\in H(\mathbb{C}^n,\rho]$, $\rho > 0$, and

$$\mathfrak{L}^*_{f_1f_2} = \mathfrak{L}^*_{f_1} + \mathfrak{L}^*_{f_2},\quad \forall f_2\in H(\mathbb{C}^n,\rho],$$

then the function f_1 is of c.r.g.

Theorem 3.2.4. If $f_1 \in H(\mathbb{C}^n,\rho]$, $\rho > 0$, and at the point $z^0 \neq 0$ the equality

$$\mathfrak{L}^*_{f_1 f_2}(z^0) = \mathfrak{L}^*_{f_1}(z^0) + \mathfrak{L}^*_{f_2}(z^0), \ \forall f_2 \in H(\mathbb{C}^n,\rho],$$

holds, then the function $f_1(z)$ is of c.r.g. on the ray l_{z^0}.

As in the case of subharmonic functions, **Theorem 3.2.3** is a consequence of **Theorems 3.2.4** and 2.3.2. Further analysis of the proof of **Theorem 2.4.4** shows that to **Theorem 3.2.4** it is sufficient to construct for a given sequence $r_j \uparrow \infty$, $r_{j+1} > 2r_j$, $\forall j$, an entire function $f_2 \in H(\mathbb{C}^n,\rho]$ satisfying for some sufficient small η and $\gamma = \gamma(\eta) > 0$ the condition

$$\ln|f_2(z)| \le (1-\gamma)\mathfrak{L}^*_{f_2}(z), \ \forall z \in \mathfrak{A}_1 \setminus \mathfrak{A}_2,$$

where

$$\mathfrak{A}_1 = \bigcup_{R \ge 0} \left\{ z \in \mathbb{C}^n : \left| z - \frac{z^0}{|z^0|} R \right| < \eta R \right\},$$

$$\mathfrak{A}_2 = \bigcup_{j=1}^{\infty} \left\{ z \in \mathbb{C}^n : \left| z - \frac{z^0}{|z^0|} r_j \right| < 4r_j \right\}.$$

Without loss of generality we may assumed that $z^0 = (1,0,\dots,0)$. Then it is easy to see that any function $f(z_1) \in H(\mathbb{C}^n,\rho]$ which satisfies the following conditions for sufficiently small $\eta_1 > 0$ and $\gamma_1 = \gamma_1(\eta_1) > 0$ may be taken as the function $f_2(z)$ sought for:

i) $\ln|f(z_1)| \le |z_1|^\rho + o(|z_1|^\rho)$ as $z_1 \to \infty$, $z_1 \in \mathfrak{A}'_1 = \{z_1 \in \mathbb{C} : |\arg z| < < \arcsin \eta\}$;

ii) $\ln|f(z_1)| \le (1-\gamma_1)|z_1|^\rho$ for $z_2 \in \mathfrak{A}'_1 \setminus \mathfrak{A}'_2$, $|z_1| > R_0$, where $\mathfrak{A}'_2 = \bigcup_j \{z_1 \in \mathbb{C} : |z_1 - r_j| < \eta r_j\}$;

iii) $h_f(0) = 1$.

To construct such a function we use the Mittag-Leffler function

$$E_\rho(w) = \sum_{m=0}^{\infty} \frac{w^m}{\Gamma\left(\frac{w}{\rho}+1\right)}, \ w \in \mathbb{C}.$$

It is known (see, for example, M.Džrbasjan [1]) that $E_\rho(w)\in H(\mathbb{C},\rho]$, and if $|w| \to \infty$, then

$$E_\rho(w) = \begin{cases} \rho e^{w^\rho} + O\left(\frac{1}{|w|}\right), & |\arg w| \le \frac{\pi}{2\rho}, \ \rho \ge \frac{1}{2} \\ O\left(\frac{1}{|w|}\right), & |\arg w| \ge \frac{\pi}{2\rho}, \ \rho \ge \frac{1}{2} \\ \rho e^{w^\rho}(1/+o(1)), & 0 < \rho \le \frac{1}{2} \end{cases} \tag{3.2.1}$$

We look for the function $f(w)$, $w\in\mathbb{C}$, in the form

$$f(w) = \kappa(w)E_\rho(w) - F(w),$$

where the function $\kappa(w)$ is fixed and the function $F(w)$ is a solution of the equation $\frac{\partial F}{\partial \bar{w}} = \frac{\partial \kappa}{\partial \bar{w}} E_\rho$ satisfying some special bounds. The last equality ensures $\bar{\partial} f = 0$, and therefore guarantees the analyticity of f.

Let $0 < \eta_2 < \eta_3 < \eta_4 < \eta_1 < 1$. We denote by $\alpha_1(w)$ and $\alpha_2(w)$ two non-negative functions in $C^\infty(\mathbb{C})$ such that $\alpha_1(w) = 1$ if $|w| < \eta_3$, $\alpha_1(w) = 0$ if $|w| > \eta_4$, $\alpha_2(w) = 0$ if $|w| < \eta_2$ and $|w| > \eta_1$, and $\alpha_2(w) = 1$ if $\eta_3 < |w| < \eta_4$. We set

$$\kappa(w) = \sum_{j=1}^{\infty} \alpha_1\left(\frac{w-r_j}{r_j}\right)$$

and

$$\Psi(w) = \frac{1}{1+\gamma_2}|w|^\rho\left(1+\gamma_2\sum_{j=1}^{\infty}\alpha_2\left(\frac{w-r_j}{r_j}\right)\right) + \ln(1+|w|^2).$$

It follows from the proof of **Theorem 2.4.4** that the function $\Psi(w)$ is subharmonic in $\mathbb{C}$ for sufficiently small $\gamma_2 = \gamma_2(\eta_1)$. We show that

$$\int_{\mathbb{C}} \left|\frac{\partial \kappa}{\partial \bar{w}}\right|^2 |E_\rho(w)|^2 e^{-2\Psi(w)} d\omega_w = c < \infty .$$

Indeed

$$\int_{\mathbb{C}} \left|\frac{\partial \kappa}{\partial \bar{w}}\right|^2 |E_\rho(w)|^2 e^{-2\Psi(w)} d\omega_w =$$

$$= \sum_{j=1}^{\infty} \int_{\eta_3 r_j<|w-r_j|<\eta_4 r_j} \left|\frac{\partial \kappa}{\partial \bar{w}}\right|^2 |E_\rho(w)|^2 e^{-2\Psi(w)} d\omega_w \le$$

$$\le \text{const} \sum_{j=1}^{\infty} \frac{1}{r_j} \int_{\eta_3 r_j < |w-r_j| < \eta_4 r_j} (1+|w|^2)^{-2} d\omega_w < \infty .$$

Now, using a well-known theorem of Hörmander on the solvability of $\bar{\partial}$-problem in weighted L^2-spaces[17], we choose the function F(w) as a solution of the equation $\frac{\partial F}{\partial \bar{w}} = E_\rho \frac{\partial \kappa}{\partial \bar{w}}$ such that

$$\int_{\mathbb{C}} |F(w)|^2 e^{-2\Psi(w)} (1+|w|^2)^{-2} d\omega_w \le c. \tag{3.2.2}$$

Now we shall show that the entire function f(w) constructed from F(z) as above belongs to the space $H(\mathbb{C}^n,\rho]$ and satisfies all the conditions i), ii), iii). First of all we note that

$$|f(w)|^2 \le \frac{1}{\pi} \int_{U(w,1)} |f(w')|^2 d\omega_{w'} \le$$

$$\le \frac{2}{\pi} \int_{U(w,1)} \kappa^2 |E_\rho(w')|^2 d\omega_{w'} + \frac{2}{\pi} \int_{U(w,1)} |F(w')|^2 d\omega_{w'} \le$$

$$\le \frac{2}{\pi} \int_{U(w,1)} \kappa^2 |E_\rho(w')|^2 d\omega_{w'} +$$

$$+ \frac{2}{\pi} \sup_{w' \in U(w,1)} \{exp(2\Psi(w')+2ln(1+|w'|^2))\} \times$$

$$\times \int_{U(w,1)} |F(w')|^2 e^{-2\Psi(w')} (1+|w'|^2)^2 d\omega_{w'} \le$$

---

[17] Hörmander's theorem can be formulated as follows.

Let Ψ be an arbitrary plurisubharmonic function in $\mathbb{C}^n$. Then for every exterior differential form g of type (p,q+1) such that

$$\int_{\mathbb{C}^n} |g|^2 e^{-2\Psi} d\omega_z < \infty,$$

there exists a form u of type (p,q) such that $\bar{\partial} u = g$ and

$$\int_{\mathbb{C}^n} |u|^2 e^{-2\Psi} (1+|z|)^2)^{-2} d\omega_z \le \int_{\mathbb{C}^n} |g|^2 e^{-2\Psi} d\omega_z .$$

$$\le 2 \sup_{w'\in U(w,1)} \{\kappa^2(w')|E_\rho(w')|^2)\} +$$

$$+ 2c \sup_{w'\in U(w,1)} exp\{2\Psi(w')+2ln(1+|w'|^2\}. \qquad (3.2.3)$$

From this and (3.2.1) it follows that $f\in H(\mathbb{C},\rho]$,

$$ln|f(w)| \le |w|^\rho + o(|w|^\rho), \; w \to \infty,$$

while for $w\in\mathfrak{A}'_1\setminus\mathfrak{A}'_2$, $0 < \gamma_1 < \gamma_2$, $|w| > R_0=R_0(\gamma_1)$, the inequality

$$ln|f(z)| \le \text{const} + 2ln(1+|w|) + \frac{1}{1+\gamma_2}(1+|w|)^\rho \le (1-\gamma_1)|w|^\rho \qquad (3.2.4)$$

is valid. Thus f(w) satisfies the conditions i) and ii). It remains to verify iii). We estimate f(w) at the points $z = w_j$. Since $\frac{\partial\kappa}{\partial\bar{w}} = 0$ for $|w-r_j| < \eta_3 r_j$, the function F(w) is analytic in the disc $|w-r_j| < \eta_3 r_j$ (3.2.2) implies

$$ln|F(r_j)| \le r_j^\rho(1-\gamma_1)$$

for $0 < \gamma_1 < \gamma_2$, $r_j > R_0(\gamma_1)$. From here, taking into account the equality $f(r_j) = E_\rho(r_j) - F(r_j) = \rho exp(r_j^\rho) - F(r_j)$, we conclude that $h_f(0) \ge 1$. At the same time it follows from i) that $h_f(0) \le 1$. Thus $h_f(0) = 1$ and therefore condition iii) is satisfied.

This finished the proof.

§3 Entire functions with prescribed behaviour at infinity

One of the most interesting and traditional kinds of problems in the theory of entire functions is the problem on the construction of functions with prescribed asymptotic properties. For example, the problem on the construction of an entire function with prescribed indicator is of this kind. In the case of one variable such a problem is solved by constructing a Weierstrass canonical product, and hence reduces to the proper choice of zeros of product. This choice can be made using the relationship between the density of the zero distribution of an entire f.c.r.g. and its indicator (this relationship was considered in **Chapter 1**). In the case of several variables a similar construction of a function with prescribed indicator is impossible, because the support of the divisor of an entire function is not a discrete set. Instead, in the multidimensional case Hörmander's theorems on the solvability of the $\bar{\partial}$-problem in weighted spaces are usually applied. However, this way is also applicable in the one-dimensional situation. The construction of the function f(w) in the proof of **Theorem 3.2.2** is an example of this.

The problem of constructing an entire function in $\mathbb{C}^n$, $n > 1$, with prescribed indicator was solved by C.Kiselman [1] (for the case $\rho = 1$) and A.Martineau [1] (for arbitrary $\rho > 0$). However, the constructed functions did not *a priori* have additional properties. Therefore the problem of constructing a function of c.r.g. with prescribed indicator is of undoubted interest. Taking into account properties of the indicator, this problem may be regarded as a problem of approximating in $\mathbb{C}^n$ a positively homogeneous plurisubharmonic function by logarithm of the modulus of an entire function. The problem of approximating an arbitrary plurisubharmonic function in this manner is more general.

The following result, given recently by R.Sigurdsson [1], concerns this problems.

Theorem 3.3.1. Let a plurisubharmonic function u(z) in $\mathbb{C}^n = \mathbb{R}^{2n}$ be of at most normal type with respect to the order $\rho > 0$ (i.e. $u \in PSH(\mathbb{C}^n, \rho] \overset{\text{def}}{=} PSH(\mathbb{C}^n) \cap SH(\mathbb{C}^n, \rho]$). Then there exists a function $f(z) \in H(\mathbb{C}^n, \rho]$ such that: if $t \to \infty$, then

$$u^{[t]}(z) - \frac{\ln|f(tz)|}{t^\rho} \to 0$$

in $L^1_{loc}(\mathbb{C}^n)$.

The proof of this theorem is based on some lemmas. For the statement of these lemmas we need to introduce a special operator R_δ on $L^1_{loc}(\mathbb{C}^n)$.

Let $A = \{a_{i,j}\}$ be a complex (nxn)-matrix. We consider such matrices as points in the space $\mathbb{C}^{n^2}$. The Euclidean volume element of this space is denoted by $d\omega_A$. We denote by $\alpha(w)$, $w \in \mathbb{C}$, a nonnegative function in $C^\infty(\mathbb{C})$ that depends on $|w|$ only, for which $supp\alpha \subset U$, and such that

$$\int_{\mathbb{C}} \alpha(w)d\omega_w = 1.$$

We set

$$\tilde{\alpha}(A) = \prod_{i,j=1}^{n} \alpha(a_{i,j})$$

and

$$\tilde{\alpha}_\delta(A) = \delta^{-2n^2}\tilde{\alpha}\left(\frac{A-I}{\delta}\right),$$

where I is the unit (nxn)-matrix.

The operator R_δ mentioned above is defined on functions $v \in L^1_{loc}(\mathbb{C}^n)$ as follows:

$$R_\delta v(z) = \int_{\mathbb{C}^{n^2}} v(z+\delta Az)\tilde{\alpha}(A)d\omega_A = \int_{\mathbb{C}^{n^2}} v(Az)\tilde{\alpha}_\delta(A)d\omega_A.$$

This operator is clearly linear. We show that it is also continuous. For this it suffices to prove that for any domain $G \subset\subset \mathbb{C}^n$ there is a constant $c_G < \infty$ such that

$$\|R_\delta v\|_{L^1(G)} \le c_G\|v\|_{L^1(G)}, \forall v \in L^1_{loc}(\mathbb{C}^n).$$

Without loss of generality we may assume that the domain G does not contain the points $z = (z_1,\dots,z_n)$ with $z_1 = 0$. In this case we let us set

$$A' = \begin{pmatrix} a_{1,2} & \cdots & a_{1,n} \\ \cdots & \cdots & \cdots \\ a_{n,2} & \cdots & a_{n,n} \end{pmatrix},$$

$$\zeta_j = a_{j,1}z_1 + \sum_{j=2}^{n} a_{j,i}z_i, \quad \zeta = (\zeta_1,\dots,\zeta_n).$$

Then: $Az = \zeta$, $d\omega_A = |z_1|^{-2n}d\omega_\zeta d\omega_{A'}$, $\tilde{\alpha}_\delta(A) = \alpha_\delta^{(1)}(A')\alpha_\delta^{(2)}(A',\zeta,z)$, where

$$\alpha_\delta^{(1)}(A') = \frac{1}{\delta^{2n^2-2n}} \prod_{i\neq 1, i\neq j} \alpha\left(\frac{a_{i,j}}{\delta}\right) \prod_{j=2} \alpha\left(\frac{a_{j,j}-1}{\delta}\right),$$

$$\alpha_\delta^{(2)}(A',\zeta,z) = \frac{1}{\delta^{2n}} \alpha\left(\frac{1}{\delta z_1}\left(\zeta_1 - \sum_{i=2}^{n} a_{1,i}z_i - z_1\right)\right) \prod_{j=2} \alpha\left(\frac{1}{\delta z_1}\left(\zeta_j - \sum_{i=2}^{n} a_{1,i}z_i\right)\right).$$

Taking into account these relations, we make the substitution $A \to (A'.\zeta)$ in the integral defining $R_\delta v(z)$. We have

$$R_\delta v(z) = |z_1|^{-2n} \int_{\mathbb{C}^{n^2-n}} \alpha_\delta^{(1)}(A')d\omega_{A'} \int_{\mathbb{C}^n} v(\zeta)\alpha_\delta^{(2)}(A',\zeta,z)d\omega_\zeta =$$

$$= |z_1|^{-2n} \int_{\mathbb{C}^{n^2-n}} \alpha_\delta^{(1)}(A')d\omega_{A'} \int_{B_r} v(\zeta)\alpha_\delta^{(2)}(A',\zeta,z)d\omega_\zeta, \qquad (3.3.1)$$

where $r = n(n+1)|z|$. Hence, for $v_1 \in L^1_{loc}(\mathbb{C}^n)$,

$$\int_G |R_\delta v_1(z)|d\omega_z \le \frac{1}{\delta^{2n}}\left(\max_w \alpha(w)\right)^n \int_G |z_1|^{-2n} d\omega_z \int_{B_{r'}} |v_1(\zeta)|d\omega_\zeta,$$

where

$$r' = \sup\{|\zeta| : \alpha_\delta^{(2)}(A',\zeta,z) \ne 0,\ z\in G,\ |a_{j,i}| \le 1\} \le n(n+1)d,$$

$$d = \sup\{|z| : z\in G\}$$

and therefore the operator R_δ is continuous in $L^1_{loc}(\mathbb{C}^n)$.

In the following lemma the properties of $R_\delta v$ as a function of v and δ are discussed.

Lemma 3.3.1. Let $v\in PSH(\mathbb{C}^n)$. Then $R_\delta v \in PSH(\mathbb{C}^n) \cap C^\infty(\mathbb{C}^n \setminus \{0\})$ and $R_\delta v \downarrow v$ as $\delta \downarrow 0$.

Proof. For every fixed A the function $v(z+\delta Az)$ is plurisubharmonic in z. Thus it is obvious that $R_\delta v$ is a plurisubharmonic function. Note that for each fixed z the function $v(z+\delta Az)$ is plurisubharmonic in $A\in\mathbb{C}^{n^2}$. This implies in a standard manner, taking into account the peculiarities of the construction of $\tilde{\alpha}_\delta(A)$ and properties of averages of subharmonic functions, that if $\delta\downarrow 0$, then the functions $R_\delta v(z)$

converge monotonically decreasing to the function $v(Az)|_{A=I}$, i.e. to the function $v(z)$. Finally, since $v\in L^1_{loc}(\mathbb{C}^n)$ and for any fixed A', ζ the function $\alpha^{(2)}_\delta(A',\zeta,z)\in C^\infty(\mathbb{C}^n_{(z)}\setminus\{0\})$, it follows from the representation (3.3.1) that $R_\delta v\in C^\infty(\mathbb{C}^n\setminus\{0\})$.

This finishes the proof.

Let us give some properties of families of functions of the form $\{R_\delta v\}_{v\in M}$.

Lemma 3.3.2. Let a family M of functions $v\in PSH(\mathbb{C}^n,\rho]$ be such that

$$\sup_{u\in M}\mathfrak{N}_{|u|}(0,r)\le cr^\rho,\quad \forall r>1,$$

where c is some constant. Then

1) for each fixed r, as $\delta\to 0$

$$\|R_\delta v-v\|_{L^1(B_r)}\to 0$$

uniformly with respect to $v\in M$.

2) for each $k\in\mathbb{Z}^n_+$ and $k'\in\mathbb{Z}^n_+$ there is a positive constant $c_{k,k'}$ such that

$$\left|\frac{\partial^{\|k\|+\|k'\|}}{\partial z^k\partial\bar z^{k'}}R_\delta v(z)\right|\le\frac{c_{k,k'}|z|^{\rho-\|k\|-\|k'\|}}{\delta^{2n+\|k\|+\|k'\|}},\quad\forall z\in\mathbb{C}^n\setminus\{0\},\ v\in M.\tag{3.3.2}$$

Proof. Let $v\in L^1_{loc}(\mathbb{C}^n)$. We then choose $v_\varepsilon\in C(\bar B_{2r})$ such that $\|v_\varepsilon-v\|_{L^1(B_{2r})}<\varepsilon$. Then for $\delta\le\frac{1}{n^2}$ and any $A\in supp\tilde\alpha$,

$$\int_{B_r}|v(z+\delta Az)-v(z)|d\omega_z\le\int_{B_r}|v(z+\delta Az)-v_\varepsilon(z+\delta Az)|d\omega_z+$$

$$+\int_{B_r}|v_\varepsilon(z+\delta Az)-v_\varepsilon(z)|d\omega_z+\int_{B_r}|v_\varepsilon(z)-v(z)|d\omega_z\le$$

$$\le\det{}^{-1}(I+\delta A)\|v_\varepsilon-v\|_{L^1(B_{2r})}+$$

$$+ \max_{|z|<r,\ |\Delta z|<\delta n^2 r} |v_\varepsilon(z+\Delta z)-v_\varepsilon(z)| \int_{B_r} d\omega + \|v_\varepsilon - v\|_{L^1(B_r)}.$$

Therefore there is $\delta^0 > 0$, independent of A, such that

$$\int_{B_r} |v(z+\delta Az)-v(z)|d\omega_z < 3\varepsilon, \ \forall\delta<\delta^0, \ A \in supp\ \tilde{\alpha}.$$

Taking into account the inequality

$$\|R_\delta v-v\|_{L^1(B_r)} = \int_{B_r} \left| \int_{\mathbb{C}^{n^2}} v(z+\delta Az)\tilde{\alpha}(A)d\omega_A - v(z)\int_{\mathbb{C}^{n^2}} \tilde{\alpha}(A)d\omega_A \right| d\omega_z \le$$

$$\le \int_{\mathbb{C}^{n^2}} \tilde{\alpha}(A)d\omega_A \int_{B_r} |v(z+\delta Az) - v(z)|d\omega_z,$$

we conclude that

$$\lim_{\delta\to 0} \|v-R_\delta v\|_{L^1(B_r)} = 0, \ \forall v\in L^1_{loc}(\mathbb{C}^n).$$

Note that the operator R_δ is linear and continuous, and that by the **Corollary** to **Lemma 2.1.4** M is relatively compact in $L^1_{loc}(\mathbb{C}^n)$. Hence a form of Banach-Steinhaus theorem (see L.Schwartz [1]) may be applied, and therefore the last limit is uniform with respect to the functions $v\in M$.

To prove the estimates (3.3.2) and (3.3.2') we use the representation (3.3.1). Without loss of generality we may assume that $|z_1| > \frac{1}{\sqrt{n}}|z|$. Then

$$\frac{\partial^{\|k\|+\|k'\|}}{\partial z^k \partial \bar{z}^{k'}} R_\delta v(z) =$$

$$= \int_{\mathbb{C}^{n^2-n}} \alpha_\delta^{(1)}(A')d\omega_{A'} \int_{|\zeta|<n(n+1)|z|} v(\zeta)\frac{\partial^{\|k\|+\|k'\|}}{\partial z^k \partial \bar{z}^{k'}}(|z_1|^{-2n}\alpha_\delta^{(2)}(A',\zeta,z))d\omega_\zeta,$$

and therefore

$$\left| \frac{\partial^{\|k\|+\|k'\|}}{\partial z^{k} \partial \bar{z}^{k'}} R_{\delta} v(z) \right| \leq \int_{|\zeta| < n(n+1)|z|} |v(\zeta)| d\omega_{\zeta} \times \tag{3.3.3}$$

$$\times \max\left\{ \left| \frac{\partial^{\|k\|+\|k'\|}}{\partial z^{k} \partial \bar{z}^{k'}} (|z_1|^{-2n} \alpha_{\delta}^{(2)}(A', \zeta, z)) \right| : |\zeta| < n(n+1)|z|, |a_{i,j}| \leq (1+\delta)|\zeta| \right\} .$$

It can be immediately verified that

$$\frac{\partial^{\|k\|+\|k'\|}}{\partial z^{k} \partial \bar{z}^{k'}} (|z_1|^{-2n} \alpha_{\delta}^{(2)}(A', \zeta, z)) =$$

$$= \sum_{\substack{p+q=2n+\|k\|+\|k'\| \\ l \leq 2n+\|k\|+\|k'\|}} z_1^{-p} \bar{z}_1^{-q} \delta^{l} \beta_{p,q,l}(A', \zeta, z), \tag{3.3.4}$$

where the $\beta_{p,q,l}(A', \zeta, z)$ are functions which are bounded on $\mathbb{C}^{n^2-n}_{(A')} \times \mathbb{C}^{n}_{(\zeta)} \times \mathbb{C}^{n}_{(z)}$.

Note that the condition of the lemma implies

$$\sup_{v \in M} \int_{|\zeta| < n(n+1)|z|} |v(\zeta)| d\omega_{\zeta} \leq \text{const} \cdot |z|^{2n+\rho}. \tag{3.3.5}$$

This and (3.3.3) and (3.3.4), taking into account the assumption $|z_1| > \frac{1}{\sqrt{n}}|z|$, imply that

$$\left| \frac{\partial^{\|k\|+\|k'\|}}{\partial z^{k} \partial \bar{z}^{k'}} R_{\delta} v(z) \right| \leq \text{const} \cdot \frac{|z|^{\rho-\|k\|-\|k'\|}}{\delta^{2n+\|k\|+\|k'\|}} , \quad \forall v \in M, \ 0<\delta<1, \ z \neq 0.$$

This finishes the proof.

Remark. It is obvious that any set $\{u^{[t]}\}_{t>1}$, where $u \in PSH(\mathbb{C}^n, \rho]$, can be taken as the set M in **Lemma 3.3.2**.

In the theory of functions of several complex variables we often encounter the need to of turn a function which is not plurisubharmonic, in general, into a plurisubharmonic function by the addition of a term with positive Levi form. Besides, this must be done in such a manner that the asymptotic properties of the function do not essentially change. In the present situation this "correction" will be done using the following lemma.

Lemma 3.3.3. Let the positive function $\gamma(r)\in C(R^+)$ be such that $\lim_{r\to\infty} \gamma(r) = 0$ and let $\rho > 0$.

Then there exists a function $\Phi\in PSH(\mathbb{C}^n,\rho]\cap C^\infty(\mathbb{C}^n\setminus\{0\})$ such that

1)

$$\sum_{i,j} \frac{\partial^2\Phi}{\partial z_i\partial \bar z_j} w_i\bar w_j \geq \gamma(|z|)|z|^{\rho-2}|w|^2, \ \forall w\in\mathbb{C}^2, \ z \neq 0; \tag{3.3.6}$$

2)

$$\lim_{z\to\infty} \frac{\Phi(z)}{|z|^\rho} = 0; \tag{3.3.7}$$

3) for any $k\in Z_+^n$ and $k'\in Z_+^n$, $\|k\|+\|k'\| \leq 2$, there exists a constant $c_{k,k'}$ such that

$$\left| \frac{\partial^{\|k\|+\|k'\|}}{\partial z^k\partial \bar z^{k'}} \Phi(z)\right| \leq c_{k,k'} |z|^{\rho-\|k\|-\|k'\|}, \ \forall z\neq 0. \tag{3.3.8}$$

Proof. We look for the function $\Phi(z)$ in the form

$$\Phi(z) = \alpha(|z|^2)|z|^\rho,$$

where the function $\alpha(t)\in C^\infty(\mathbb{R}_+)$ is positive and convex, and satisfies the condition $\alpha(t)\downarrow 0$ as $t\to\infty$. For the function $\Phi(z)$ we obtain

$$\sum_{i,j} \frac{\partial^2\Phi}{\partial z_i\partial \bar z_j} w_i\bar w_j =$$

$$= \left\{\alpha''(|z|^2)|z|^\rho + \rho|z|^{\rho-2}\alpha'(|z|^2) + \frac{\rho}{2}\left(\frac{\rho}{2} - 1\right)|z|^{\rho-4}\alpha(|z|^2)\right\}\cdot|\langle z,w\rangle|^2 +$$

$$+ \left(|z|^\rho\alpha'(|z|^2) + \frac{\rho}{2}|z|^{\rho-2}\alpha(|z|^2)\right)|w|^2 \geq$$

$$\geq \left\{\rho|z|^{\rho-2}\alpha'(|z|^2) - \frac{\rho}{2}\left(\frac{\rho}{2} - 1\right)^+|z|^{\rho-4}\alpha(|z|^2)\right\}|z|^{\rho-2}|w|^2 +$$

$$+ \left(|z|^\rho\alpha'(|z|^2) + \frac{\rho}{2}\cdot|z|^{\rho-2}\alpha(|z|^2)\right)|w|^2 =$$

$$= \left\{(1+\rho)|z|^2\alpha'(|z|^2) + \left(\frac{\rho}{2} - \left(\frac{\rho}{2}\left(1 - \frac{\rho}{2}\right)\right)\right)^+\alpha(|z|^2)\right\}|z|^{\rho-2}|w|^2.$$

Thus, inequality (3.3.6), i.e. condition 1), will be satisfied if the function $\alpha(t)$ is such that

$$(1+\rho)t\alpha'(t) + \left(\frac{\rho}{2} - \left(\frac{\rho}{2}\left(1 - \frac{\rho}{2}\right)\right)\right)^+\alpha(t) \geq \gamma(t)$$

or, equivalently, if

$$t\alpha'(t) + a\alpha(t) \geq \frac{\gamma(t)}{1+\rho}, \tag{3.3.9}$$

where $a = (1+\rho)^{-1}\left(\frac{\rho}{2} - \left(\frac{\rho}{2}\left(1 - \frac{\rho}{2}\right)\right)\right)^+$. Note that $0 < a < 1$.

Let $\gamma_1(t)$ be a function in $C^\infty(\mathbb{R}_+)$ satisfying the conditions: $\gamma_1(t) \geq (1+\rho)^{-1}\gamma(t)$ on $\mathbb{R}_+$, $\gamma_1(t)\downarrow 0$ as $t \to \infty$, $\gamma_1''(t) \geq 0$, $\forall t \geq 0$. The existence of such a function is obvious. Now we show that the function

$$\alpha(t) = t^{-a}\int_0^t \gamma_1(s)s^{a-1}ds,$$

which is a solution of the equation

$$t\alpha'(t) + a\alpha(t) = \gamma_1(t), \tag{3.3.10}$$

is as desired. Indeed, condition (3.3.9) is obviously satisfied, since (3.3.10) takes place and $\gamma_1(t) \geq \frac{\gamma(t)}{1+\rho}$. Then

$$\alpha'(t) = -\frac{a}{t^{a+1}}\int_0^t \gamma_1(s)s^{a-1}ds + \frac{\gamma_1(t)}{t} \leq -\frac{a\gamma_1(t)}{t^{a+1}}\int_0^t s^{a-1}ds + \frac{\gamma_1(t)}{t} = 0$$

and therefore $\alpha(t)$ is monotonically decreasing on $\mathbb{R}_+$. It is also obvious that $\lim_{t\to\infty} \alpha(t) = 0$.

We verify that $\alpha(t)$ is a convex function. We have

$$\alpha''(t) = \frac{(a+1)a}{t^{a+2}}\int_0^t \gamma_1(s)s^{a-1}ds + \frac{\gamma_1(t)}{t^2}(-2a+a-1) + \frac{\gamma_1'(t)}{t} =$$

$$= \frac{(a+1)a}{t^{a+2}}\left\{\frac{t^a}{a}\gamma_1(t) - \frac{1}{a}\int_0^t \gamma_1'(s)s^a ds\right\} - \frac{a+1}{t^2}\gamma_1(t) + \frac{\gamma_1'(t)}{t} =$$

$$= \frac{\gamma_1'(t)}{t} - \frac{(a+1)}{t^{a+2}}\int_0^t \gamma_1'(s)s^a ds. \qquad (3.3.11)$$

Since $\gamma_1(t)$ is a convex function, the function $\gamma_1'(t)$ is monotonically nondecreasing. Hence, by (3.3.11) it follows that

$$\alpha''(t) \geq \frac{\gamma_1'(t)}{t} - \frac{(a+1)}{t^{a+2}}\gamma_1'(t)\int_0^t s^a ds = 0.$$

Thus the required function $\alpha(t)$ has been constructed. Simultaneously we have constructed a function $\Phi(z)\in C^\infty(\mathbb{C}^n\setminus\{0\})$ satisfying the conditions 1) and 2). By 1) it follows that $\Phi\in PSH(\mathbb{C}^n\setminus\{0\})$. Since $\phi(0) = 0 < \Phi(z)$, $\forall z\in\mathbb{C}^n\setminus\{0\}$, we have $\Phi\in PSH(\mathbb{C}^n)$.

In order that condition 3) be satisfied it is evidently sufficient to require that $sup\ t^m\alpha^{(m)}(t) < \infty$. In turn, this will be true if the function $\gamma_1(t)$ is such that $\sup_{0<t<\infty} t^m\gamma_1^{(m)}(t) < \infty$, $\forall m\leq 3$. It is clear that this additional condition on $\gamma_1(t)$ does not lead to loss of generality.

The proof is finished.

We now show that we can approximate (in some sense "sufficient good")an arbitrary function $u\in PSH(\mathbb{C}^n,\rho]$ by an infinitely differentiable plurisubharmonic function.

Lemma 3.3.4. Let $u\in PSH(\mathbb{C}^n,\rho]$, $\rho > 0$, and let κ be a positive number. Then there exists a function $v\in PSH(\mathbb{C}^n,\rho]\cap C^\infty(\mathbb{C}^n\setminus\{0\})$ satisfying the conditions

i) $u^{[t]} - v^{[t]} \to 0$ in $L^1_{loc}(\mathbb{C}^n)$ as $t \to \infty$

ii) for any $k\in\mathbb{Z}^n_+$ and $k'\in\mathbb{Z}^n_+$, $\|k\|+\|k'\| \leq 2$, there exists a constant $c_{k,k'}$ such that

$$\left|\frac{\partial^{\|k\|+\|k'\|}}{\partial z^k\partial\bar{z}^{k'}} v(z)\right| \leq$$

$$\le c_{k,k'}|z|^{\rho-\|k\|-\|k'\|}\left(1+(ln(1+|z|))^{2n+\|k\|+\|k'\|}\right),\ \forall z\in\mathbb{C}^n\setminus\{0\} \quad (3.3.12)$$

iii)

$$\sum_{i,j}\frac{\partial^2 v}{\partial z_i\partial\bar{z}_j}w_i\bar{w}_j \ge \frac{|z|^{\rho-2}|w|^2}{(1+|z|)^{\kappa}},\ \forall w\in\mathbb{C}^n,\ z\in\mathbb{C}^n\setminus\{0\}. \quad (3.3.13)$$

Proof. Taking into account the properties of the operator R_δ mentioned in **Lemmas 3.3.1** and **3.3.2**, it is natural to look for the function v(z) in the form

$$v(z) = \sum_{m=0}\varphi_m(z)R_{\delta_m}u(z),$$

where $\{\varphi_m\}$ is a partition of unity in $\mathbb{C}^n$ and $\delta_m\downarrow 0$ as $m\to\infty$. This function belongs, of cause, to the space $C^\infty(\mathbb{C}^n\setminus\{0\})$. However, it need not obliged be plurisubharmonic. That is why we must add to it a plurisubharmonic function $\Phi(z)$ which would not essentially affect the asymptotic properties of the sum $\sum\varphi_m R_{\delta_m}u$, but at the same time would compensate for the possible negativity of the Levi form of this sum. To construct a partition of unity that would ensure that i) is satisfied we consider a sequence $\{\delta_m\}_{m=0}^{\infty}$ and a sequence $\sigma_m\uparrow\infty$, $\sigma_0 = 1$, such that

$$\lim_{m\to\infty}\delta_m^{2n+2}ln\sigma_m = \infty \quad (3.3.14)$$

and

$$\delta_m \ge (ln(1+\sigma_1\sigma_2\cdot\ldots\cdot\sigma_m))^{-1},\ \forall m. \quad (3.3.15)$$

These clearly exist.

Then we let $\eta(t)$ be some function in $C^\infty(\mathbb{R})$ satisfying the conditions: $0\le\eta(t)\le 1$ for $-\infty<t<\infty$, $\eta(t)=1$ if $t<\frac{1}{3}$, and $\eta(t) = 0$ if $t>\frac{2}{3}$. Now we set

$$\varphi_0(z) = \eta\left(\frac{ln|z|}{ln\sigma_1}\right),$$

$$\varphi_m(z) = \eta\left(\frac{1}{ln\sigma_{m+1}}\,ln\frac{|z|}{\sigma_0\sigma_1\ldots\sigma_m}\right) - \eta\left(\frac{1}{ln\sigma_m}\,ln\frac{|z|}{\sigma_0\sigma_1\ldots\sigma_{m-1}}\right),\ m\ge 1.$$

It is clear that $\varphi_m \in C^\infty(\mathbb{C}^n)$. We verify directly that

a) $0 \le \varphi_m \le 1$;

b) $\sum_{m=0}^{\infty} \varphi_m \equiv 1$;

c) $supp\varphi_m \subset \{ z: \sigma_1 \dots \sigma_m^{1/3} \le |z| \le \sigma_1 \dots \sigma_m \sigma_{m+1}^{2/3} \}$;

$supp\varphi_0 \subset \{ z: |z| < \sigma_1^{2/3} \}$;

d) $supp\varphi_m \cap supp\varphi_{m+j} = \varnothing$, $\forall j \ge 2$, $m \ge 0$;

e) $\varphi_m(z) = 1$ for $\sigma_1 \dots \sigma_m^{2/3} \le |z| \le \sigma_1 \dots \sigma_m \sigma_{m+1}^{1/3}$, $m \ge 1$,

and $\varphi_0(z) = 1$ for $|z| < \sigma_1^{2/3}$;

f) $\varphi_m(z) + \varphi_{m+1}(z) = 1$ for $\sigma_1 \dots \sigma_{m-1}^{1/3} \le |z| \le \sigma_1 \dots \sigma_{m+1} \sigma_{m+2}^{2/3}$, $m \ge 1$,

and $\varphi_0(z) + \varphi_1(z) = 1$ for $|z| < \sigma_1 \sigma_2^{1/3}$;

g) $\forall\ k \in \mathbb{Z}_+^n,\ k' \in \mathbb{Z}_+^n,\ \exists c_{k,k'} < \infty$:

$$\left| \frac{\partial^{\|k\|+\|k'\|}}{\partial z^k \partial \bar{z}^{k'}} \varphi_m(z) \right| \le c_{k,k'} |z|^{-\|k\|-\|k'\|} (ln\sigma_{m+1})^{-1}, \quad m \ge 0.$$

From these properties of the functions φ_m it also follows that for any compact set $K \subset \mathbb{C}^n \setminus \{0\}$ there is a number m_0 such that if $m > m_0$, then

$$K \subset \{ z: \varphi_m(\sigma_1 \dots \sigma_m z) = 1 \}; \tag{3.3.16}$$

$$\varphi_j(tz) = 0 \tag{3.3.17}$$

for $z \in K$, $\sigma_1 \dots \sigma_m < t < \sigma_1 \dots \sigma_{m+1}$, $j \ne m$, $j \ne m+1$;

$$\varphi_m(tz) + \varphi_{m+1}(tz) = 1 \tag{3.3.18}$$

for $z \in K$, $\sigma_1 \dots \sigma_m < t < \sigma_1 \dots \sigma_{m+1}$.

Put

$$u_m(z) = R_{\delta_m} u(z)$$

and

$$v_1(z) = \sum_{m=0}^{\infty} \varphi_m(z)u_m(z).$$

In accordance with **Lemma 3.3.2** and the **Remark** following it the functions $u_m^{[t]}(z)$, which can clearly be represented in the form $u_m^{[t]} = R_{\delta_m} u^{[t]}$, satisfy the condition: for any $R > 0$,

$$u_m^{[t]} - u^{[t]} \to 0$$

in the space $L^1(B_R)$, uniformly with respect to $t \geq 1$ as $m \to \infty$.

At the same time, from (3.3.17) and (3.3.18) we infer that for arbitrary r and ε, $r > \varepsilon > 0$, all $m > m_0$ and $\sigma_1 \ldots \sigma_m < t < \sigma_1 \ldots \sigma_{m+1}$ the following holds:

$$\|v_1^{[t]}(z) - u^{[t]}(z)\|_{L^1(B_r \setminus B_\varepsilon)} = \int_{B_r \setminus B_\varepsilon} |v_1^{[t]}(z) - u^{[t]}(z)| d\omega_z =$$

$$= \int_{B_r \setminus B_\varepsilon} \left| \sum_{j=0}^{\infty} \varphi_j(tz)(u_j^{[t]}(z) - u^{[t]}(z)) \right| d\omega_z =$$

$$= \int_{B_r \setminus B_\varepsilon} \left| \sum_{j=m}^{m+1} \varphi_j(tz)(u_j(z)^{[t]}(z) - u^{[t]}(z)) \right| d\omega_z \leq$$

$$\leq \|u_m^{[t]}(z) - u^{[t]}(z)\|_{L^1(B_r \setminus B_\varepsilon)} + \|u_{m+1}^{[t]}(z) - u^{[t]}(z)\|_{L^1(B_r \setminus B_\varepsilon)}.$$

Hence

$$\lim_{t \to \infty} \|v_1^{[t]}(z) - u^{[t]}(z)\|_{L^1(B_r \setminus B_\varepsilon)} = 0. \qquad (3.3.19)$$

To construct a "correcting" function $\Phi(z)$, we note that, as follows from property f) of the functions φ_m, if $\beta_m \leq |z| \leq \beta_{m+1}$, where $\beta_0 = 0$, $\beta_m = \sigma_1 \ldots \sigma_m$, $m = 1,2,\ldots$, then $v_1 = u_m\varphi_m + \varphi_{m+1}u_{m+1}$. Therefore, taking into account the estimates of the derivatives of φ_m and $R_\delta u$ mentioned above (see g) and **Lemma 3.3.2**, we conclude that the Levi form of v_1 can for $\beta_m \leq |z| \leq \beta_{m+1}$ be estimated as follows:

$$\sum_{i,j} \frac{\partial^2 v}{\partial z_i \partial \bar{z}_j} w_i \bar{w}_j =$$

$$= \sum_{i,j} \sum_{l=m}^{m+1} \left\{ \varphi_l \frac{\partial^2 u_l}{\partial z_i \partial \bar{z}_j} + 2Re \frac{\partial \varphi_l}{\partial z_i} \cdot \frac{\partial u_l}{\partial \bar{z}_j} + u_l \frac{\partial^2 \varphi_l}{\partial z_i \partial \bar{z}_j} \right\} w_i \bar{w}_j \ge$$

$$\ge -c|w|^2 \sum_{l=m}^{m+1} \left\{ \frac{|z|^{\rho-1}}{|z| \delta_l^{2n+1} \ln \sigma_{l+1}} + \frac{|z|^{\rho}}{|z|^2 \delta_l^{2n+1} \ln \sigma_{l+1}} \right\}, \tag{3.3.20}$$

where c is a positive constant.

We define the function $\gamma_1(t)$ on $\mathbb{R}_+$ by the equality

$$\gamma_1(t) = \frac{2c}{\delta_m^{2n+1} \ln \sigma_{m+1}} + \frac{2c}{\delta_{m+1}^{2n+1} \ln \sigma_{m+2}}$$

for $\beta_m \le t \le \beta_{m+1}$, $m = 0,1,\dots$. Then from (3.3.20) it follows that

$$\sum_{i,j} \frac{\partial^2 v}{\partial z_i \partial \bar{z}_j} w_i \bar{w}_j \ge -\gamma_1(|z|)|z|^{\rho-2}|w|^2. \tag{3.3.21}$$

Note that from (3.3.14) it follows that the condition

$$\lim_{t \to \infty} \gamma_1(t) = 0$$

is satisfied. Hence the condition

$$\lim_{t \to \infty} (\gamma_1(t) + (1+t)^{-\kappa}) = 0$$

is satisfied as well. We set $\gamma(t) = \gamma_1(t) + (1+t)^{-\kappa}$, and let us construct from $\gamma(t)$ the function $\Phi(z)$ (see **Lemma 3.3.3**) satisfying the conditions (3.3.6)-(3.3.8). Then for the Levi form of the function $v(z) = v_1(z) + \Phi(z)$ we have the estimate

$$\sum_{i,j} \frac{\partial^2 v}{\partial z_i \partial \bar{z}_j} w_i \bar{w}_j \ge -|z|^{\rho-2}|w|^2\gamma_1(|z|) + |w|^2|z|^{\rho-2}\gamma(|z|) =$$

$$= \frac{|z|^{\rho-2}|w|^2}{(1+|z|)^{\kappa}}, \quad \forall z \in \mathbb{C}^n \setminus \{0\},\ w \in \mathbb{C}^n.$$

Thus, the function $v \in PSH(\mathbb{C}^n \setminus \{0\})$ satisfies iii). The fact that v(z) is a plurisubharmonic function in a neighbourhood of the origin follows from the fact that the functions $u_0(z)$ and $\Phi(z)$ are both plurisubharmonic.

Further, since $\Phi(z)$ is of minimal type with respect to the order ρ (satisfies (3.3.7)), we have

$$\lim_{t\to\infty}\int_{B_r}|\Phi^{[t]}(z)|d\omega_z = 0,\ \forall r>0.$$

From this and (3.3.19) it follows that

$$\lim_{t\to\infty}\int_{B_r\setminus B_\varepsilon}|v^{[t]}(z) - u^{[t]}(z)|d\omega_z = 0. \tag{3.3.22}$$

We show that $v\in PSH(\mathbb{C}^n,\rho]$. Since $u\in PSH(\mathbb{C}^n,\rho]$,

$$\sup_{t>1}\int_{B_r}|u^{[t]}(z)|d\omega_z < \infty.$$

From this inequality and (3.3.22) it follows that

$$\sup_{t>1}\int_{B_r\setminus B_\varepsilon}|v^{[t]}(z)|d\omega_z = c^{(r,\varepsilon)} < \infty.$$

Taking into account that the $v^{[t]}$ are subharmonic, we conclude that

$$\sup_{|z|=1,t\ge1} v^{[t]}(z) \le \sup_{|z|=1,t\ge1}\mathfrak{M}_{v^{[t]}}(z,1/2) \le \frac{2^{2n}}{V_{2n}}\,c^{(3/2,1/2)},$$

and hence $v\in PSH(\mathbb{C}^n,\rho]$.

The fact that $u(z)$ and $v(z)$ belong to the space $PSH(\mathbb{C}^n,\rho]$ also implies (see §1.4 of **Chapter 2**) the existence of constants $c(u)$ and $c(v)$ such that for any $\varepsilon > 0$ and all t larger than some $t_0 = t_0(u,v,\varepsilon)$ the following inequalities hold:

$$\int_{B_\varepsilon}|v^{[t]}(z)|d\omega_z \le c(v)\varepsilon^\rho,$$

$$\int_{B_\varepsilon}|u^{[t]}(z)|d\omega_z \le c(u)\varepsilon^\rho.$$

From this and (3.3.22) we conclude that

$$\lim_{t\to\infty} \|v^{[t]} - u^{[t]}\|_{L^1(B_r)} = 0.$$

Hence condition i) is valid. To complete the proof it remains to verify condition ii).

Similarly to estimating the Levi form of v(z) for $\beta_m \le |z| \le \beta_{m+1}$, we have

$$\frac{\partial^{\|k\|+\|k'\|}v}{\partial z^k \partial \bar{z}^{k'}} = \frac{\partial^{\|k\|+\|k'\|}}{\partial z^k \partial \bar{z}^{k'}} (\varphi_m u_m + \varphi_{m+1}u_{m+1} + \Phi) =$$

$$= \sum_{j=m}^{m+1} \sum_{\substack{0\le q\le k\\ 0\le q'\le k'}} c_{q,q'} \frac{\partial^{\|q\|+\|q'\|}\varphi_j}{\partial z^q \partial \bar{z}^{q'}} \cdot \frac{\partial^{\|k-q\|+\|k'-q'\|}u_j}{\partial z^{k-q} \partial \bar{z}^{k'-q'}} + \frac{\partial^{\|k\|+\|k'\|}\Phi}{\partial z^k \partial \bar{z}^{k'}},$$

where $q = (q_1,\dots,q_n)$ and the $c_{q,q'}$ are positive constants.

From here taking into account the above given estimates of the derivatives of the functions $\varphi_j(z)$, $u_j(z)$ and $\Phi(z)$, we obtain

$$\frac{\partial^{\|k\|+\|k'\|}v}{\partial z^k \partial \bar{z}^{k'}} \le \text{const}\cdot |z|^{\rho-\|k\|-\|k'\|}\gamma_2(|z|), \ z \ne 0, \qquad (3.3.23)$$

where $\gamma_2(t) = 1 + \delta_{m+1}^{-2n-\|k\|-\|k'\|}$ if $\beta_m < t \le \beta_{m+1}$. Since the numbers δ_m and σ_m have been chosen in such a way that the inequality (3.3.15) is valid, we have $\delta_{m+1}^{-1} \le \ln(1+\beta_m) \le \ln(1+t)$ if $t \ge \beta_m$. Hence $\gamma_2(t) \le 1 + (\ln(1+t))^{2n+\|k\|+\|k'\|}$. In accordance with (3.3.23) this implies that ii) is satisfied.

The proof of the lemma is finished.

An estimate of the point set on which the average of a subharmonic function essentially differs from the function value is obtained in the following lemma.

Lemma 3.3.5. Let Ω and Ω_1 be open sets in $\mathbb{C}^n$, $\Omega_1 \subset\subset \Omega$. Let also $u \in SH(\mathbb{C}^n,\rho]$, $\|u\|_{L^1(\Omega)} < \infty$ and

$$X(u,\delta) \overset{\text{def}}{=} \{\, z\in\Omega_1 : \mathfrak{M}_u(z,\varepsilon) - u(z) \le \varepsilon\|u\|_{L^1(\Omega)},\ \forall\varepsilon\in(0,\delta)\,\}.$$

Then there exists a positive constant c, depending on Ω and Ω_1 only, such that

$$mes_{2n}(\Omega_1\backslash X(u,\delta)) \le c\delta,\ \forall\delta \le dist(\Omega_1,\partial\Omega).$$

Proof. Take a function $\psi\in D(\Omega)$ that equals 1 in some neighbourhood of the set $\bar{\Omega}_1$. Applying Taylor's formula we obtain

$$\sup_{0<\varepsilon<A,\, z\in\mathbb{C}^n} |\mathfrak{N}_\psi(z,\varepsilon)-\psi(z)| = c' = c'(A) < \infty.$$

It can be verified immediately that for $0 < \varepsilon < dist(\partial\Omega,\Omega_1)$

$$\int_\Omega \psi(z)\mathfrak{N}_u(z,\varepsilon)d\omega_z = \int_\Omega \mathfrak{N}_\psi(z,\varepsilon)u(z)d\omega_z.$$

Taking all this into account as well as the monotonicity of $\mathfrak{N}_u(z,\varepsilon)$ with respect to ε for $\varepsilon \le A = dist(supp\psi,\partial\Omega)$, we obtain

$$\int_{\Omega_1} (\mathfrak{N}_u(z,\varepsilon)-u(z))d\omega_z \le \int_\Omega \psi(z)(\mathfrak{N}_u(z,\varepsilon)-u(z))d\omega_z =$$

$$= \int_\Omega (\mathfrak{N}_\psi(z,\varepsilon)u(z)-\ \psi(z)u(z))d\omega_z \le c'\varepsilon^2\|u\|_{L^1(\Omega)}.$$

Using the obvious inequality

$$\int_{\Omega_1} (\mathfrak{N}_u(z,\varepsilon)-u(z))d\omega_z \ge \frac{\varepsilon}{2}\|u\|_{L^1(\Omega)}\cdot mes_{2n}\{\ z\in\Omega_1:\mathfrak{N}_u(z,\varepsilon)-u(z)\ge \frac{\varepsilon}{2}\|u\|_{L^1(\Omega)}\}$$

it follows that

$$mes_{2n}\{\ z\in\Omega_1:\ \mathfrak{N}_u(z,\varepsilon)-u(z) \ge \frac{\varepsilon}{2}\|u\|_{L^1(\Omega)}\} \le 2c'\varepsilon.$$

Note that in accordance with the definition of the set $X(u,\delta)$ at every point $z\in\Omega_1\backslash X(u,\delta)$ the inequality

$$\mathfrak{N}_u(z,\varepsilon)-u(z) > \varepsilon\|u\|_{L^1(\Omega)}. \tag{3.3.24}$$

is valid, for some $\varepsilon=\varepsilon(z)\in(0,\delta)$. We choose a nonnegative integer $l = l(z)$ such that $2^{-l-1}\delta \le \varepsilon(z) \le 2^{-l}\delta$. Then (3.3.24) implies that

$$\mathfrak{N}_u(z,2^{-l}\delta) -u(z) \ge \delta 2^{-l-1}\|u\|_{L^1(\Omega)}$$

at this point z. Hence

$$mes_{2n}(\Omega_1\setminus X(u,\delta)) \le \sum_{l=0}^{\infty} mes_{2n}\{ z\in\Omega_1 : \mathfrak{N}_u(z,2^{-l}\delta)-u(z) \ge 2^{-l-1}\delta\|u\|_{L^1(\Omega)} \} \le$$

$$\le 2c' \sum_{l=0}^{\infty} 2^{-l}\delta = 4c'\delta.$$

This finishes the proof.

The existence of an estimate for the integral of a differentiable function does not imply, in general, any estimate of its value at individual points. But such a point estimate becomes possible if we impose an additional restriction on the gradient of the function. A statement of such a type is given, in the form needed by us in the following lemma.

Lemma 3.3.6. Let κ be a positive number, $u(z)$ a continuously differentiable function in $\mathbb{C}^n = \mathbb{R}^{2n}$, and $\psi(z)$ a real-valued, measurable function in $\mathbb{C}^n$. Let κ, $u(z)$, $\psi(z)$ be such that

$$\int_{\mathbb{C}^n} |u(z)|^2(1+|z|^2)^{-\mu}e^{-2\psi(z)}d\omega_z \overset{\text{def}}{=} c_1 < \infty$$

and

$$\sup_{z\in\mathbb{C}^n}\{ |\bar\partial u|e^{-\psi(z)} \} \overset{\text{def}}{=} c_2 < \infty .$$

Then there is a constant $c > 0$ such that

$$|u(z)| \le c(1+|z|)^{\mu}e^{M_\psi(z,1)},$$

where $M_\psi(z,1) = \sup_{|\zeta|\le 1} \psi(z+\zeta)$.

Proof. It is well-known (see, for example, E.Čirka [1]) that if a domain $G\subset\mathbb{C}^n$ has piecewise smooth boundary, then for any continuously differentiable function f on $\bar G$ the Martinelli-Bochner holds: i.e.

$$f(z) = \int_{\partial G} f(\zeta)\omega_{MB}(\zeta-z) - \int_G \bar\partial f\wedge\omega_{MB}(\zeta-z),$$

where

$$\omega_{MB}(\zeta-z) =$$

$$= \frac{(n-1)!}{(2\pi i)^n}\sum_{l=1}^{n}\frac{(-1)^{l-1}}{|\zeta-z|^{2n}}(\bar\zeta_l-\bar z_l)d\bar\zeta_1\wedge\ldots\wedge d\bar\zeta_{l-1}\wedge d\bar\zeta_{l+1}\wedge d\bar\zeta_n\wedge d\zeta_1\wedge\ldots\wedge d\zeta_n.$$

Applying this formula to the case $G = B_1(z)$ and $f(\zeta) = u(\zeta)\chi(\zeta-z)$, where $\chi(\zeta)\in D(\bar{B}_1)$ and $\chi(0) = 1$ for $|\zeta| < \frac{1}{2}$, we obtain

$$u(z) = - \int_{B_1(z)} \bar{\partial}(u(\zeta)\chi(\zeta-z))\wedge\omega_{MB}(\zeta-z) =$$

$$= - \frac{(n-1)!}{\pi^n} \int_{B_1(z)} \sum_{l=1}^{n} \frac{\bar{\zeta}_l-\bar{z}_l}{|z-\zeta|^{2n}} \frac{\partial}{\partial\bar{\zeta}_l} (u(\zeta)\chi(\zeta-z))d\omega_\zeta =$$

$$= - \frac{(n-1)!}{\pi^n} \int_{B_1(z)} \sum_{l=1}^{n} \frac{\bar{\zeta}_l-\bar{z}_l}{|z-\zeta|^{2n}} \frac{\partial u(\zeta)}{\partial\bar{\zeta}_l} \chi(\zeta-z)d\omega_\zeta +$$

$$- \frac{(n-1)!}{\pi^n} \int_{B_1(z)\setminus B_{1/2}(z)} \sum_{l=1}^{n} \frac{\bar{\zeta}_l-\bar{z}_l}{|z-\zeta|^{2n}} u(\zeta) \frac{\partial\chi(z-\zeta)}{\partial\bar{\zeta}_l} d\omega_\zeta = I_1 + I_2 .$$

We estimate I_1:

$$|I_1| \le \frac{(n-1)!}{\pi^n} \|\chi\|_\infty c_2 e^{M_\psi(z,1)} \int_{B_1} \frac{n}{|\zeta|^{2n-1}} d\omega_\zeta = c_3 e^{M_\psi(z,1)} .$$

Then we estimate I_2, and obtain

$$|I_2| \le \frac{(n-1)!}{\pi^n} (1/2)^{2n-1} \|\mathrm{grad}\ \chi\|_\infty \int_{1/2<|z|<1} |u(\zeta)|d\omega_\zeta \le$$

$$\le \frac{(n-1)!}{\pi^n} (1/2)^{2n-1} \|\mathrm{grad}\ \chi\|_\infty \left(\int_{|\zeta|<1} |u|^2(1+|\zeta|^2)^{-\mu} e^{-2\psi(\zeta)} d\omega_\zeta \right)^{1/2} \times$$

$$\times \left(\int_{|\zeta|<1} (1+|\zeta|^2)^{\mu} e^{2\psi(\zeta)} d\omega_\zeta \right)^{1/2} \le$$

$$\le \frac{(n-1)!}{\pi^n} (1/2)^{2n-1} \|\mathrm{grad}\ \chi\|_\infty c_1 V_{2n} e^{M_\psi(z,1)} 2^{\mu}(1+|z|)^{\mu} =$$

$$= c_4(1+|z|)^{\mu} e^{M_\psi(z,1)} .$$

Hence

$$|u(z)| = |I_1 + I_2| \le e^{M_\psi(z,1)} (c_3+c_4(1+|z|)^{\mu}) \le c(1+|z|)^{\mu} e^{M_\psi(z,1)} .$$

This finishes the proof.

Now we construct the function f(z) looked for.

We set

$$X = \{\ z\in\mathbb{C}^n:\ 1/2 < |z| < 4\ \},$$

$$X' = \{\ z\in\mathbb{C}^n:\ 1 < |z| < 2\ \},$$

$$s_j = 2^j,\ \ j\in\mathbb{Z}\ .$$

Then we choose positive numbers κ and κ' such that $\kappa+2\kappa' < \min\{\rho,1\}$, $\kappa < \kappa'$. For the function v(z) figuring in **Lemma 3.3.4** and for given integers j and l we consider finite sets $W \subset X'$ satisfying the conditions:

i) $|z'-z''| > 8s_j^{\kappa'}$ for $z'\in W$, $z''\in W$, $z' \neq z''$;

ii) $\operatorname{dist}(W,\partial X') > 4s_j^{-\kappa'}$;

iii) $\mathfrak{M}_{v^{[s_j]}}(z,\varepsilon) - v^{[s_j]}(z) < \varepsilon\|v^{[s_j]}\|_{L^1(x)}$, $\forall\varepsilon\in(0,2^{-l})$, $z\in W$.

By **Lemma 3.3.5**, for sufficiently large l there exists a nonempty set W. Moreover, by i) the number of elements of any W is bounded by a constant that is independent of W. We denote by $Y_{j,l}$ some set W having maximal number of elements. Since the number l does not figure in the conditions i) and ii), for all l larger than some l_0 this maximal number of elements is the same and $Y_{j,l}$ may be chosen such that $Y_{j,l}=Y_{j,l_0} \overset{\text{def}}{=} Y_j$, $\forall l\geq l_0$. Consider the set $\bigcup_j s_jY_j$. This set is countable. By arbitrarily listing its points we obtain a sequence $\{z^{(m)}\}_{m=1}^{\infty}$. We set

$$\Omega_m = \{\ z\in\mathbb{C}^n: |z-z^{(m)}| \leq 2|z^{(m)}|^{1-\kappa'}\},$$

$$\Omega_m^{(1)} = \{\ z\in\mathbb{C}^n: |z-z^{(m)}| \leq |z^{(m)}|^{1-\kappa'}\},$$

$$\Omega_m^{(2)} = \left\{z\in\mathbb{C}^n:\ \tfrac{1}{2}|z^{(m)}|^{1-\kappa'} \leq |z-z^{(m)}| \leq |z^{(m)}|^{1-\kappa'}\right\}.$$

It is obvious that $\Omega_m^{(2)}\subset \Omega_m^{(1)}\subset \Omega_m$. Note also that for $s_j \leq |z^{(m)}| \leq$

s_{j+1} or, equivalently, for $z^{(m)} \in s_j Y_j$ the inclusion

$$\Omega_m \subset \{ z \in \mathbb{C}^n : s_j < |z| < s_{j+1} \} \tag{3.3.25}$$

is valid. Indeed, by ii) it follows that in this case

$$\left| \frac{z^{(m)}}{s_j} - 1 \right| \geq 4s_j^{-\kappa'}, \quad \left| \frac{z^{(m)}}{s_j} - 2 \right| \geq 4s_j^{-\kappa'},$$

and therefore

$$|z^{(m)} - s_j| \geq 4s_j^{1-\kappa'} \geq 2|z^{(m)}|^{1-\kappa'} \cdot 2^{\kappa'},$$

$$|z^{(m)} - s_{j+1}| \geq 4s_j^{1-\kappa'} \geq 2^{\kappa'} \cdot 2|z^{(m)}|^{1-\kappa'}.$$

From this we conclude that the required inclusion is valid.

Now we show that $\Omega_m \cap \Omega_{m'} = \emptyset$ for $m \neq m'$. In the case $z^{(m)} \in s_j Y_j$, $z^{(m')} \in s_{j'} Y_{j'}$, $j \neq j'$, this follows immediately from (3.3.25). In the case $z^{(m)} \in s_j Y_j$, $z^{(m')} \in s_j Y_j$, using i) we obtain

$$\operatorname{dist}(\Omega_m, \Omega_{m'}) \geq |z^{(m)} - z^{(m')}| - 2(|z^{(m)}|^{1-\kappa'} + |z^{(m')}|^{1-\kappa'}) >$$

$$> 8s_j^{1-\kappa'} - 4(2s_j)^{1-\kappa'} = 8s_j^{1-\kappa'}(1 - 2^{-\kappa'}) > 0$$

and therefore $\Omega_m \cap \Omega_{m'} = \emptyset$. We set

$$h_m(z) = v(z^{(m)}) + 2 \sum_{i=1}^{n} \frac{\partial v}{\partial z_i} \Big|_{z=z^{(m)}} (z - z_i^{(m)}) +$$

$$+ \sum_{i,j=1}^{n} \frac{\partial^2 v}{\partial z_i \partial z_j} \Big|_{z=z^{(m)}} (z_i - z_i^{(m)})(z_j - z_j^{(m)}).$$

In accordance with Taylor's formula for the function

$$Q_m(z) = v(z) - Re h_m(z) - \sum_{i,j=1}^{n} \frac{\partial^2 v}{\partial z_i \partial z_j} \Big|_{z=z^{(m)}} (z_i - z_i^{(m)})(z_j - z_j^{(m)}),$$

we find that if $|z - z^{(m)}| < |z^{(m)}|$, then

$$|Q_m(z)| \leq$$

$$\le c|z-z^{(m)}|^3 sup\left\{\left|\frac{\partial^{\|k\|+\|k'\|}v(\zeta)}{\partial\zeta^k\partial\bar{\zeta}^{k'}}\right| : \|k\|+\|k'\|=3,\ |\zeta-z^{(m)}| \le |z-z^{(m)}|\right\}$$

for some constant c independent of z. Note that for all m larger than some m_0 and all $|\zeta| \le 2|z|$ we have $|\zeta-z^{(m)}| \le |z-z^{(m)}|$, $z\in\Omega_m$. Now, taking into account (3.3.12), we find that

$$|Q_m(z)| \le c_1(ln(1+|z|))^{2n+3}|z|^{\rho-3}|z-z^{(m)}|,\ \forall z\in\Omega_m,$$

where c_1 is some constant independent of m.

Using (3.3.13) we arrive at the inequality

$$v(z) - Reh_m(z) \ge$$

$$\ge |z|^{\rho-2}|z-z^{(m)}|^2 \left\{\frac{1}{(1+|z|)^{\kappa}} - c_1(ln(1+|z|))^{2n+3}\frac{|z-z^{(m)}|}{|z|(1-\kappa)}\right\}. \qquad (3.3.26)$$

It is easy to see that for $\kappa < \kappa'$,

$$\lim_{m\to\infty}\sup_{z\in\Omega_m}\left\{(ln(1+|z|))^{2n+3}\frac{|z-z^{(m)}|}{|z|}\right\} = 0.$$

Hence, by (3.3.26) it follows that for all m larger than some m_1 the following estimate holds:

$$v(z) - Reh_m(z) \ge \frac{1}{2}|z|^{\rho-2-\kappa}|z-z^{(m)}|^2,\ \forall z\in\Omega_m . \qquad (3.3.27)$$

Since $|z-z^{(m)}| \ge \frac{1}{4}|z|^{1-\kappa'}$, $\forall z\in\Omega_m^{(2)}$, it follows that

$$Reh_m(z) \le v(z) - \frac{1}{32}|z|^{\rho-\kappa-2\kappa'},\ \forall z\in\Omega_m^{(2)},\ m \ge m_1. \qquad (3.3.28)$$

Then, and this has been repeatedly, we take some function $\chi\in C^\infty(\mathbb{C}^n)$ satisfying the conditions: $supp\chi \subset B_1$; $\chi(z) = 1$, $\forall z\in B_{1/2}$, and set

$$\chi_m(z) = \chi(z-z^{(m)})|z|^{\kappa'-1},$$

$$\tilde{\chi}_m(z) = \chi_m\left(\frac{1}{2}z\right).$$

It is evident that $supp\tilde{\chi}_m \subset \Omega_m$, $supp\chi_m \subset \Omega_m^{(1)}$, $supp\bar{\partial}\chi_m \subset \Omega_m^{(2)}$, $\tilde{\chi}_m = 1$ if $z\in supp\chi_m$ and

$$\sup_{m,z} |\bar{\partial}\chi_m(z)| = c_3 < \infty . \tag{3.3.29}$$

We construct the function f(z) whose existence is stated in the theorem to be proved in the form

$$f = \sum_{m=0}^{\infty} \chi_m e^{h_m} - U . \tag{3.3.30}$$

Since f(z) must be an entire function, the function U in (3.3.30) must satisfy the following condition:

$$\bar{\partial}U = \sum_{m=0}^{\infty} \bar{\partial}\chi_m e^{h_m} \overset{def}{=} g .$$

To solve this $\bar{\partial}$-problem with the aid of Hörmander's theorem mentioned and used in the previous paragraph, we have to find a plurisubharmonic function $\psi(z)$ such that

$$\int_{\mathbb{C}^n} |g|^2 e^{-2\psi(z)} d\omega_z < \infty.$$

We set

$$\psi(z) = \frac{n+1}{2} \ln(1+|z|^2) + v(z) - c_2|z|^{\rho-\kappa-2\kappa'} ,$$

where c_2 is a positive constant.

It can be directly verified that

$$\left|\frac{\partial^2}{\partial z_i \partial \bar{z}_j}\left(|z|^{\rho-\kappa-2\kappa'} \sum_{m=0}^{\infty} \tilde{\chi}_m(z) \right)\right| \le c_4|z|^{\rho-\kappa-2\kappa'-2}, \ \forall z\in\mathbb{C}^n, \ i,j=1,\dots n,$$

where c_4 is a constant. From this and the estimate (3.3.12) of the Levi form of v(z) given in **Lemma 3.3.4** we conclude that the function $\psi(z)$ belongs to $PSH(\mathbb{C}^n,\rho]$ for c_2 sufficient small. We estimate $|g|$. Taking into account (3.3.28), (3.3.29), the inclusion $supp\bar{\partial}\tilde{\chi}_m \subset \Omega_m^{(2)}$ and assuming $c_2 < \frac{1}{32}$, we obtain

$$|g| \le c_3 e^{Reh_m} \le c_3 exp\left\{v(z) - \frac{1}{32}|z|^{\rho-\kappa-2\kappa'}\right\} \le$$

$$\le c_3 e^{\psi(z)} (\ln(1+|z|^2))^{-(n+1)/2} , \ \forall z\in\Omega_m^{(2)}, \ m \ge m_1,$$

and

$$|g(z)| = 0, \ \forall z \notin \bigcup_{m=0}^{\infty} \Omega_m^{(2)}.$$

Hence

$$|g|^2 \le (\ln(1+|z|^2))^{-(n+1)} e^{2\psi(z)}, \tag{3.3.31}$$

and therefore

$$\int_{\mathbb{C}^n} |g|^2 e^{-2\psi(z)} d\omega_z < \infty. \tag{3.3.32}$$

In accordance with Hörmander's theorem it follows that the equation $\bar{\partial}U = g$ has a solution $U(z) \in C^\infty(\mathbb{C}^n)$ satisfying the condition

$$\int_{\mathbb{C}^n} |U|^2 e^{2\psi(z)} (1+|z|^2)^{-2} d\omega_z = c_5 < \infty . \tag{3.3.33}$$

Substituting this solution into (3.3.30) we obtain the entire function $f(z)$.

For brevity we set $\Phi(z) = \ln|f(z)|$. First of all we show that $\Phi^{[s_j]} - v^{[s_j]} \to 0$ in $L^1_{loc}(\mathbb{C}^n)$ as $j \to \infty$. To prove this it is obviously sufficient to show that every subsequence of the sequence $\{s_j\}$ contains a subsequence $\{t_l\}$ such that $\Phi^{[t_l]} - v^{[t_l]} \to 0$ in $L^1_{loc}(\mathbb{C}^n)$ as $l \to \infty$. The existence of a subsequence $\{t_l\}$ satisfying the conditions: $\exists \mathcal{D}'\text{-}lim\ \Phi^{[t_l]} = \Phi_1 \in Fr\Phi$ and $\exists \mathcal{D}'\text{-}lim\ v^{[t_l]} = v_1 \in Frv$, follows from the **Corollary** to **Lemma 2.1.4**. We prove that $v_1 = \Phi_1$.

By (3.3.27), (3.3.30) and (3.3.31) it follows that for the function $f(z)$ we have the inequality

$$\int_{\mathbb{C}^n} |f|^2 (1+|z|^2)^{-2} e^{-2\psi(z)} d\omega_z < \infty.$$

From this and (3.3.31), in accordance with **Lemma 3.3.6** it follows that

$$|f(z)| \le c_5 (1+|z|)^{-2} exp\left\{ \sup_{\zeta \in B_1} \psi(z+\zeta) \right\} \le$$
$$\le c_5 (1+|z|)^{-2} exp\left\{ \sup_{\zeta \in B_1} v(z+\zeta) \right\}, \ \forall z \in \mathbb{C}^n.$$

Hence

$$\overline{\lim_{l\to\infty}}\ \Phi^{[t_l]}(z) \le \overline{\lim_{l\to\infty}}\ \sup_{\zeta\in B_1} v^{[t_l]}\left(z+\frac{\zeta}{t_l}\right) \le$$

$$\le \overline{\lim_{l\to\infty}}\ \sup_{\zeta\in B_1} v^{[t_l]}(z+\varepsilon\zeta), \forall z\in\mathbb{C}^n,\ \varepsilon > 0. \qquad (3.3.34)$$

Since $\mathcal{D}'\text{-}\lim_{l\to\infty} v^{[t_l]}(z+\varepsilon\zeta) = v_1(z+\varepsilon\zeta)$ we have (see point ii) of **Lemma 2.1.4**) the inequality

$$\lim_{l\to\infty} v^{[t_l]}(z+\varepsilon\zeta) \le v_1(z+\varepsilon\zeta).$$

Take some $A > v_1(z)$ and choose $\varepsilon>0$ so small that $v_1(z+\varepsilon\zeta) < A$, $\forall\zeta\in B_1$. Then

$$\overline{\lim_{l\to\infty}}\ v^{[t_l]}(z+\varepsilon\zeta) \le A,\ \forall\varepsilon>0,$$

and by **Hartogs' Lemma** it follows that

$$\overline{\lim_{l\to\infty}}\ \sup_{\zeta\in B_1} v^{[t_l]}(z+\varepsilon_1\zeta) \le A,\ 0 < \varepsilon_1 < \varepsilon.$$

From this and (3.3.34) we conclude that

$$\overline{\lim_{l\to\infty}}\ \Phi^{[t_l]}(z) \le v_1(z)$$

and by **Lemma 2.1.4**,

$$\Phi_1(z) = reg\ \overline{\lim_{l\to\infty}}\ \Phi^{[t_l]}(z).$$

Then we have

$$\Phi_1(z) \le v_1(z),\ \forall z\in\mathbb{C}^n.$$

Now we show that $\Phi_1(z) \ge v_1(z)$, $\forall z\in\mathbb{C}^n$. We set $t_l = s_{j_l}$ and denote by F_p the set of limit points of the sequence $\{\zeta_l\}_{l=1}^{\infty}$, where $\zeta_l\in Y_{j_l,p}$. It is obvious that for each p the set F_p is closed, and correspondingly $X\backslash F_p$ is open. Let K be some compact set in $X'\backslash F_p$. Then there exist numbers d, $0 < d < \text{dist}(K,\partial X')$, and $l_0\in\mathbb{N}$ such that $\text{dist}(Y_{j_l,p},K) > d$, $\forall l>l_0$. We choose $l > l_0$ such that $8t_l^{-\kappa} < d$. Note that $Y_{j_l,p}$ has the property that the number of its elements is maximal (as it should be according to its definition). Therefore there is no point in K at which point the condition iii) holds with $l = p$ and $s_j = t_l$. Hence **Lemma 3.3.5** may be applied here. From this lemma it follows that $mes_{2n}K < c\cdot 2^{-p}$, where c is a positive constant independent of p. From this, in turn, we find that $mes_{2n}(X'\setminus F_p) < c2^{-p}$, and hence

$$mes_{2n}\left(X'\setminus\bigcup_p F_p\right) = mes_{2n}\left(\bigcap_p (X'\setminus F_p)\right) = 0.$$

Let $\tilde{z} \in \bigcup_p F_p$. Then for some p the point $\tilde{z}$ is the limit of a point sequence $\tilde{\zeta}_\nu \in F_{j_l,p}$, $l = l(\nu)$, $\nu = 1, 2, \ldots$. Thus

$$v_1(z) \le \mathfrak{M}_{v_1}(\tilde{z},\varepsilon) = \lim_{\nu\to\infty} \mathfrak{M}_{v_1}(\tilde{\zeta}_\nu,\varepsilon) \qquad (3.3.35)$$

(we recall that the function $\mathfrak{M}(\zeta,r)$ is continuous in both ζ and r). Since $v^{[t_l]} \xrightarrow{\mathcal{D}'} v_1$ as $l \to \infty$, we have $v^{[t_l]} \to v_1$ in the space $L^1_{loc}(\mathbb{C}^n)$ (**Lemma 2.1.4**). Therefore $\forall\delta>0\ \exists\nu_\delta$:

$$\mathfrak{M}_{v_1}(\tilde{\zeta}_\nu,\varepsilon) \le \mathfrak{M}_{v^{[t_{l(\nu)}]}}(\tilde{\zeta}_\nu,\varepsilon) + \delta, \ \forall\nu \ge \nu_\delta. \qquad (3.3.36)$$

Note that $\tilde{\zeta}_\nu \in Y_{l(\nu),p}$ and $t_{l(\nu)} = s_{j_{l(\nu)}}$. Using the condition iii) of the definition of $Y_{j,p}$ we obtain

$$\mathfrak{M}_{v_1}(\tilde{\zeta}_\nu,\varepsilon) \le v^{[t]}(\tilde{\zeta}_\nu) + \varepsilon \int_X | v^{[t_{l(\nu)}]}(\zeta)|d\omega_\zeta \ . \qquad (3.3.37)$$

The function $v_1 \in SH(\mathbb{C}^n,\rho]$, hence

$$\sup_{t>1} \int_X |v^{[t_{l(\nu)}]}(\zeta)|d\omega_\zeta = c_6 < \infty.$$

Now we note that the point $t_{l(\nu)}\tilde{\zeta}_n$ belongs to the sequence $\{z^{(m)}\}$. Taking this into account we estimate the value

$$|v(z^{(m)}) - ln|f(z^{(m)})||.$$

The definitions of the functions v and f imply that

$$|v(z^{(m)}) - ln|f(z^{(m)})|| = |ln|1-U(z^{(m)})e^{-v(z^{(m)})}|| \ . \qquad (3.3.38)$$

Since the conditions (3.3.32) and (3.3.33) are satisfied, the function $U(z)$ can be estimated, using **Lemma 3.3.6**, as follows:

$$|U(z)| \le c_7(1+|z|)^2 exp\{M_\psi(z,1)\}.$$

Estimating the function $M_\psi(z,1) = \sup_{|\zeta|<1} \psi(z+\zeta)$ using Taylor's formula and the estimate of the derivative, mentioned above, we find that for some positive constants c_8, c_9,

$$M_\psi(z^{(m)},1) \le v(z^{(m)}) + c_8|z^{(m)}|^{\rho-1}(ln(1+|z^{(m)}|))^{2n+1} - c_9|z^{(m)}|^{\rho-\kappa-2\kappa'}.$$

Therefore

$$|U(z^{(m)})|e^{-v(z^{(m)})} \le$$

$$\le c_7(1+|z|)^2 exp\left\{ c_8|z^{(m)}|^{\rho-1}(ln(1+|z^{(m)}|))^{2n+1} - c_9|z^{(m)}|^{\rho-\kappa-2\kappa'}\right\},$$

and for $\kappa+2\kappa' < \min\{\rho,1\}$,

$$\lim_{m\to\infty} |U(z^{(m)})|e^{-v(z^{(m)})} = 0.$$

By (3.3.38) we thus conclude that

$$\lim_{m\to\infty} (v(z^{(m)}) - ln|f(z^{(m)})|) = 0,$$

and, as a consequence,

$$\lim_{\nu\to\infty} \left| v^{[t_{1(\nu)}]}(\tilde{\zeta}_\nu) - \Phi^{[t_{1(\nu)}]}(\tilde{\zeta}_\nu)\right| = 0 \qquad (3.3.39)$$

(recall that $t_{1(\nu)}\tilde{\zeta}_\nu \in \{z^{(m)}\}$). Now we note that since $\Phi^{[t_{1(\nu)}]} \to \Phi_1$ in $L^1_{loc}(\mathbb{C}^n)$ as $\nu \to \infty$, we have

$$\overline{\lim_{\nu\to\infty}}\, \Phi^{[t_{1(\nu)}]}(\zeta) \le \Phi_1(\zeta), \ \forall\zeta\in\mathbb{C}^n.$$

Hence

$$\overline{\lim_{\nu\to\infty}}\, \Phi^{[t_{1(\nu)}]}(\tilde{\zeta}_\nu) \le \Phi_1(\tilde{z}) .$$

Comparing this inequality with (3.3.35), (3.3.36), (3.3.37), and (3.3.39), we find that

$$v_1(\tilde{z}) \le \Phi_1(\tilde{z}) + c_{10}\varepsilon + \delta.$$

Since ε and δ are arbitrary numbers, it follows that $v_1(\tilde{z}) \le \Phi_1(\tilde{z})$, $\forall \tilde{z} \in \bigcup_p F_p$. Since $mes_{2n}(X' \setminus \bigcup_p F_p) = 0$, we have $v_1(z) \le \Phi_1(z)$, $\forall z \in X'$. In combination with the reverse inequality proved above, this leads to

$$v_1(z) = \Phi_1(z), \ \forall z \in X' . \tag{3.3.40}$$

Since

$$v_1\Big|_{s_j X'} = \left(\mathcal{D}' - \lim_{l \to \infty} v^{[t_l s_j]}\right)\Big|_{X'}, \ \forall j \in \mathbb{Z},$$

and

$$\Phi_1\Big|_{s_j X'} = \left(\mathcal{D}' - \lim_{l \to \infty} \Phi^{[t_l s_j]}\right)\Big|_{X'}, \ \forall j \in \mathbb{Z},$$

from (3.3.40) it obviously follows that

$$v_1(z) = \Phi_1(z), \ \forall z \in \bigcup_j s_j X'.$$

Taking into account the equality $mes_{2n}\left(\mathbb{C}^n \setminus \bigcup_{j=-\infty}^{\infty} s_j X\right) = 0$, we find that

$$v_1(z) = \Phi_1(z), \ \forall z \in \mathbb{C}^n.$$

So, by virtue of the above, we have proved that as $j \to \infty$,

$$\Phi^{[s_j]} - v^{[s_j]} \to 0 \text{ in } L^1_{loc}(\mathbb{C}^n). \tag{3.3.41}$$

Now we consider an arbitrary sequence $t_l \uparrow \infty$ as $j \to \infty$. We define numbers $j(l) \in \mathbb{Z}_+$ for which $s_{j(l)} \le t_l \le s_{j(l)+1}$. Then the sequence $t_l / s_{j(l)}$ contains a convergent subsequence. By (3.3.41) and the obvious equality

$$v^{[t_l]} - \Phi^{[t_l]} = (v - \Phi)^{[t_l]} = \left(v^{[s_{j(l)}]} - \Phi^{[s_{j(l)}]}\right)^{[t_l / s_{j(l)}]}$$

this fact implies that the sequence of functions $\left(v^{[t_l]} - \Phi^{[t_l]}\right)$ contains a subsequence converging to zero. Since the sequence $\{t_l\}$ is arbitrary, we further conclude that $\left(v^{[t]} - \Phi^{[t]} \right) \to 0$ in $L^1_{loc}(\mathbb{C}^n)$ as $t \to \infty$. Recall that $v^{[t]} - u^{[t]} \to 0$ in $L^1_{loc}(\mathbb{C}^n)$ as $t \to \infty$, where u is the function given in the condition of the theorem. Thus,

$$\Phi^{[t]} - u^{[t]} \to 0 \text{ in } L^1_{loc}(\mathbb{C}^n) \text{ as } t \to \infty .$$

The proof is finished

The following theorem is a consequence of **Theorems 3.3.1** and **2.2.1**.

Theorem 3.3.2. Let $u(z)$ be a function which is plurisubharmonic in $\mathbb{C}^n$ and positively homogeneous of degree $\rho > 0$. Then there is an entire function $f \in H(\mathbb{C}^n, \rho]$ of c.r.g. such that $\mathfrak{L}_f(z) = u(z)$, $\forall z \in \mathbb{C}^n$.

Notes

Apparently, entire functions in $\mathbb{C}^n$ that are of c.r.g. on certain special sets of complex rays were first considered by L.Ronkin [2] in connection with the theorem on the addition of Polya-Plancherel indicators. Entire functions in $\mathbb{C}^n$ that are of c.r.g. on almost every ray l_λ were introduced by L.Gruman [2] (see also P.Agranovič, L.Ronkin [4]). **Theorem 3.1.2** was obtained by P.Agranovič and L.Ronkin [1], [2]. **Theorem 3.1.3** is due to S.Favorov [5]. Functions of c.r.g. in a distinguished variable were introduced by P.Agranovič and L.Ronkin [4], [1], [2]. In particular, these papers contain **Theorem 3.1.4**. **Theorem 3.2.3** is due to S.Favorov [3]. The similar **Theorem 3.2.4** was first published in Lelong-Gruman [1], were it is ascribed to S.Favorov. **Theorem 3.3.1** is due to R.Sigurdsson [1].

CHAPTER 4

FUNCTIONS OF COMPLETELY REGULAR GROWTH IN THE HALF-PLANE OR A CONE

§1 Preliminary information on functions holomorphic in a half-plane

Before proceeding to the account of the necessary notions and facts, we note that everything presented here, as well as in §4.2 and §4.3, can be obviously reformulated for an arbitrary sector

$$Y(\theta_1,\theta_2) = Y(\infty;\theta_1,\theta_2) = \{z\in\mathbb{C}: \theta_1 < argz < \theta_2\}.$$

The half-plane is considered for simplicity of notation only. Besides, all results (with some changes in the proofs can be generalized to subharmonic functions in a half-plane (see also §4.4, where we give the corresponding theory in the case of arbitrary dimension).

1. The order $\hat{\rho}_f$ and the class $H(\mathbb{C}^+,\rho]$. We set

$$\mathbb{C}^+ = \{z\in\mathbb{C}^+: Rez > 0\},$$

$$U_r^+ = \{z\in\mathbb{C}^+: |z| < r\},$$

$$S_r^+ = \{z\in\mathbb{C}^+: |z| = r\},$$

$$U_{r,R}^+ = \{z\in\mathbb{C}^+: r < |z| < R\},$$

$$\hat{M}_f(r) = sup\ \{|f(z)|: z\in S_r^+\}.$$

We introduce the notion of order $\hat{\rho}_f$ of a function $f\in H(\mathbb{C}^+)$. Taking into account the problems of constructing the theory of f.c.r.g. in $\mathbb{C}^+$, it is natural to proceed from the requirement that an indicator $h_f(\theta)$ with respect to the order $\hat{\rho}_f$ should give essential information on the asymptotic behaviour of the function f. If we would proceed by direct analogy with the case of entire functions and define $\hat{\rho}_f$ by the equality

$$\hat{\rho}_f = \overline{\lim_{r\to\infty}} \frac{\ln^+\ln^+\hat{M}_f(r)}{\ln r}, \tag{4.1.1}$$

then in the general case we are unable to satisfy this requirement. For example, if we calculate by (4.1.1) the order $\hat{\rho}_f$ in $\mathbb{C}^+$ of the function $f(z) = e^{iz}$, then we obtain $\hat{\rho}_f = 0$ and in this case the notion of the indicator becomes of no use. At the same time $\ln|f(re^{i\theta})| = -r\cos\theta$, so it is clear that we should have $\hat{\rho}_f = 1$ and, correspondingly, $h_f(\theta) = -\cos\theta$. There are different definitions of $\hat{\rho}_f$, for such that this situation ($\hat{\rho}_f = 0$ for $f(z) = e^{iz}$, etc.) is impossible (see N.Govorov [1], E.Titchmarsh [1], A.Rashkovskiĭ [1]). We use the following definition.

A number $\hat{\rho}_f$, $0 \le \hat{\rho}_f \le \infty$, is called the order of a function $f \in H(\mathbb{C}^+)$ (order in the half-plane) if it is defined by

$$\hat{\rho}_f = \max\left(\overline{\lim_{r\to\infty}} \frac{\ln^+\ln^+\hat{M}_f(r)}{\ln r}, \overline{\lim_{r\to\infty}} \frac{1}{\ln r} \ln^+\int_0^\pi |\ln|f(re^{i\theta})||\sin\theta d\theta \right) \tag{4.1.2}$$

The class $H(\mathbb{C}^+,\rho]$ is defined by analogy with $H(\mathbb{C},\rho]$. Namely, a function $f(z)$ in $H(\mathbb{C}^+)$ is said to belong to the class $H(\mathbb{C}^+,\rho]$ if

$$\overline{\lim_{r\to\infty}} r^{-\rho}\ln^+\hat{M}_f(r) < \infty \tag{4.1.3}$$

and

$$\overline{\lim_{r\to\infty}} r^{-\rho}\int_0^\pi |\ln|f(re^{i\theta})||\sin\theta d\theta < \infty. \tag{4.1.4}$$

Below we shall show that the number defined by (4.1.1) and (4.1.2) are equal even if one of them is larger than 1 (see the **Remark** to **Lemma 4.1.3**).

2. Boundary measure, generalized Carleman's formula, estimates. Jensen's formula and the representation of a function by a Weierstrass product were basic instruments in the study of entire functions of c.r.g. For functions in $\mathbb{C}^+$, the role of Jensen's formula is played by Carleman's formula, which is contained in the following lemma (for a proof of the lemma see, for example, B.Levin [1], Gol'dberg-Ostrovskiĭ [1]).

Lemma. Let $0 < \lambda < R < \infty$, $f \in H(\overline{U}^+_{\lambda,R})$, and let $z_k = r_k e^{i\theta_k}$, $k = 1, 2,$

... , be the zeros of f in $U^+_{\lambda,R}$, counted with their multiplicities. If λ is such that $f(z) \neq 0$, $\forall z \in \bar{S}^+$ [18] then

$$\sum_{k:\lambda<|z_k|<R} \left(\frac{1}{r_k} - \frac{r_k}{R^2} \right) \sin\theta_k = \frac{1}{\pi R} \int_0^\pi \ln|f(Re^{i\theta})| \sin\theta d\theta +$$

$$+ \frac{1}{2\pi} \int_\lambda^R \left(\frac{1}{x^2} - \frac{1}{R^2} \right) \ln|f(x)f(-x)| dx +$$

$$- Im \frac{1}{2\pi} \int_0^\pi \left(\frac{\lambda e^{i\theta}}{R^2} - \frac{e^{i\theta}}{\lambda} \right) \ln f(\lambda e^{i\theta}) dx, \quad (4.1.5)$$

where in the last term on the right $\ln f$ is some continuous branch of $Ln f$ on S^+_λ.

We give some statements that can be proved using Carleman's formula and which are needed later on. Without special stipulation, in the sequel we denote by $c_1(A,B,\dots)$, $c_2(A,B,\dots)$ quantities that depend on listed parameters only and that are locally bounded in their domains.

First of all we prove a lemma on the existence of some weak limit value of $\ln|f|$, called the boundary measure of $\ln|f|$ and denoted by $\mu_{f,\partial}$.

Lemma 4.1.1. Let a function $f \not\equiv 0$ be holomorphic in the semidisc U^+_R and let $\sup\{|f(z)|: z \in U^+_R\} = A < \infty$. Then the functions $\ln|f(x+ih)|$, $h > 0$, for $h \to 0$, converge, as functionals on $C_0([-R,R]) = \{\varphi \in C(-R,R): supp\varphi \subset (-R,R)\}$, to some real-valued measure $\mu_{f,\partial}$.

Proof. Since the function f is bounded on U^+_r, it is well-known that for almost all $\lambda \in (-R,R)$ the angular limits

$$\lim_{\substack{z \to \lambda \\ \delta < \arg(z-\lambda) < \pi-\delta}} f(z) \overset{\text{def}}{=} f^*(\lambda), \ \delta > 0,$$

exist, and since $f \not\equiv 0$, we have $f^*(\lambda) \neq 0$ for almost all $\lambda \in (-R,R)$. We denote by E the set of $\lambda \in (-R,R)$ for which the angular limit $f^*(\lambda)$

---

[18] The statement of the lemma remains true when there are zeros of function f on S^+_λ, but in this case the definition of $\ln f$ needs modification.

exists and does not vanish. Then by E_0 we denote the set of those $\lambda \in E$ such that f does not vanish on the semicircle $S^+_{|\lambda|}$. Like E, the set E_0 differs from (-R,R) by a set of measure zero. For every $\lambda \in E_0$ there clearly exists an $h = h_0(\lambda)$ such that the functions $f_h(z) = f(z+ih)$ do not vanish on $S^+_{|\lambda|}$ when $0 \le h \le h_0$. Therefore the functions $\ln f_h\big|_{S^+_{|\lambda|}}$ in Carleman's formula may be regarded as being continuous in $h \in [0,h_0]$. Hence

$$\sup_{0 \le h \le h_0} |\ln|f_h(z)|| = d < \infty,$$

and

$$\lim_{h \to 0} \int_0^\pi \ln f_h(|\lambda|e^{i\theta})\cdot\left(\frac{|\lambda|e^{i\theta}}{R^2} - \frac{e^{-i\theta}}{|\lambda|}\right)d\theta =$$

$$= \int_0^\pi \ln f(|\lambda|e^{i\theta})\cdot\left(\frac{|\lambda|e^{i\theta}}{R^2} - \frac{e^{-i\theta}}{|\lambda|}\right)d\theta. \qquad (4.1.6)$$

It is also clear that

$$\lim_{h \to 0} \int_0^\pi \ln|f_h(|\lambda|e^{i\theta})|\sin\theta d\theta = \int_0^\pi \ln|f(|\lambda|e^{i\theta})|\sin\theta d\theta, \ \forall \lambda \in E. \qquad (4.1.7)$$

We denote by $z_k = r_k e^{i\theta_k}$ the zeros of f(z), counted with their multiplicities. We set $z_k(h) = z_k - ih = r_k(h)e^{i\theta_k(h)}$, $h \ge 0$. From Carleman's formula, applied to f_h, it follows that for $0 < \lambda_1 < \lambda_2 < R$, $0 \le h < R - \lambda_2$,

$$\sum_{k: z_k(h) \in U^+_{\lambda_1,\lambda_2}} \left(\frac{1}{r_k(h)} - \frac{r_k(h)}{R^2}\right)\sin\theta_k(h) < \infty.$$

It is easy to see that

$$\lim_{h \to 0} \sum_{k: z_k(h) \in U^+_{\lambda_1,\lambda_2}} \left(\frac{1}{r_k(h)} - \frac{r_k(h)}{R^2}\right)\sin\theta_k(h) =$$

$$= \sum_{k:\lambda_1<|z_k|<\lambda_2} \left(\frac{1}{r_k} - \frac{r_k}{R^2} \right) sin\theta_k, \lambda_1 \in E_0, \lambda_2 \in E_0. \qquad (4.1.8)$$

From this and (4.1.6), (4.1.7) using Carleman's formula we conclude that $\forall\lambda_1 \in E_0$, $\forall\lambda_2 \in E_0$, $0 < \lambda_1 < \lambda_2$,

$$\exists \lim_{h\to 0} \int_{\lambda_1'}^{\lambda_2} \left(\frac{1}{x^2} - \frac{1}{\lambda_2^2} \right) \cdot ln|f_h(x)f_h(-x)|dx.$$

In turn it follows from this statement that for $0 < \lambda_1 < \lambda_1' < \lambda_2$, $\lambda_1 \in E_0$, $\lambda_2 \in E_0$, $\lambda_1' \in E_0$,

$$\exists \lim_{h\to 0} \int_{\lambda_1}^{\lambda_2} \left(\frac{1}{x^2} - \frac{1}{\lambda_2^2} \right) \cdot ln|f_h(x)f_h(-x)|dx.$$

By varying λ_2 we can prove the existence of the limit

$$\lim_{h\to 0} \int_{\lambda_1}^{\lambda_1'} ln|f_h(x)f_h(-x)|dx, \ \forall\lambda_1 \in E_0, \ \forall\lambda_1' \in E_0. \qquad (4.1.9)$$

Let $a_1 \in E$, $a_2 \in E$, $-R < a_1 < a_2 < R$. We choose $a_1' \in E$, $a_2' \in E$, $a_1 < a_1' < a_2' < a_2$ so that on the semicircle $\left\{z\in C^+: \left|z - \left(a_2' + \frac{1}{2}(a_1'-a_1)\right)\right| = a_2' - \frac{1}{2}(a_1+a_1')\right\}$, $\{z\in C^+: \left|z - \left(a_2' + \frac{1}{2}(a_1'-a_1)\right)\right| = a_1' - \frac{1}{2}(a_1'-a_1)\right\}$, $\left\{z\in C^+: \left|z - \frac{1}{2}(a_2'+a_1')\right| = a_2' - \frac{1}{2}(a_2'+a_1')\right\}$, $\left\{z\in C^+: \left|z - \frac{1}{2}(a_2'+a_1')\right| = a_1 + \frac{1}{2}(a_2'+a_1')\right\}$, the functions $f(z)$ would not vanish (the possibility of such a choice is obvious because of the countability of the zero set of the function f and the fact that E and $(-R,R)$ differ on a set of measure zero). Then the statement on the existence of the limit (4.1.9) with $\lambda_1 = \frac{1}{2}(a_2'-a_1')$, $\lambda_1' = \frac{1}{2}(a_1+a_2')-a_1$ and $\lambda_1 = \frac{1}{2}(a_1'-a_1)$, $\lambda_1' = a_2' + \frac{1}{2}(a_1+a_1')$ can be applied to the functions $f^*(z) = f\left(z-\frac{1}{2}(a_1'+a_2')\right)$ and $f^{**}(z) = f\left(z-\frac{1}{2}(a_1'+a_1)\right)$, respectively. Since

$$\int_{a_1}^{2a_2'-a_1} ln|f_h(x)|dx =$$

$$= \int_{\frac{1}{2}(a_2'-a_1')}^{\frac{1}{2}(a_1+a_2')-a_1} ln|f_h^*(x)f_h^*(-x)|dx + \int_{\frac{1}{2}(a_1'-a_1)}^{a_2' - \frac{1}{2}(a_1+a_1')} ln|f_h^*(x)f_h^*(-x)|dx,$$

$\exists \lim_{h\to 0} \int_{a_1}^{2a_2'-a_1} ln|f_h(x)|dx$ and therefore

$$\exists \lim_{h\to 0} \int_{a_1}^{a_2} ln|f_h(x)|dx, \ \forall a_1 \in E, \ a_2 \in E. \tag{4.1.10}$$

We estimate the integral

$$\int_{\lambda_1}^{\lambda_2} |ln|f_h(x)||dx.$$

For this we represent it as follows:

$$\int |ln|f_h(x)||dx = \int ln^+|f_h(x)|dx + \int ln^+\left|\frac{1}{f_h(x)}\right|dx.$$

The estimate of the first term is obvious. The second term is estimated by Carleman's formula. Let us assume that $\lambda_1 > 0$ and $\lambda_1 \in E$. This clearly does not lead to loss of generality. Note also that

$$\left(\frac{1}{r_k(h)} - \frac{r_k(h)}{\lambda_2^2} \right) sin\theta_k(h) \geq 0$$

when $r_k(h) < \lambda_2$ and

$$\int ln|f_h|dx = \int ln^+|f_h|dx - \int ln^+\left|\frac{1}{f_h}\right|dx.$$

Then for any $\lambda_3 \in (\lambda_2, R)$ and $h \in (0, h(\lambda_1))$, $h+\lambda_3 < R$, we have

$$\int_{\lambda_1}^{\lambda_2} ln^+\left|\frac{1}{f_h(x)}\right|dx \leq \left(\frac{1}{\lambda_2^2} - \frac{1}{\lambda_3^2} \right)^{-1} \int_{\lambda_1}^{\lambda_2} \left(\frac{1}{x^2} - \frac{1}{\lambda_3^2} \right) ln^+\left|\frac{1}{f_h(x)}\right|dx \leq$$

$$\leq \left(\frac{1}{\lambda_2^2}-\frac{1}{\lambda_3^2}\right)^{-1}\left\{\int_{\lambda_1}^{\lambda_3}\left(\frac{1}{x^2}-\frac{1}{\lambda_3^2}\right)ln^+|f_h(x)|dx + \int_{\lambda_1}^{\lambda_3}\left(\frac{1}{x^2}-\frac{1}{\lambda_3^2}\right)ln|f_h(-x)|dx + \right.$$

$$\left. + \frac{2}{\lambda_3}\int_0^{\pi} ln|f(\lambda_3 e^{i\theta}+h)|d\theta - Im\int_0^{\pi}\left(\frac{\lambda_1 e^{i\theta}}{\lambda_3^2} - \frac{e^{-i\theta}}{\lambda_1}\right)lnf(\lambda_1 e^{i\theta}+ih)d\theta\right\} \leq$$

$$\leq \left(\frac{1}{\lambda_2^2}-\frac{1}{\lambda_3^2}\right)^{-1}\int_{\lambda_1}^{\lambda_3}\left(\frac{1}{x^2}-\frac{1}{\lambda_3^2}\right)ln^+Adx + \int_{\lambda_1}^{\lambda_3}\left(\frac{1}{x^2}-\frac{1}{\lambda_3^2}\right)lnAdx +$$

$$+ \pi d\left(\frac{\lambda_1}{\lambda_3^2}+\frac{1}{\lambda_1}\right) = c(A,\lambda_1,\lambda_2,\lambda_3) < \infty.$$

It is clear that the restriction $h < h_0$ is inessential here; only the requirement $h \leq h_1$, where h_1 satisfies the condition $\{x+ih\colon \lambda_1 \leq x \leq \lambda_3\} \subset U_R^+$, $\forall h$, $0 < h \leq h_1$ is important. Thus,

$$\sup_{0<h\leq h_1}\int_{\lambda_1}^{\lambda_2} |ln|f_h(x)||dx < \infty. \tag{4.1.11}$$

We now finish the proof of the lemma as follows.

Let $\varphi \in C^1((-R,R))$ and $supp\varphi \subset (-R^*,R^*)$, where $0 < R^* < R$. Then

$$\int_{-R}^{R}\varphi(x)ln|f_h(x)|dx = -\int_{-R^*}^{R^*}\varphi'(x)\int_{-R^*}^{x}ln|f_h(t)|dt$$

and by (4.1.10) and (4.1.11) the limit of the right-hand side exists. Hence

$$\exists \lim_{h\to 0}\int_{-R}^{R}\varphi(x)ln|f_h(x)|dx. \tag{4.1.12}$$

The functions $\varphi \in C_0((-R,R)) \cap C^1((-R,R))$ are dense in $C_0((-R,R))$. Therefore the limit (4.1.12) exists for any function $\varphi \in C_0((-R,R))$, and it defines on $(-R,R)$ a real-valued measure, denoted by $\mu_{f,\partial}$.

This finishes the proof.

Remark. It is obvious that if the function f figuring in this lemma is holomorphic on $\bar{U}^+_R$, then $d\mu_{f,\partial} = \ln|f(x)|dx$.

Now that the notion of boundary measure has been introduced and a condition for its existence has been given, the assumptions on f(z) in Carleman's formula may be relaxed. We obtain the following statement.

Lemma 4.1.2 (generalized Carleman's formula). Let $0 < \lambda' < R' < \infty$, $f \in H(U^+_{\lambda',R'})$, $\sup\{|f(z)|: z \in U^+_{\lambda',R'}\} < \infty$ and let $z_k = r_k e^{i\theta_k}$, $k=1,2,\ldots$, be the zeros of f, counted with multiplicities. Then if λ and R, $\lambda' < \lambda < R < R'$, are such that $\mu_{f,\partial}(\{\pm\lambda\}) = 0$, $\mu_{f,\partial}(\{\pm R\}) = 0$, $f(z) \neq 0$, $\forall z \in S^+_\lambda$, and if angular limits $f^*(\pm R)$ exist at the points $\pm R + 0 \cdot i$ and are non-zero, then

$$\sum_{k:\lambda<r_k<R} \left(\frac{1}{r_k} - \frac{r_k}{R^2} \right) \sin\theta_k =$$

$$= \frac{1}{\pi R} \int_0^\pi \ln|f(Re^{i\theta})|\sin\theta d\theta + \frac{1}{2\pi} \left\{ \int_{-R}^{-\lambda} + \int_\lambda^R \left(\frac{1}{x^2} - \frac{1}{R^2} \right) d\mu_{f,\partial}(x) \right\} +$$

$$- \operatorname{Im} \frac{1}{2\pi} \int_0^\pi \left(\frac{\lambda e^{i\theta}}{R^2} - \frac{e^{-i\theta}}{\lambda} \right) \ln f(\lambda e^{i\theta}) d\theta. \qquad (4.1.13)$$

To prove this lemma it is sufficient to write formula (4.1.5), i.e. Carleman's formula, for the function $f_h(z)$ and then using (4.1.6)-(4.1.8) and **Lemmas 4.1.1, 2.1.1**, pass to the limit as $h \to \infty$.

When $f \in H(\mathbb{C}^+, \rho]$, a number of integral estimates of $\ln|f(z)|$ can be obtained with the aid of the generalized Carleman's formula. Namely, the following lemma holds.

Lemma 4.1.3. Let $f \in H(\mathbb{C}^+)$, $z_k = r_k e^{i\theta_k}$ be the zeros of f(z) counted with multiplicities, and let the inequalities

$$\ln|f(z)| \le A|z|^\rho, \ \forall z \in \mathbb{C}^+ \setminus U_\lambda, \qquad (4.1.14)$$

$$\frac{1}{\pi} \int_0^\pi |\ln|f(re^{i\theta})||\sin\theta d\theta \le Br^\rho, \ \forall r \ge \lambda, \qquad (4.1.15)$$

be valid for some $A > 0$, $B > 0$, $\rho > 0$, $\lambda > 0$. Let also $f(z) \neq 0$, $\forall z \in S_\lambda^+$, $\sup_{z\in S_\lambda^+} |\ln f(z)| = d < \infty$, $\mu_{f,\partial}(\{\pm\lambda\}) = 0$ and $\exists f^*(\pm\lambda) \neq 0$. Then the following inequalities are true:

$$\int_\lambda^r x^\alpha d|\mu_{f,\partial}|(x)dt \le c_1 r^{\rho+\alpha+1}, \ \forall r \ge \lambda, \tag{4.1.16}$$

where $\alpha \ge -1$, $c_1 = c_1(\rho, A, B, d, \lambda)$;

$$\int_{U^+_{r,R}} |\ln|f(z)|||z|^{-2} \sin(\arg z) d\omega \le c_2 (R-r) R^{\rho-1}, \forall r, R, \ \lambda < r < R, \tag{4.1.17}$$

where $c_2 = c_2(\rho, A, B, d, \lambda)$;

$$\int_\lambda^r |\ln|f(te^{i\theta})|| t^\alpha \le c_3 r^{\rho+\alpha+1}, \ \forall r \ge \lambda, \tag{4.1.18}$$

where $\alpha \ge -1$, $c_3 = c_3(\rho, A, B, d, \lambda)$;

$$\int_{(Y(=,\delta)\setminus U_\lambda)\cap U_r} \ln|f(z)|||z|^{-2} d\omega_z \le$$

$$\le \delta c_4 r^\rho, \ \forall r \ge \lambda, \ 0 < \delta < \frac{\pi}{2}, \tag{4.1.19}$$

where $c_4 = c_4(A, B, \rho, d, \lambda)$;

$$\sum_{k:\lambda<r_k<R} \sin\theta_k \le c_5 R^\rho, \ \forall R \ge \lambda, \tag{4.1.20}$$

where $c_5 = c_5(A, B, \lambda, \rho)$.

Proof. We represent the measure $\mu_{f,\partial}$ in standard manner as $\mu_{f,\partial} = \mu^+_{f,\partial} - \mu^-_{f,\partial}$, and choose numbers r' and r'', $\lambda < r' < r''$, satisfying the conditions such that the generalized Carleman's formula is valid in the semi-annulus $U^+_{\lambda,r'}$ and $U^+_{\lambda,r''}$. Using this formula we obtain

$$\frac{1}{2\pi} \int_{\lambda<|x|<r''} \left(\frac{1}{x^2} - \frac{1}{r''^2} \right) d\mu_{f,\partial}(x) - \frac{1}{2\pi} \int_{\lambda<|x|<r'} \left(\frac{1}{x^2} - \frac{1}{r'^2} \right) d\mu_{f,\partial}(x) +$$

$$+ \frac{1}{\pi r''} \int_0^{\pi} ln|f(r''e^{i\theta})|sin\theta d\theta - \frac{1}{\pi r'} \int_0^{\pi} ln|f(r'e^{i\theta})|sin\theta d\theta +$$

$$+ Im \frac{1}{2\pi} \int_0^{\pi} lnf(\lambda e^{i\theta})\left(\frac{\lambda}{r'^2} - \frac{\lambda}{r''^2} \right)e^{i\theta}d\theta =$$

$$= \sum_{\lambda<r_k<r'} \left(\frac{1}{r'^2} - \frac{1}{r''^2} \right)r_k sin\theta_k + \sum_{r'\le r_k<r''} \left(\frac{1}{r_k} - \frac{r_k}{r''^2} \right)sin\theta_k \ge 0.$$

From this, by the equality

$$\frac{1}{2\pi} \int_{\lambda<|x|<r''} \left(\frac{1}{x^2} - \frac{1}{r''^2} \right)d\mu_{f,\partial}(x) - \frac{1}{2\pi} \int_{\lambda<|x|<r'} \left(\frac{1}{x^2} - \frac{1}{r'^2} \right)d\mu_{f,\partial}(x) =$$

$$= \frac{1}{2\pi} \int_{r'<|x|<r''} \left(\frac{1}{x^2} - \frac{1}{r''^2} \right)d\mu_{f,\partial}(x) + \frac{1}{2\pi} \left(\frac{1}{r'^2} - \frac{1}{r''^2}\right) \int_{\lambda<|x|<r'} d\mu_{f,\partial}(x),$$

it follows that

$$\frac{1}{2\pi} \int_{r'}^{r''} \left(\frac{1}{x^2} - \frac{1}{r''^2} \right)d\mu^-_{f,\partial}(x) \le \frac{1}{2\pi} \int_{r'}^{r''} \left(\frac{1}{x^2} - \frac{1}{r''^2} \right)d\mu^+_{f,\partial}(x) +$$

$$\frac{1}{2\pi} \int_{-r''}^{-r'} \left(\frac{1}{x^2} - \frac{1}{r''^2} \right)d\mu_{f,\partial}(x) + \frac{1}{2\pi} \left(\frac{1}{r'^2} - \frac{1}{r''^2}\right) \int_{\lambda<|x|<r'} d\mu_{f,\partial}(x) +$$

$$+ \frac{1}{\pi r''} \int_0^{\pi} ln|f(r''e^{i\theta})|sin\theta d\theta - \frac{1}{\pi r'} \int_0^{\pi} ln|f(r'e^{i\theta})|sin\theta d\theta +$$

$$+ \frac{\lambda}{2\pi} \left(\frac{1}{r'^2} - \frac{1}{r''^2} \right) Im \int_0^{\pi} lnf(\lambda e^{i\theta})d\theta = I.$$

Hence

$$\frac{1}{2\pi}\int_{r'}^{r''}\left(\frac{1}{x^2}-\frac{1}{r''^2}\right)d|\mu_{f,\partial}|(x) \le I + \frac{1}{2\pi}\int_{r'}^{r''}\left(\frac{1}{x^2}-\frac{1}{r''^2}\right)d\mu^+_{f,\partial}(x).$$

Then, taking into account the estimates (4.1.14), (4.1.15), we find

$$\frac{1}{2\pi}\int_{r'}^{r''}\left(\frac{1}{x^2}-\frac{1}{r''^2}\right)d|\mu_{f,\partial}|(x) \le \frac{3A}{\pi|\rho^2-1|}(r'^{\rho-1}+r''^{\rho-1}) +$$

$$+\frac{2}{\pi(\rho+1)}r'^{\rho-1} + B'r^{\rho-1} + Br''^{\rho-1} + \frac{\lambda d}{2r'^2} =$$

$$= c_{1,1}r'^{\rho-1} + c_{1,2}r''^{\rho-1} + \frac{\lambda d}{2r'^2} \le c_{1,3}r'^{\rho-1} + c_{1,2}r''^{\rho-1} \tag{4.1.21}$$

for $\rho \ne 1$ and

$$\frac{1}{2\pi}\int_{r'}^{r''}\left(\frac{1}{x^2}-\frac{1}{r''^2}\right)d|\mu_{f,\partial}|(x) \le$$

$$\le \frac{3A}{2\pi}\ln\frac{r''}{r'} + \frac{2}{\rho+1} + Br'^{\rho-1} + Br''^{\rho-1} + \frac{\lambda}{2}\frac{d}{r''^2} \tag{4.1.21'}$$

for $\rho = 1$.

Since the right hand sides of the inequalities (4.1.21) and (4.1.21') are continuous in r' and r'', the initial restrictions on r' and r'' may be omitted. We estimate the integral $\int_\lambda^r x^\alpha d|\mu_{f,\partial}|(x)$, using these inequalities. Assuming $2^{N-1}\lambda \le r \le 2^N\lambda$, $\lambda > 1$, we have for $\rho \ne 1$,

$$I \le \int_\lambda^{2^N\lambda} x^\alpha d|\mu_{f,\partial}|(x) = \sum_{k=0}^{N-1}\int_{2^k\lambda}^{2^{k+1}\lambda} x^\alpha d|\mu_{f,\partial}|(x) \le$$

$$\le \frac{4}{3}\sum_{k=0}^{N-1}\lambda^2 2^{2k}(2^{k+1}\lambda)^\alpha\int_{2^k\lambda}^{2^{k+1}\lambda}\left(\frac{1}{x^2}-\frac{1}{(2^{k+2}\lambda)^2}\right)d|\mu_{f,\partial}|(x) \le$$

$$\le \frac{1}{3}\sum_{k=0}^{N-1}(2\lambda)^{\alpha+2}2^{2k(\alpha+2)}\,(c_{1,3}(2^k\lambda)^{\rho-1}+c_{1,2}(2^{k+2}\lambda)^{\rho-1})\le$$

$$\le \text{const}\cdot\sum_{k=0}^{N-1}2^{k(\alpha+\rho+1)}\le \text{const}\cdot 2^{(N-1)(\alpha+\rho+1)}\sum_{k=0}^{\infty}2^{-k\rho}\le \text{const}\cdot r^{\alpha+\rho+1}.$$

Thus we have proved inequality (4.1.16) in case $\rho \neq 1$. If $\rho = 1$ then inequality (4.1.16) can be obtained similarly, using (4.1.21') instead of (4.1.21).

Inequality (4.1.17) follows from (4.1.15) in an obvious manner.

Now we prove (4.1.18) under the assumption $\theta < \frac{\pi}{2}$, which involves no loss of generality. Consider the function $F(z) = f(z^{\gamma}e^{i\theta})$, $0 < \arg z < \pi$, where $\gamma = 1 - \theta/\pi$. It is clear that $F\in H(\mathbb{C}^+)$ and $\ln|F(z)| \le A|z|^{\rho\gamma}$. It can be checked in a standard manner that

$$\sin\frac{\psi-\theta}{\gamma} \le \frac{1}{\gamma}\cdot\sin\psi,\ \forall\psi\in(0,\pi),$$

and hence

$$\frac{1}{\pi}\int_0^{\pi}|\ln|F(re^{i\varphi})||\sin\varphi d\varphi =$$

$$= \frac{1}{\gamma}\,\frac{1}{\pi}\int_0^{\pi}|\ln|f(r^{\gamma}e^{i\psi})||\sin\frac{\psi-\theta}{\gamma}\,d\psi \le \frac{1}{\gamma^2}\,Br^{\theta\rho} \le \frac{B}{4}\,r^{\gamma\rho}.$$

Applying the estimate (4.1.16) to F and taking into account that in this situation $d\mu_{f,\partial}(x) = \ln|F(x)|dx$, we obtain

$$\int_{\lambda}^{r} s^{\beta}|\ln|F(s)||ds \le c_1 r^{\rho\gamma+\beta+1},\ \beta \ge -1.$$

This implies

$$\int_{\lambda}^{r} t^{\alpha}|\ln|f(te^{i\theta})||dt = \gamma\int_{\lambda^{1/\gamma}}^{r^{1/\gamma}} s^{\alpha\gamma+\gamma-1}|\ln|f(s^{\gamma}e^{i\theta})||ds =$$

$$= \gamma\int_{\lambda^{1/\gamma}}^{r^{1/\gamma}} s^{\alpha\gamma+\gamma-1}|\ln|F(s)||ds \le c_1(r^{1/\gamma})^{\rho\gamma+\alpha\gamma+\gamma} = \gamma c_1 r^{\rho+\alpha+1}.$$

Thus inequality (4.1.18) is proved.

Inequality (4.1.19) is an obvious consequence of (4.1.18).

To prove inequality (4.1.20) for $\rho \ge 1$ we note that

$$\sum_{k:\lambda<r_k<R} \sin\theta_k \le \frac{4}{3} R \sum_{k:\lambda<r_k<R} \left(\frac{1}{r_k} - \frac{r_k}{(2R)^2} \right) \sin\theta_k \le$$

$$\le \frac{4}{3} R \sum_{k:\lambda<r_k<2R} \left(\frac{1}{r_k} - \frac{r_k}{(2R)^2} \right) \sin\theta_k.$$

Now it suffices to refer to estimates (4.1.15) and (4.1.21) and the generalized Carleman's formula.

In case $\rho < 1$ we proceed as in the derivation of inequality (4.1.6). First we estimate the sum

$$\sum_{k:r'<r_k<r''} \left(\frac{1}{r_k} - \frac{r_k}{r''^2} \right) \sin\theta_k.$$

Using the generalized Carleman's formula we obtain

$$\sum_{k:r'<r_k<r''} \left(\frac{1}{r_k} - \frac{r_k}{r''^2} \right) \sin\theta_k \le$$

$$\le \sum_{k:\lambda<r_k<r''} \left(\frac{1}{r_k} - \frac{r_k}{r''^2} \right) \sin\theta_k - \sum_{k:\lambda<r_k<r'} \left(\frac{1}{r_k} - \frac{r_k}{r'^2} \right) \sin\theta_k =$$

$$= \frac{1}{\pi r''} \int_0^{\pi} \ln|f(r''e^{i\theta})| \sin\theta d\theta - \frac{1}{\pi r'} \int_0^{\pi} \ln|f(r'e^{i\theta})| \sin\theta d\theta +$$

$$+ \frac{1}{2\pi} \int_{\lambda<|x|<r'} \left(\frac{1}{r'^2} - \frac{1}{r''^2} \right) d\mu_{f,\partial}(x) + \frac{1}{2\pi} \int_{r'<|x|<r'} \left(\frac{1}{x^2} - \frac{1}{r''^2} \right) d\mu_{f,\partial}(x) +$$

$$+ \frac{1}{2\pi} \left(\frac{\lambda}{r'^2} - \frac{\lambda}{r''^2} \right) \cdot Im \int_0^{\pi} \ln f(\lambda e^{i\theta}) e^{i\theta} d\theta.$$

From this and estimates (4.1.15), (4.1.16) and (4.1.21) it follows that

$$\sum_{k:r'<r_k<r''} \left(\frac{1}{r_k} - \frac{r_k}{r''^2} \right) \sin\theta_k \le c_{5,1} r'^{\rho-1}, \quad c_{5,1} = c_{5,1}(A,B,\rho,\lambda,d).$$

Setting $r' = 2^{l-1}\lambda$, $r'' = 2^{l+1}\lambda$ we have

$$\sum_{k:2^{l-1}\lambda<r_k<2^l\lambda} \sin\theta_k \le \frac{4}{3}2^l\lambda \sum_{k:2^{l-1}\lambda<r_k<2^l\lambda} \left(\frac{1}{r_k} - \frac{r_k}{4^{l+1}\lambda^2}\right)\sin\theta_k \le$$

$$\le \frac{4}{3}2^l\lambda \sum_{k:2^{l-1}\lambda<r_k<2^{l+1}\lambda} \left(\frac{1}{r_k} - \frac{r_k}{4^{l+1}\lambda^2}\right)\sin\theta_k \le$$

$$\le \frac{4}{3}2^l\lambda c_{5,1}(2^{l+1}\lambda)^{\rho-1} = c_{5,2}(2^l)^\rho.$$

Then, assuming $2^{N-1}\lambda \le R \le 2^N\lambda$, we obtain

$$\sum_{k:\lambda<r_k<R} \sin\theta_k \le \sum_{l=1}^{N} \sum_{k:2^{l-1}\lambda<r_k<2^l\lambda} \sin\theta_k \le c_{5,2}\sum_{l=1}^{N} 2^{l\rho} \le$$

$$\le c_{5,2}(2^{N-1})^\rho \sum_{-1}^{\infty} 2^{-l\rho} = c_5 2^{(N-1)\rho} = c_5 R^\rho.$$

In case $\rho = 1$ the inequality (4.1.20) can be proved similarly. This finishes the proof of the lemma.

Remark. In case $\rho > 1$ the inequality (4.1.16) follows from (4.1.14). Indeed, by the generalized Carleman's formula we have

$$\int_0^\pi |\ln|f(re^{i\theta})||d\theta + \int_0^\pi \ln^+|f(re^{i\theta})|d\theta + \int_0^\pi \ln^+\frac{1}{|f(re^{i\theta})|}d\theta \le$$

$$\le 2\int_0^\pi \ln^+|f(re^{i\theta})|d\theta + \frac{r}{2}\int_{\lambda<|x|<r} \left(\frac{1}{x^2} - \frac{1}{r^2}\right)d\mu_{f,\partial}(x) + d\left(\frac{\lambda}{r} + \frac{r}{\lambda}\right) \le$$

$$\le 2\int_0^\pi A^+ r^\rho \sin\theta d\theta + \frac{rA^+}{2}\int_{\lambda<|x|<r} \left(\frac{1}{x^2} - \frac{1}{r^2}\right)|x|^\rho dx + d\left(\frac{\lambda}{r} + \frac{r}{\lambda}\right) =$$

$$= 2A^+ \frac{\rho^2}{\rho^2-1} r^\rho + d\left(\frac{\lambda}{r} + \frac{r}{\lambda}\right) \le c_6 r^\rho, \quad c_6 = c_6(A,d,\lambda) .$$

An obvious consequence of this implication is the equality

$$\hat{\rho}_f = \overline{\lim_{r\to\infty}} \frac{\ln\ln\hat{M}_f(r)}{\ln r}$$

when $\hat{\rho}_f \geq 1$, and the coincidence for $\rho > 1$ of the class $H(\mathbb{C}^+, \rho]$ with the set of functions $f \in H(\mathbb{C}^+)$ satisfying the condition

$$\overline{\lim_{r \to \infty}} \frac{\ln^+ \hat{M}_f(r)}{r^\rho} < \infty.$$

3. Green's formula. To give an exposition of the theory of holomorphic f.c.r.g. in the above manner it is necessary to introduce Green's formula for functions of the form $\ln|f|$, $f \in H(\mathbb{C}^+)$, next to Carleman's formula. This formula is given in **Lemma 4.1.5**, and is preceded by the following statement.

Lemma 4.1.4. Let $f \in H(\mathbb{C}^+)$ be bounded in every semidisc U_R^+, $R > 0$, and let $z_k = r_k e^{i\theta_k}$, $k = 1, 2, \ldots$, be the zeros of f counted with multiplicities. We denote by $\tilde{\Lambda}_f$ the set of points $r > 0$ such that the nonzero angular limits $f^*(\pm r)$ exist, $f(z)$ does not vanish on the semicircle S_r^+ and $\mu_{f,\partial}(\{\pm r\}) = 0$. Then for any function $\eta \in C^2([0,\pi])$ and any $r \in \tilde{\Lambda}_f$ the limit

$$\lim_{h \to 0} \int_0^\pi \eta(\theta) \sin\theta \frac{\partial}{\partial r} \ln|f_h(re^{i\theta})| d\theta \overset{\text{def}}{=} Q_f^+(\eta(\theta), r)$$

exists.

Proof. Let $\gamma(r)$ be a function satisfying the conditions: $\gamma \in C^2(0,\infty)$, $supp\gamma \subset (\lambda, 2r)$, $\lambda \in \tilde{\Lambda}_f$, $r \in \tilde{\Lambda}_f$, $\gamma(r) = 1$, $\gamma'(r) = 0$. We set $v(z) = \eta(\theta)\sin\theta\gamma(t)$, $z = te^{i\theta}$. We apply the ordinary Green's formula to the functions $v(z)$ and $\ln|f(z)|$ in the domain $U_{\lambda,r}^+$. Taking into account that $\Delta \ln|z-z_k| = 2\pi\delta(z-z_k)$, we obtain

$$2\pi \sum_{z_k \in (ih + U_{\lambda,r}^+)} v(z_k - ih) - \int_{U_{\lambda,r}^+} \ln|f_h(z)| \Delta v(z) d\omega_z =$$

$$= r\gamma(r) \int_0^\pi \eta(\theta) \sin\theta \frac{\partial}{\partial r} \ln|f_h(re^{i\theta})| d\theta +$$

$$+ \int_{\lambda < |x| < r} \eta\left(\arg \frac{x}{|x|}\right) \ln|f_h(x)| \frac{\gamma(|x|)}{|x|} dx.$$

Since f is bounded in the semidiscs U_R^+ , it follows from Carleman's formula that

$$\sum_{k:z_k\in U^+_{\lambda,r}} \sin\theta_k < \infty. \tag{4.1.22}$$

In turn, this implies that

$$\exists \lim_{h\to 0} \sum_{k:z_k\in(U^+_{\lambda,r}+ih)} v(z_k-ih) = \sum_{k:z_k\in U^+_{\lambda,r}} v(z_k).$$

Then, using the inequality

$$\int_{U^+_R} |\ln|f(z)||d\omega_z < \infty, \ \forall R < \infty$$

which follows from the estimates in **Lemma 4.1.3**, we conclude that

$$\exists \lim_{h\to 0} \int_{U^+_{\lambda,r}} \ln|f_h(z)|\Delta v(z)d\omega_z = \int_{U^+_{\lambda,r}} \ln|f(z)|\Delta v(z)d\omega_z.$$

Finally, since $\lambda\in\tilde{\Lambda}_f$, $r\in\tilde{\Lambda}_f$, we have

$$\exists \lim_{h\to 0} \int_{\lambda<|x|<r} \eta\left(\arg\frac{x}{|x|}\right)\frac{\gamma(|x|)}{|x|} \ln|f_h(x)|dx =$$

$$= \int_{\lambda<|x|<r} \eta\left(\arg\frac{x}{|x|}\right)\frac{\gamma(|x|)}{|x|} d\mu_{f,\partial}(x).$$

Combining all this we find that

$$\exists \lim_{h\to 0} \int_0^{\pi} \eta(\theta)\sin\theta \frac{\partial}{\partial r} \ln|f_h(re^{i\theta})|d\theta =$$

$$= \frac{1}{r\gamma(r)}\left\{ 2\pi \sum_{k:z_k\in U^+_{\lambda,r}} v(z_k) - \int_{U^+_{\lambda,r}} \ln|f(z)|\Delta v(z)d\omega_z + \right.$$

$$\left. - \int_{\lambda<|x|<r} \eta(\arg\frac{x}{|x|}) \frac{\gamma(|x|)}{|x|} d\mu_{f,\partial}(x) \right\} \overset{\text{def}}{=} Q^+_f(\eta(\theta),r). \tag{4.1.23}$$

The proof of the lemma is finished.

It is naturally to regard the variable (functional) Q^+_f introduced in this lemma as the integral

$$\int_0^{\pi} \eta(\theta)sin\theta \frac{\partial}{\partial r} ln|f(re^{i\theta})|d\theta,$$

which does not exist in the ordinary sense, in general. Using Q_f^+ as well as the notion of boundary measure, we extend the Green's formula for $U_{\lambda,r}^+$ to the case when one of the functions can be represented as $u = ln|f|$, $f\in U_{r,R}^+$.

Lemma 4.1.5 (Green's formula). Let $f(z)$, $z_k = r_k e^{i\theta_k}$, and let $\tilde{\Lambda}_f$ be as in **Lemma 4.1.4**. Then for any $\lambda\in\tilde{\Lambda}_f$, $R\in\tilde{\Lambda}_f$ and $\varphi\in C^2(U_{\lambda,R}^+)$ the following equality is true:

$$2\pi \sum_{k:z_k\in U_{\lambda,R}^+} \varphi(z_k)sin(argz_k) - \int_{U_{\lambda,R}^+} ln|f(z)|\Delta(sin(argz)\varphi(z))d\omega_z =$$

$$= (rQ_f^+(\varphi(re^{i\theta}),r))\Big|_{r=\lambda}^{r=R} - \left(r\int_0^{\pi} sin\theta\cdot ln|f(re^{i\theta})| \frac{\partial}{\partial r} \varphi(re^{i\theta})d\theta\right)\Big|_{r=\lambda}^{r=R} +$$

$$+ \int_{\lambda<|x|<r} \frac{\varphi(x)}{|x|} d\mu_{f,\partial}(x). \qquad (4.1.24)$$

Proof. Applying Green's formula to the functions $ln|f_h(z)|$ and $\varphi(z)sin(argz)$ in the domain $U_{\lambda,R}^+$ we obtain

$$2\pi \sum_{z_k\in(ih+U_{\lambda,r}^+)} \varphi(z_k-ih)sin(arg(z_k-ih)) +$$

$$- \int_{U_{\lambda,r}^+} ln|f_h(z)|\Delta(\varphi(z)sin(argz))d\omega_z =$$

$$= \left(r\int_0^{\pi} \varphi(re^{i\theta})sin\theta \frac{\partial}{\partial r} ln|f(re^{i\theta})|d\theta\right)\Big|_{r=\lambda}^{r=R} +$$

$$- \left(r\int_0^{\pi}\varphi'_r(re^{i\theta})sin\theta\cdot ln|f_h(re^{i\theta})|d\theta\right)\Big|_{r=\lambda}^{r=R} + \int_{\lambda<|x|<r} \frac{\varphi(x)}{|x|} ln|f_h(x)|dx.$$

Letting $h \to 0$ in this inequality and taking into account the inequalities (4.1.22), (4.1.23), **Lemmas 4.1.2, 4.1.4** and the fact that

$\lambda\in\tilde{\Lambda}_f$ and $R\in\tilde{\Lambda}_f$ we arrive at the conclusion that equality (4.1.24) is valid.

This finishes the proof.

Remark. It is obvious the the statements of **Lemmas 4.1.4** and **4.1.5** are also true in case $f\in H(\bar{U}^+_{\lambda,R})$ and $r\in(\lambda,R]$, respectively.

Green's formula (in proper notation) is valid not only for such functions as $ln|f|$, $f\in H(\mathbb{C}^+)$ but also for arbitrary subharmonic functions in $\mathbb{C}^+$. Moreover, it is true also for functions that are subharmonic in a cone of $\mathbb{R}^n$. The corresponding result will be given in §4.4. Here we consider, in addition to the case of functions $ln|f|$, only the case of positively homogeneous subharmonic functions.

Lemma 4.1.6. Let $\mathfrak{L}(z) = h(\theta)r^\rho$, $z = re^{i\theta}$, be a function which is subharmonic and bounded from above on S^+_1. Then the equality

$$\int_{U^+_1} \mathfrak{L}(z)\Delta(sin(argz)\varphi(z))d\omega_z = 2\pi\int_{U^+_1}\varphi(z)sin(argz)d\mu_{\mathfrak{L}}(z) +$$

$$- h(+0)\int_0^1 \varphi(x)x^{\rho-1}dx - h(\pi-0)\int_{-1}^0 \varphi(x)|x|^{\rho-1}dx +$$

$$+\int_0^\pi h(\theta)sin\theta\,\frac{\partial\varphi}{\partial r}\Big|_{r=1}d\theta - \rho\int_0^\pi h(\theta)sin\theta\varphi(e^{i\theta})d\theta \qquad (4.1.25)$$

is true for any function $\varphi\in C^2(\bar{U}^+_1)$.

Proof. Let us choose functions $\mathfrak{L}_j(z) = r^\rho h_j(\theta)$ in such a way that $\mathfrak{L}_j(z)\in SH(\mathbb{C}^+)$, $h_j(\theta)\in C^2((0,\pi))$, $h_j(\theta) \rightrightarrows h(\theta)$ on $(0,\pi)$ as $j\to\infty$, $h_j(\theta) \ge h_{j+1}(\theta)$, $\forall j$, and $h_j(+0) = h(+0)$, $h_j(\pi-0) = h(\pi-0)$. The possibility of such a choice is obvious, because as it is well-known (see, for example, L.Ronkin [1]) that in this case a positively homogeneous function of degree 1 is subharmonic if and only if it is convex in an the ordinary sense. The case of an arbitrary $\rho > 0$ can be reduced to the case $\rho = 1$ by the substitution $z = w^{1/\rho}$.

We apply Green's formula to the functions $\mathfrak{L}_j(z)$ and $\varphi(z)sin(argz)$ in the semidisc $U^+_{\lambda,1}$, and then pass to the limit as $\lambda\to 0$ to obtain

$$\int_{U^+_1}\mathfrak{L}_j(z)\Delta(sin(argz)\varphi(z))d\omega_z = 2\pi\int_{U^+_1}\varphi(z)sin(argz)d\mu_{\mathfrak{L}_j}(z) +$$

$$- h_j(+0) \int_0^1 \varphi(x)x^{\rho-1}dx - h_j(\pi-0) \int_{-1}^0 \varphi(x)|x|^{\rho-1}dx +$$

$$+ \int_0^\pi h_j(\theta)sin\theta \frac{\partial\varphi}{\partial r}\Big|_{r=1} d\theta - \rho \int_0^\pi h_j(\theta)sin\theta\varphi(e^{i\theta})d\theta. \qquad (4.1.26)$$

Note that

$$\Delta\mathfrak{L}(z) = r^{\rho-2}(h''(\theta) + \rho^2 h(\theta)) \geq 0$$

(all derivatives are regarded as distributions).

Therefore $h''(\theta)$ is a real measure on $(0,\pi)$.

From the positivity of the measure $(h''(\theta) + \rho^2 h(\theta))$ it follows also that the function $h'(\theta)$ is bounded below on $(0,\pi)$. Further, since

$$\int_\alpha^\beta h'(\theta)d\theta = h(\beta) - h(\alpha)$$

and $h(+0) = \lim_{\alpha\to+0} h(\alpha)$, we have

$$\underline{lim}_{\alpha\to 0}\ \alpha h'(\alpha) = 0.$$

Using this we find that

$$\int_0^\pi sin\theta h''(\theta)\eta(\theta)d\theta = - \int_0^\pi h'(\theta)(\eta(\theta)sin\theta)'d\theta, \quad \forall\eta\in C^1([0,\pi]),$$

and

$$\int_0^\pi h'(\theta)\eta(\theta)d\theta = h(\theta)\eta(\theta)\Big|_0^\pi - \int_0^\pi h(\theta)\eta'(\theta)d\theta, \quad \forall\eta\in C^1([0,\pi]).$$

It follows from these equalities that

$$\lim_{j\to\infty} \int_0^\pi (h_j'' + \rho^2 h_j)\eta sin\theta d\theta = \int_0^\pi (h'' + \rho^2 h)\eta sin\theta d\theta, \quad \forall\eta\in C^2([0,\pi]).$$

In turn, this implies the equality

$$\lim_{j\to\infty} \int_{U_1^+} \varphi(z)sin(arg z)d\mu_{\mathfrak{L}_j}(z) = \int_{U_1^+} \varphi(z)sin(arg z)d\mu_{\mathfrak{L}}(z), \quad \forall\varphi\in C^2(\bar{U}_1^+).$$

The corresponding statements for the other terms in (4.1.26) are obvious. Hence equality (4.1.25) can be obtained by limit transition as $j \to \infty$.

The proof is finished.

4. Factorization of holomorphic functions of finite order in $\mathbb{C}^+$. In connection with the existence of the nonempty boundary $\partial\mathbb{C}^+=\mathbb{R}$ of the half-plane $\mathbb{C}^+$, the factorization of functions in $\mathbb{C}^+$ differs from that of entire functions as follows: the factors are characterized not only by the zeros but by their behaviour on $\partial\mathbb{C}^+$ as well. The Blaschke product

$$\prod_k \frac{z-a_k}{z-\bar{a}_k},$$

where $\{a_k\}$ is a discrete set in U_1^+ and the Weierstrass-Nevanlinna product

$$\prod_k \frac{G_q(z/a_k)}{G_q(z/\bar{a}_k)},$$

where $\{a_k\}$ is a discrete set in $\mathbb{C}^+\backslash U_1$ and $G_q(z)$ is the prime Weierstrass factor as above, are used instead of the canonical Weierstrass product. It will be clear from the following exposition that the Weierstrass-Nevanlinna product possesses the properties of the Weierstrass and the Blaschke product. In $\mathbb{C}^+$ natural non-vanishing factors have to take part in the factorization side by side with the above factorization elements. So the functions $e^{ip(z)}$, where $p(z)$ is a polynomial with real coefficients,

$$exp\left\{\frac{1}{\pi i}\int_{-1}^{1}\frac{d\mu(t)}{t-z}\right\}$$

and

$$exp\left\{\frac{z^{q+1}}{\pi i}\int_{-1}^{1}\frac{d\mu(t)}{t^{q+1}(t-z)}\right\}$$

are the standard factors.

Before formulating the factorization theorem for holomorphic functions of finite order in $\mathbb{C}^+$ we give a number of statements concerning existence conditions for and properties of these factorization elements. First we consider the factors that can be prescribed by their zeros.

Lemma 4.1.7. Let $z_k = r_k e^{i\theta_k}$, $0 < \theta_k < \pi$, $k = 1,2,\dots$, be a point sequence such that $0 < r_k < 1$ and

$$\sum_k r_k \sin\theta_k < \infty. \tag{4.1.27}$$

Then the Blaschke product

$$f(z) = \prod_k \frac{z-z_k}{z-\bar{z}_k}$$

converges absolutely in $\mathbb{C}^+$, is a holomorphic function in $\mathbb{C}^+$ and satisfies the conditions: $\mu_{f,\partial} = 0$, $\hat{\rho}_f = 0$.

Proof. The first three statements of lemma are well-known properties of the Blaschke product in the half-plane[19] (see, for example, K.Hoffman [1]). To prove $\hat{\rho}_f = 0$ we note that for $z \to \infty$,

$$\ln \frac{z-z_k}{z-\bar{z}_k} \sim 2i \frac{r_k \sin\theta_k}{z-\bar{z}_k}.$$

Hence

$$|\ln f(z)| \le \frac{c}{|z|-1} \sum_k r_k \sin\theta_k.$$

This immediately implies $\hat{\rho}_f = 0$.

The proof is finished.

Lemma 4.1.8. Let $z_k = r_k e^{i\theta_k}$, $r_k > 1$, $0 < \theta_k < \pi$, $k = 1, 2, \dots$, be a point sequence such that the function

$$\hat{n}(r) \overset{\text{def}}{=} \sum_{k:r_k<r} \sin\theta_k$$

is finite for any r and, moreover, the inequality

$$\hat{n}(r) \le cr^{\rho}, \ \forall r > 0, \tag{4.1.28}$$

is true for some $c > 0$ and $\rho > 0$. Then the Weierstrass-Nevanlinna product

---

[19] Recall also that condition (4.1.27) is necessary and sufficient for the existence of a function $f \in H(\mathbb{C}^+)$ which is bounded in $\mathbb{C}^+$ and whose divisor is the set $Z_f = \{z_k\}_{k=1}^{\infty}$. The corresponding condition for a discrete set $\{z_k\}_k \subset \mathbb{C}^+ \setminus U_1$ is: $\sum_k r_k^{-1} \sin\theta_k < \infty$.

$$f(z) = \prod_k D_q(z, z_k),$$

where $q = [\rho]$,

$$D_q(z,\zeta) = \frac{G_q(z/\zeta)}{G_q(z/\bar{\zeta})}$$

and

$$G_q(z) = (1-z)exp\left\{z + \frac{1}{2} z^2 + \dots + \frac{1}{q} z^q\right\},$$

converges absolutely in $\mathbb{C}^+$, is a holomorphic function in $\mathbb{C}^+$ and satisfies the conditions $\mu_{f,\partial} = 0$, $\hat{\rho}_f \le \rho$.

Proof. First of all we note that under these conditions

$$\sum_k r_k^{-q-1} sin\theta_k < \infty. \tag{4.1.29}$$

Indeed,

$$\sum_k r_k^{-q-1} sin\theta_k = \int_1^\infty t^{-q-1} d\hat{n}(t) = \left.\frac{\hat{n}(t)}{t^{q+1}}\right|_1^\infty + \frac{1}{q+2} \int_1^\infty \frac{\hat{n}(t)}{t^{q+2}} dt.$$

From this and (4.1.28) we obtain (4.1.29).

From rather quite simple calculations, omitted here, we can obtain the following statements:

$$|lnD_q(z,z_k)| \le 4|z/z_k|^{q+1} sin\theta_k \text{ for } |z| \le \frac{1}{2} r_k\ , \tag{4.1.30}$$

$$|lnD_q(z,z_k)| \le 2^{q+1} |z/z_k|^q sin\theta_k \text{ for } |z| \ge \frac{1}{2} r_k\ , \tag{4.1.31}$$

and

$$\left| ln\left\{ \frac{\bar{z}_k}{z_k} exp\left(\sum_{m=1}^q \frac{1}{m} ((z/z_k)^m - (z/\bar{z}_k)^m) \right) \right\} \right| \overset{def}{=} |ln\tilde{D}_q(z,z_k)| \le$$

$$\le 2\left(\frac{\pi}{2} + q|z|^q\right) sin\theta_k\ ,\ \forall z_k, z. \tag{4.1.32}$$

Now we take an arbitrary $R > 0$ and represent the product $f = \prod_k D_q(z,z_k)$ for $|z| < R/2$ in the form

$$f = f_1^{(R)} f_2^{(R)} f_3^{(R)},$$

where

$$f_1^{(R)} = \prod_{k:r_k \geq R} D_q(z,z_k),$$

$$f_2^{(R)} = \prod_{k:r_k < R} D_q(z,z_k),$$

and

$$f_3^{(R)} = \prod_{k:r_k < R} \tilde{D}_q(z,z_k).$$

Inequality (4.1.30) implies the absolute and uniform convergence of the product $f_1^{(R)}$. It is obvious that condition (4.1.27) is satisfied by the product $f_2^{(R)}$, hence it converges and is a holomorphic function in $\mathbb{C}^+$. The absolute and uniform convergence of the product $f_3^{(R)}$ is a consequence of the estimate (4.1.32) and the fact that $\hat{n}(t) < \infty$. Thus, the product f converges and is a holomorphic function in $U^+_{R/2}$ and, since R is arbitrary, the statement is true in the whole half-plane $\mathbb{C}^+$.

We show that $\mu_{f,\partial} = 0$. For this we use the representation $f = f_1^{(R)} f_2^{(R)} f_3^{(R)}$ again.

Note that as $y \to 0$

$$\ln|D_q(x+iy,z_k)| \rightrightarrows 0, \quad -R/2 \leq x \leq R/2,$$

$$\ln|\tilde{D}_q(x+iy,z_k)| \rightrightarrows 0, \quad -R/2 \leq x \leq R/2.$$

In accordance with the dominated convergence theorem, which can be applied by the estimates (4.1.30) and (4.1.31), it follows that if $y \to 0$, then

$$\ln|f_1^{(R)}(x+iy)| \to 0 \text{ and } \ln|f_2^{(R)}(x+iy)| \to 0$$

uniformly with respect to $x \in [-R/2,R/2]$. At the same time the function $f_3^{(R)}$, satisfies, as a Blaschke product in $\mathbb{C}^+$ the condition

$$\lim_{y \to 0} \int_{-a}^{a} |\ln|f_3^{(R)}(x+iy)||dx = 0, \ \forall a > 0.$$

Taking into account that $f = f_1^{(R)} f_2^{(R)} f_3^{(R)}$ we conclude that

$$\lim_{y \to 0} \int_{-R/2}^{R/2} |ln|f(x+iy)||dx = 0.$$

Hence $\mu_{f,\partial} = 0$.

Now we estimate the function $\hat{M}_f(r)$. It follows from (4.1.30) and (4.1.31) that

$$ln|f(z)| \le 4|z|^{q+1} \int_{2|z|}^{\infty} \frac{d\hat{n}(t)}{t^{q+1}} + 2^q|z|^q \int_0^{2|z|} \frac{d\hat{n}(t)}{t^q} \le$$

$$\le \hat{n}(2|z|) + \frac{2^q|z|^q}{q+1} \int_0^{2|z|} \frac{\hat{n}(t)}{t^{q+1}} dt + \frac{4|z|^{q+1}}{q+2} \int_{2|z|}^{\infty} \frac{\hat{n}(t)}{t^{q+2}} dt.$$

Taking into account (4.1.28), this gives

$$\overline{lim}_{r \to \infty} \frac{ln^+ ln^+ \hat{M}_f(r)}{lnr} \le \rho.$$

According to the **Remark** following **Lemma 4.1.3**, for $\rho \ge 1$this inequality implies:

$$\overline{lim}_{r \to \infty} \frac{1}{lnr} ln^+ \int_0^{\pi} |ln|f(re^{i\theta})||sin\theta d\theta \le \rho.$$

Therefore, in this case $\hat{\rho}_f \le \rho$. To obtain the required estimate when $\rho < 1$ and $q = [\rho] = 0$ we calculate, as above the integral

$$\int_0^{\pi} |ln|D_0(re^{i\theta},z_k)||sin\theta d\theta = - \int_0^{\pi} ln\left| \frac{1 - z/z_k}{1 - z/\bar{z}_k} \right| sin\theta d\theta .$$

Using Carleman's formula for $r_k < r$ we find:

$$\int_0^{\pi} ln\left| \frac{1 - z/z_k}{1 - z/\bar{z}_k} \right| sin\theta d\theta =$$

$$= \pi\left(\frac{r}{r_k} - \frac{r_k}{r}\right)\sin\theta_k + \frac{1}{2} r\, Im \lim_{\lambda \to 0} \int_0^{\pi} \ln D_0(\lambda e^{i\theta})\left(\frac{\lambda e^{i\theta}}{r^2} - \frac{e^{i\theta}}{\lambda}\right)d\theta =$$

$$= \pi\left(\frac{r}{r_k} - \frac{r_k}{r}\right)\sin\theta_k + \frac{\pi}{2} r\, Im\, Res_{z=0} \frac{\ln D_0(z)}{z^2} =$$

$$= \pi\left(\frac{r}{r_k} - \frac{r_k}{r}\right)\sin\theta_k - \frac{\pi r}{r_k}\sin\theta_k = -\pi\frac{r_k}{r}\sin\theta_k.$$

Similarly, if $r_k > r$, then

$$\int_0^{\pi} \ln\left|\frac{1 - z/z_k}{1 - z/\bar{z}_k}\right|\sin\theta d\theta = -\frac{\pi r}{r_k}\sin\theta_k.$$

Therefore, in this situation

$$\int_0^{\pi} |\ln|f(re^{i\theta})||\sin\theta d\theta = \frac{\pi}{r}\cdot \sum_{k:r_k<r} r_k \sin\theta_k + \pi r\cdot \sum_{k:r_k\geq r} \frac{\sin\theta_k}{r_k} =$$

$$= \frac{\pi}{r}\int_0^r t d\hat{n}(t) + \pi r \int_r^{\infty} \frac{d\hat{n}(t)}{t} =$$

$$= -\frac{\pi}{r}\int_0^r \hat{n}(t)dt + \pi r\int_r^{\infty} \frac{\hat{n}(t)}{t^2}dt \leq \pi r \int_r^{\infty}\frac{\hat{n}(t)}{t^2}dt.$$

It can be easily seen that the above implies that

$$\overline{\lim_{r\to\infty}} \frac{1}{\ln r} \ln^+\left(\int_0^{\pi} |\ln|f(re^{i\theta})||\sin\theta d\theta\right) \leq \rho,$$

hence $\hat{\rho}_f \leq \rho$.

This finishes the proof.

Now we consider the factors that are prescribed by their boundary measures.

Lemma 4.1.9. Let μ be a real measure on $\mathbb{R}$, *supp*$\mu \subset\subset \mathbb{R}$. Then the function

$$f(z) = exp\left\{ \frac{1}{\pi i} \int_{-\infty}^{\infty} \frac{d\mu(t)}{t-z} \right\}$$

is holomorphic in $\mathbb{C}^+$ and satisfies the conditions $\mu_{f,\partial} = \mu$, $\hat{\rho}_f = 0$.

Proof. Since $supp\mu \subset\subset \mathbb{R}$, $\int_{\mathbb{R}} \frac{d\mu(t)}{t-z}$ converges absolutely and uniformly on every compact subset of $\mathbb{C}^+$. Therefore the function f(z) in the lemma is holomorphic in $\mathbb{C}^+$. Further, since

$$ln|f(z)| = \frac{1}{\pi} \int_{\mathbb{R}} Im \frac{1}{t-z} d\mu(t),$$

and the right-hand integral is the Poisson integral of the measure μ, the equality $\mu_{f,\partial} = \mu$ is an obvious consequence (or even paraphrase) of the well-known statement on the boundary properties of the Poisson integral (see, for example, E.Stein [1]). The equality $\hat{\rho}_f = 0$ follows from the fact that

$$\left| \frac{1}{\pi} \int_{\mathbb{R}} Im \frac{1}{t-z} d\mu(t) \right| = O\left(\frac{1}{|z|}\right).$$

This finishes the proof.

Lemma 4.1.10. Let μ be a real measure on $\mathbb{R}$ vanishing on [-1,1] and such that the inequalities

$$\|\mu\|(t) \overset{def}{=} |\mu|([-t,t]) \le c_1 t^{\rho+1}, \ \forall t \ge 1,$$

and

$$d\mu \le c_2 t^{\rho} dt$$

are valid for some $c_1 > 0$, $c_2 > 0$ and $\rho > 0$. The last inequality should be understood as an inequality between the corresponding measures. Then the function

$$f(z) = exp\left\{ \frac{z^{q+1}}{\pi i} \int_{-\infty}^{\infty} \frac{d\mu(t)}{t^{q+1}(t-z)} \right\},$$

where $q = [\rho]$, is holomorphic in $\mathbb{C}^+$ and satisfies the conditions $\mu_{f,\partial} = \mu$, $\hat{\rho}_f \le \rho$.

Proof. It follows from the condition $\|\mu\| \le c_1 t^{\rho+1}$ that the integral prescribing the function f(z) converges absolutely and uniformly on every compact set in $\mathbb{C}^+$. Hence $f \in H(\mathbb{C}^+)$. To prove the equality $\mu_{f,\partial} = \mu$

we fix an arbitrary $R > 0$ and represent $f(z)$ in the form $f = f_1 f_2$, where

$$f_1(z) = exp\left\{ \frac{z^{q+1}}{\pi i} \int_{|t|\le R} \frac{d\mu(t)}{t^{q+1}(t-z)} \right\},$$

$$f_2(z) = exp\left\{ \frac{z^{q+1}}{\pi i} \int_{|t|>R} \frac{d\mu(t)}{t^{q+1}(t-z)} \right\}.$$

Then

$$ln|f_1(z)| = Im\left\{ \frac{z^{q+1}}{\pi} \int_{-R}^{R} \frac{d\mu(t)}{t^{q+1}(t-z)} \right\} =$$

$$= Im\left\{ \frac{1}{\pi} \int_{-R}^{R} \left(\frac{1}{t-z} - \frac{1}{t} - \frac{z}{t^2} - \ldots - \frac{z^q}{t^{q+1}} \right) d\mu(t) \right\} =$$

$$= Im\left\{ \frac{1}{\pi} \int_{-R}^{R} \frac{d\mu(t)}{t-z} \right\} - \frac{1}{\pi} \sum_{m=1}^{q} r^m sin m\theta \int_{-R}^{R} \frac{d\mu(t)}{t^{m+1}}, \quad z = re^{i\theta},$$

and we find that the measures $\mu_{f,\partial}$ and μ coincide on the interval $[-R,R]$, as a consequence of the above-mentioned property of the Poisson integral. At the same time it can be easily seen that $ln|f_2(x+iy)| \to 0$ as $y \to 0$, uniformly with respect to $x \in [-R',R']$, $0 < R' < R$. Consequently, $\mu_{f,\partial} = \mu$ on $[-R',R']$ and hence this is true on all of $\mathbb{R}$.

Now we estimate $\hat{M}_f(r)$. We have:

$$ln|f(z)| = Im\left\{ \frac{z^{q+1}}{\pi} \int_{|t|>1} \frac{d\mu(t)}{t^{q+1}(t-z)} \right\} =$$

$$= Im\left\{ \frac{1}{\pi} \int_{|t|\le \frac{1}{2}|z|} + \frac{1}{\pi} \int_{\frac{1}{2}|z|<|t|<2|z|} + \right.$$

$$\left. + \frac{1}{\pi} \int_{|t|\ge 2|z|} \frac{z^{q+1} d\mu(t)}{t^{q+1}(t-z)} \right\} = I_1 + I_2 + I_3. \qquad (4.1.33)$$

Further

$$|I_1| \le \frac{2}{\pi}|z|^q \int_1^{\frac{1}{2}|z|} \frac{d\|\mu\|(t)}{t^{q+1}} = \frac{2}{\pi}|z|^q \left\{ \frac{\|\mu\|\left(\frac{1}{2}|z|\right)}{\left(\frac{1}{2}|z|\right)^{q+1}} + (q+1)\int_0^{\frac{1}{2}|z|} \frac{d\|\mu\|(t)}{t^{q+2}} \right\} \le$$

$$\le \frac{2}{\pi}|z|^q \left\{ c_1\left(\frac{1}{2}|z|^{\rho-q}\right) + (q+1)c_1 \int_0^{\frac{1}{2}|z|} t^{\rho-q-1}dt \right\} \le$$

$$\le \begin{cases} c_3|z|^\rho & \text{for } \rho \ne q \\ c_4|z|^\rho ln|z| & \text{for } \rho = q \end{cases} \tag{4.1.34}$$

Similarly,

$$|I_3| \le \frac{2}{\pi}|z|^{q+1} \int_{2|z|}^{\infty} \frac{d\|\mu\|(t)}{t^{q+2}} \le \frac{2}{\pi}|z|^{q+1}(q+2) \int_{2|z|}^{\infty} \frac{\|\mu\|(t)}{t^{q+3}} dt \le$$

$$= \frac{2}{\pi}|z|^{q+1} \frac{q+2}{q+1-\rho} c_1|z|^{\rho-q-1} = c_5|z|^\rho. \tag{4.1.35}$$

Finally,

$$I_2 = \frac{1}{\pi} Im \int_{\frac{1}{2}|z|<|t|<2|z|} \left(\frac{1}{t-z} - \frac{1}{t} - \frac{z}{t^2} - \ldots - \frac{z^q}{t^{q+1}}\right) d\mu(t) =$$

$$= \frac{1}{\pi} \int_{\frac{1}{2}|z|<|t|<2|z|} Im \frac{1}{t-z} d\mu(t) - \frac{1}{\pi}\sum_{m=0}^{q} r^m sinm\theta \int_{\frac{1}{2}|z|<|t|<2|z|} \frac{d\mu(t)}{t^{m+1}} \le$$

$$\le \frac{1}{\pi} \int_{\frac{1}{2}|z|<|t|<2|z|} c_2 t^\rho Im \frac{1}{t-z} d\mu(t) + \frac{(q+1)2^{q+1}}{\pi} \frac{1}{|z|} \|\mu(2|z|)\| \le$$

$$\le \frac{1}{\pi} c_2(2|z|)^\rho \int_{-\infty}^{\infty} Im \frac{1}{t-z} dt + \frac{q+1}{\pi} 2^{q+2+\rho} c_1|z|^\rho.$$

It follows from these estimates that

$$\overline{\lim_{r\to\infty}} \frac{\ln^+\ln^+\hat{M}_f(r)}{\ln r} \le \rho. \tag{4.1.36}$$

To estimate the integral

$$\int_0^\pi |\ln|f(re^{i\theta})||\sin\theta d\theta,$$

we represent it in the form

$$\int_0^\pi |\ln|f(re^{i\theta})||\sin\theta d\theta =$$

$$= \int_0^\pi |I_1(re^{i\theta})|\sin\theta d\theta + \int_0^\pi |I_2(re^{i\theta})|\sin\theta d\theta + \int_0^\pi |I_3(re^{i\theta})|\sin\theta d\theta,$$

where $I_1(z)$, $I_2(z)$ and $I_3(z)$ are as in equality (4.1.33). It follows from the estimates (4.1.34) and (4.1.35) that

$$\overline{\lim_{r\to\infty}} (\ln r)^{-1}\ln\left(\int_0^\pi |I_j(re^{i\theta})|\sin\theta d\theta\right) \le \rho,\ j = 1,\ 3.$$

Further,

$$\int_0^\pi |I_2(re^{i\theta})|\sin\theta d\theta =$$

$$= \frac{1}{\pi}\int_0^\pi \sin\theta|\ Im \int_{\frac{1}{2}|z|<|t|<2|z|} \left(\frac{1}{t-z} - \frac{1}{t} - \frac{z}{t^2} - \ldots - \frac{z^q}{t^{q+1}}\right)d\mu(t)|d\theta \le$$

$$\le \frac{1}{\pi}\int_{\frac{1}{2}r<|t|<2r} d\mu(t)\int_0^\pi \frac{r\sin^2\theta}{t^2+r^2-2rt\cos\theta}\, d\theta + \frac{q+1}{\pi}\, 2^{q+2}\, \frac{\|\mu\|(2r)}{r}.$$

The inner integral on the right can be computed in a standard way using residues. Thus we obtain that

$$\int_0^\pi \frac{r\sin^2\theta}{t^2+r^2-2rt\cos\theta}\, d\theta = \begin{cases} \frac{\pi}{2r} & \text{for } \frac{|t|}{r} < 1 \\ \frac{\pi r}{2t^2} & \text{for } \frac{r}{|t|} < 1 \end{cases}.$$

Therefore

$$\frac{1}{\pi}\int_0^{\pi} |I_2(re^{i\theta})|\sin\theta d\theta \le \frac{2}{r}\int_{r/2}^{2r} d\|\mu\|(t) + \frac{q+1}{\pi} 2^{q+2+\rho}\|\mu\|(2r) \le$$

$$\le \frac{2}{r}\|\mu\|(2r) + \frac{q+1}{\pi r} 2^{q+2}\|\mu\|(2r) \le \text{const}\cdot r^{\rho}.$$

Combining these estimates for I_1, I_2 and I_3, we conclude that

$$\overline{\lim_{r\to\infty}} \frac{1}{\ln r}\int_0^{\pi} |\ln|f(re^{i\theta})||\sin\theta d\theta \le \rho.$$

By this and (4.1.36) it follows that $\hat{\rho}_f \le \rho$.

Now we pass directly to the main problem of this section, i.e. the factorization of an arbitrary function of finite order.

Theorem 4.1.1. Let $f \in H(\mathbb{C}^+,\rho]$, $\rho > 0$, and let $z_1, z_2, \dots$ be the zeros of $f(z)$, counted with multiplicities, let $\mu_{f,\partial}$ be the boundary measure of $\ln|f|$, and $q = [\rho]$.

Then the following representation holds:

$$f(z) = e^{P(z)} \exp\left\{ \frac{1}{\pi i}\int_{-1}^{1} \frac{d\mu_{f,\partial}(t)}{t-z} + \frac{1}{\pi i}\int_{|t|\ge 1} \frac{z^{q+1} d\mu_{f,\partial}(t)}{t^{q+1}(t-z)} \right\} \times$$

$$\times \prod_{|z_k|\le 1} \frac{z-z_k}{z-\bar{z}_k} \cdot \prod_{|z_k|>1} D_q(z,z_k), \tag{4.1.37}$$

where $P(z)$ is a polynomial with real coefficients of degree at most ρ, the factors $D_q(z,z_k)$ are as in **Lemma 4.1.3**, and the products and integrals in this representation converge absolutely.

Proof. In accordance with the estimates of **Lemma 4.1.4**, the zeros z_k and the measure $\mu_{f,\partial}$ satisfy the conditions of **Lemmas 4.1.7 - 4.1.10**. Hence the products and integrals in the representation (4.1.37) converge absolutely, and the functions

$$F_1 = \prod_{|z_k|\le 1} \frac{z-z_k}{z-\bar{z}_k},$$

$$F_2 = \prod_{|z_k|>1} D_q(z,z_k),$$

$$F_3 = exp\left\{ \frac{1}{\pi i} \int_{|t|<1} \frac{d\mu_{f,\partial}(t)}{t-z} \right\},$$

and

$$F_4 = exp\left\{ + \frac{1}{\pi i} \int_{|t|\ge 1} \frac{z^{q+1} d\mu_{f,\partial}(t)}{t^{q+1}(t-z)} \right\}$$

are holomorphic in $\mathbb{C}^+$.

Consider the function

$$g = f\cdot(F_1F_2F_3F_4)^{-1}.$$

It is holomorphic in $\mathbb{C}^+$ and does not vanish there. Therefore $g(z) = e^{P(z)}$, where $P(z)\in H(\mathbb{C}^+)$. It follows from the properties of the functions F_j established by **Lemmas 4.1.7 - 4.1.10** that the function $g(z)$ satisfies the condition

$$\mathcal{D}'-\lim_{y\to 0} Re P(x+iy) = \mathcal{D}'-\lim_{y\to 0} ln|g(x+iy)| = 0.$$

By the symmetry principle, we conclude that the harmonic function $Re P(z)$ can be harmonically continued to the whole plane. Correspondingly, $P(z)$ can be continued to $\mathbb{C}$ as an entire function. We estimate the growth of $P(z)$. It follows from **Lemmas 4.1.7 - 4.1.10** that the order (in $\mathbb{C}^+$) of the functions F_1, F_2, F_3, F_4 is at most than ρ. Now taking into account the estimate (4.1.19) of **Lemma 4.1.3**, we conclude that the inequality

$$\int_{U^+_{1,r}} |ln|g(z)||\,|z|^{-2} d\omega_z \le const\cdot r^{\rho+\varepsilon}$$

is valid for any $\varepsilon > 0$ and all r sufficiently large.

Since the function $g(z)$ can be continued to $\mathbb{C}^+$ in accordance with the symmetry principle, the corresponding estimate is true in an annulus, i.e.

$$\int_{U_{1,r}} |ln|g(z)||\,|z|^{-2} d\omega_z \le const\cdot r^{\rho+\varepsilon}.$$

In view of the monotonicity of $\mathfrak{M}_{ln|f|}(0,t)$, this estimate implies

$$\mathfrak{M}_{ln|g|}(0,r/e) \le \frac{1}{2\pi}\int_0^{2\pi} d\theta \int_{r/e}^{r} ln|g(te^{i\theta})| \frac{dt}{t} \le \mathrm{const}\cdot r^{\rho+\varepsilon}.$$

Hence the order ρ_g of the entire function g(z) does not exceed ρ. Therefore P(z) is a polynomial of degree at most ρ.

This finishes the proof of the theorem.

Remark. All the statements of **Theorem 4.1.1** remain valid if the number q figuring in it is not defined by the equality $q = [\rho]$, but, as in the case of entire functions, by the equality

$$q = min\left\{\nu\in\mathbb{Z}_+ : \int^{\infty} \frac{d\|\tau_f\|(t)}{t^{\nu+1}} < \infty\right\}.$$

This modification of the formulation does not require any essential change in the proof.

§2 Functions of completely regular growth in $\mathbb{C}^+$

1. Regularity of growth in $\mathbb{C}^+$ and weak convergence. Let $f\in H(\mathbb{C}^+,\rho]$. Its indicators $h_f(\theta)$ and $\mathfrak{L}_f(z)$ are defined as in the case of entire functions, i.e.

$$h_f(\theta) = \overline{lim}_{r\to\infty}\ r^{-\rho}ln|f(re^{i\theta})|,\ 0 < \theta < \pi,$$

$$\mathfrak{L}_f(z) = \overline{lim}_{t\to\infty}\ t^{-\rho}ln|f(tz)| = h_f(\theta)r^{\rho},\ re^{i\theta} = z\in\mathbb{C}^+ .$$

Note that $h_f(\theta)$ is a trigonometrically convex function, while $\mathfrak{L}_f(z)$ is one.

First we give the definition of regularity of growth in the closed half-plane $\overline{\mathbb{C}}^+$.

A function $f\in H(\mathbb{C}^+,\rho]$ is said to be of c.r.g. in $\overline{\mathbb{C}}^+$ (the closed half-plane) if there is a set $E \subset \mathbb{R}_+$ of relative measure zero such that

$$r^{-\rho}ln|f(re^{i\theta})| \rightrightarrows h_f(\theta)$$

in the interval $0 < \theta < \pi$ as $r \to \infty$, $r\notin E$.

Similarly, regularity of growth in an arbitrary closed sector

$\overline{Y}(\theta_1,\theta_2) = \overline{Y}(\infty,\theta_1,\theta_2) = \{z\in\mathbb{C}:\ \theta_1 \le \arg z \le \theta_2\}$ can be defined. But first it is necessary to introduce the function class $H(Y(\theta_1,\theta_2),\rho]$, which is the analogue of $H(\mathbb{C}^+,\rho]$. The simplest way of doing this is to use the definition of $H(\mathbb{C}^+,\rho]$. Namely, we say that a function $f(z)$ holomorphic in a sector belongs to the class $H(Y(\theta_1,\theta_2),\rho]$ if $\varphi(w) = f(w^{\frac{1}{\theta_2-\theta_1}} e^{-\frac{i\theta_1}{\theta_2-\theta_1}})$ belongs to the class $H(\mathbb{C}^+,\rho/(\theta_2-\theta_1)]$.

A function $f\in H(Y(\theta_1,\theta_2),\rho]$ is said to be of c.r.g. in $\overline{Y}(\theta_1,\theta_2)$ if there exist a set $E \subset \mathbb{R}_+$ of relative measure zero such that

$$r^{-\rho} \ln|f(re^{i\theta})| \rightrightarrows h_f(\theta)$$

in the interval $\theta_1 < \theta < \theta_2$ as $r \to \infty$, $r\notin E$.

It can be directly verified that the mapping $x \to t = x^a$, $a > 0$, transforms a set E of relative measure zero into a set relative measure zero. Therefore $f(z)$ is a function of c.r.g. in $\overline{Y}(\theta_1\theta_2)$ if and only if the function $\varphi(w)$ constructed from $f(z)$ as above is of c.r.g. in $\overline{\mathbb{C}}^+$.

Now we define regularity of growth in the open half-plane.

A function $f\in H(\mathbb{C}^+,\rho]$ is said to be of c.r.g. in $\mathbb{C}^+$ (the open half-plane) if for any θ_1 and θ_2 there is a set $E_{\theta_1,\theta_2} \subset\subset \mathbb{R}_+$ of relative measure zero such that

$$r^{-\rho} \ln|f(re^{i\theta})| \rightrightarrows h_f(\theta) \tag{4.2.1}$$

in the interval $\theta_1 < \theta < \theta_2$ as $r \to \infty$, $r\notin E_{\theta_1,\theta_2}$.

In other words, $f\in H(\mathbb{C}^+,\rho]$ is a function of c.r.g. in $\mathbb{C}^+$ if it is of c.r.g. in any closed sector $\overline{Y}(\theta_1,\theta_2)$, $0 < \theta_1 < \theta_2 < \pi$.

Similarly to the case of entire functions, regularity of growth in $\mathbb{C}^+$ can be characterized in terms of weak convergence of the functions $t^{-\rho}\ln|f(tz)|$, $t \to \infty$. First we show that regularity of growth of $f(z)$ in $\mathbb{C}^+$ implies some weak convergence of the functions $t^{-\rho}\ln|f(tz)|$, and that this convergence is formally stronger than convergence in $\mathcal{D}'(\mathbb{C}^+)$.

Theorem 4.2.1. Let $f\in H(\mathbb{C}^+,\rho]$ be a function of c.r.g. in $\mathbb{C}^+$. Then for any $\lambda > 0$ the functions

$$v_{\lambda,t}(z) = \begin{cases} t^{-\rho} \ln|f(tz)| & \text{for } |z| > \frac{\lambda}{t} \\ 0 & \text{for } |z| \le \frac{\lambda}{t} \end{cases}$$

satisfy the condition

$$\exists \lim_{t \to \infty} \int_{U_r^+} |z|^{-2} v_{\lambda,t}(z)\varphi(z)d\omega_z =$$

$$= \int_{U_r^+} \mathfrak{L}_f(z)|z|^{-2}\varphi(z)d\omega_z, \ \forall \varphi \in C(\overline{U}_r^+), \ r > 0. \qquad (4.2.2)$$

Note that the set $\{z \in Y(\theta_1,\theta_2): |z| \in E_{\theta_1,\theta_2}\}$ outside which, in accordance with the definition of c.r.g., "large" deviations of $\ln|f(z)|$ from $\mathfrak{L}_f(z)$ are not admitted, is a C_0^2-set. Therefore **Theorem 4.2.1** is a particular case of the following theorem.

Theorem 4.2.2. Let $f \in H(\mathbb{C}^+, \rho]$ and suppose that in each sector $Y(\theta_1,\theta_2)$, $0 < \theta_1 < < \theta_2 < \pi$, there is a C_0^α-set E_{θ_1,θ_2}, $\alpha \in (0,2]$, such that

$$\lim_{\substack{z \to \infty \\ z \in Y(\theta_1,\theta_2) \setminus E_{\theta_1,\theta_2}}} (|z|^{-\rho} \ln|f(z)| - \mathfrak{L}_f(z)) = 0.$$

Then for any $\lambda > 0$ the functions $v_{\lambda,t}(z)$ defined in **Theorem 4.2.1** satisfy the condition (4.2.2).

Proof. Let $\delta_1 \in \left(0, \frac{\pi}{2}\right)$, $\delta_2 \in (0,1)$, and divide the semidisc U_r^+ as follows:

$$\Omega_1 = U_{\delta_2}^+, \ \Omega_2 = \{z \in Y(0,\delta_1): \delta_2 \le |z| \le r\},$$

$$\Omega_3 = \{z \in Y(\pi-\delta_1,\pi): \delta_2 \le |z| \le r\}, \ \Omega_4 = \{z \in \overline{Y}(\delta_1,\pi-\delta_1): \delta_2 \le |z| \le r\}.$$

Then, assuming for the simplicity of exposition that $\|\varphi\|_\infty \le 1$, where $\varphi \in C(\overline{U}_r^+)$, we have

$$\left| \int_{U_r^+} (v_{\lambda,t}(z) - \mathfrak{L}_f(z))|z|^{-2}\varphi(z)d\omega_z \right| \le \left(\left(\int_{\Omega_1} + \int_{\Omega_2} + \int_{\Omega_3} \right) |z|^{-2}|v_{\lambda,t}(z)|d\omega_z \right) +$$

$$+ \left(\left(\int_{\Omega_1} + \int_{\Omega_2} + \int_{\Omega_3}\right) |z|^{-2} |\mathcal{L}_f(z)| d\omega_z\right) + \int_{\Omega_4} |z|^{-2} |v_{\lambda,t}(z) - \mathcal{L}_f(z)| d\omega_z =$$

$$= \frac{1}{t^{\rho+2}} \left(\left(\int_{t\Omega_1 \setminus U_\lambda} + \int_{t\Omega_2 \setminus U_\lambda} + \int_{t\Omega_3 \setminus U_\lambda}\right) |z|^{-2} | \ln|f(z)| | d\omega_z\right) +$$

$$+ \left(\left(\int_{\Omega_1} + \int_{\Omega_2} + \int_{\Omega_3}\right) |z|^{-2} |\mathcal{L}_f(z)| d\omega_z\right) + \int_{\Omega_4} |z|^{-2} |v_{\lambda,t}(z) - \mathcal{L}_f(z)| d\omega_z = \sum_{j=1}^{7} I_j .$$

Since f belongs to $H(\mathbb{C}^+, \rho]$, the estimate (4.1.19) of **Lemma 4.1.3** is valid for it. This estimate implies that $I_1 \le c_1 \delta_2^\rho$, $I_2 \le c_2 \delta_1$, $I_3 \le c_2 \delta_1$, where c_1 and c_2 are constants depending on only.

The corresponding estimates of the indicator $\mathcal{L}_f(\theta)$, i.e. the estimates $I_4 \le c_3 \delta_2^\rho$, $I_5 \le c_4 \delta_1$, $I_6 \le c_4 \delta_1$, are obvious, because $\mathcal{L}_f(re^{i\theta}) = r^\rho h_f(\theta)$ and because the boundedness above and the trigonometric convexity of $h_f(\theta)$ imply $\inf h_f(\theta) > -\infty$. Now we consider I_7. It follows from condition (4.2.1) that the functions $v_{\lambda,t}$ converge on Ω_4 to $\mathcal{L}_f(z)$ with respect to the Carleson's measure as $t \to \infty$. Note also (see (4.1.18)) that

$$\int_{U^+_{\frac{1}{2}\delta_2, 2r}} |v_{\lambda,t}(z)| d\omega_z = \frac{1}{t^{\rho+2}} \int_{U^+_{\frac{1}{2}t\delta_2, 2tr}} | \ln|f(z)| | d\omega_z \le$$

$$\le \frac{1}{t^{\rho+2}} \int_{U^+_{2tr}} | \ln|f(z)| | d\omega_z \le c_5, \ \forall t > 1.$$

Hence **Lemma 2.1.8** can be applied; it implies $\lim I_7 = 0$. Taking into account that δ_1 and δ_2 are arbitrary,we conclude that (4.2.2) holds.

This finishes the proof.

The following theorem is a formal strengthening of the theorem converse to the one above.

Theorem 4.2.3. Let $f \in H(\mathbb{C}^+, \rho]$, and let $t^{-\rho} \ln|f(tz)| \xrightarrow{\mathcal{D}'(\mathbb{C}^+)} \mathcal{L}_f(z)$ as $t \to \infty$. Then for any $\alpha \in (0,2]$ there is in each sector $Y(\theta_1, \theta_2)$, $0 < \theta_1 < \theta_2 < \pi$, a C_0^α-set E such that

$$\lim_{\substack{z \to \infty \\ z \in Y(\theta_1,\theta_2)\setminus E_{\theta_1,\theta_2}}} \left| |z|^{-\rho} ln|f(z)| - \mathfrak{L}_f(z) \right| = 0.$$

Proof. As done in **Chapter 2** for the functions $u \in SH(\mathbb{R}^n,\rho]$ we can prove that convergence on compact sets in $\mathbb{C}^+$ of the functions $t^{-\rho} ln|f(tz)|$ with respect to the Carleson's α-measure implies ordinary convergence of $|z|^{-\rho} ln|f(z)|$ as $z \to \infty$ in any sector $Y(\theta_1,\theta_2)$ outside some C_0^α-set E_{θ_1,θ_2}. Therefore the statement of the theorem is a consequence of **Lemma 2.1.6**.

The proof is finished.

The following theorem is an obvious consequence of **Theorem 4.2.3**.

Theorem 4.2.4. If $f \in H(\mathbb{C}^+,\rho]$ and

$$\mathcal{D}' - \lim_{t \to \infty} \frac{ln|f(tz)|}{t^\rho} = \mathfrak{L}_f(z),$$

then f(z) is a function of c.r.g.

Generally speaking, in **Theorems 4.2.4 and 4.2.1**, as in **Theorems 4.2.2 and 4.2.3**, different types of convergence of families of functions are considered. Comparison of these theorems shows that these types of convergence are equivalent for families of the form $\{t^{-\rho} ln|f(tz)|\}_{t>1}$, $f \in H(\mathbb{C}^+,\rho]$. More exactly, the following holds:

$$\left\{ \lim_{t \to \infty} \int_{U^+_{\lambda/t,r}} \frac{ln|f(tz)|}{t^\rho} \frac{\varphi(z)}{|z|^2} d\omega_z = \int_{U^+_r} \mathfrak{L}_f(z) \frac{\varphi(z)}{|z|^2} d\omega_z, \ \forall \varphi \in C(\bar{U}^+_r) \right\} \Leftrightarrow$$

$$\Leftrightarrow \left\{ \lim_{t \to \infty} \int_{U^+_r} \frac{ln|f(tz)|}{t^\rho} \varphi(z) d\omega_z = \int_{U^+_r} \mathfrak{L}_f(z) \varphi(z) d\omega_z, \ \forall \varphi \in C_0^\infty(U^+_r) \right\}.$$

2. Argument-boundary density. Similarly to the case of entire functions, $\mathcal{D}'(\mathbb{C}^+)$-convergence of the functions $t^{-\rho} ln|f(tz)|$, $t \to \infty$, implies that the zero set Z_f of the function f (the divisor of f) has a angular density in sector $Y(\theta_1,\theta_2)$, $0 < \theta_1 < \theta_2 < \pi$. But unlike the case of entire functions, the existence of this density of Z_f, $f \in H(\mathbb{C}^+,\rho]$, $\rho \in \mathbb{Z}$, does not imply that f(z) is of c.r.g. in $\mathbb{C}^+$. This can be explained by the fact that the distance of the zeros of the function to the origin is used in constructing the angular density, and not the distance from the boundary of $\mathbb{C}^+$, which would be more natural. Besides, the angular density leaves out of account the boundary values of f on $\partial\mathbb{C}^+$, more precisely, the boundary measure $\mu_{f,\partial}$. The necessity of considerating $\mu_{f,\partial}$ is obvious, because there are non-trivial functions $f\in H(\mathbb{C}^+,\rho]$ without zeros.

The analogue of angular density corresponding to the case under consideration is the notion of argument-boundary density introduced by N.Govorov. The definition of argument-boundary density is as follows.

Let $f\in H(\mathbb{C}^+,\rho]$, let $z_k = r_k e^{i\theta_k}$, $k = 1, 2, \ldots$, be the zeros of f(z) counted with multiplicities and $\mu_{f,\partial}$ be the boundary measure. Let E be a bounded Borel set.

We us set

$$\tau_f(E) = 0 \text{ if } E \subset U_1^+,$$

$$\tau_f(E) = \sum_{z_k\in E\setminus\partial\mathbb{C}^+} \sin\theta_k - \frac{1}{2\pi}\int_{E\cap\partial\overline{\mathbb{C}}^+} |x|^{-1}d\mu_{f,\partial}(x) \text{ if } E \subset \overline{\mathbb{C}}^+\setminus U_1,$$

and

$$\tau_f(E) = \tau_f(E\cap U_1) + \tau_f(E\setminus U_1) \text{ if } E \subset \overline{\mathbb{C}}^+.$$

Thus we relate to each function $f\in H(\mathbb{C}^+,\rho]$ a measure τ_f in $\overline{\mathbb{C}}^+$ containing information both on the divisor Z_f and on the boundary measure $\mu_{f,\partial}$. It follows from **Lemma 4.1.3** that the estimate

$$\|\tau_f\|(t) \le c_f t^\rho, \ \forall t > 0,$$

is valid for the function $\|\tau_f\|(t) \overset{\text{def}}{=} |\tau_f|(\overline{U}_t^+)$, for some constant c_f.

A function f is said to have argument-boundary density if the limit (called the argument-boundary density)

$$a_f(\theta_1,\theta_2) = \lim_{t\to\infty} t^{-\rho}\tau_f(\overline{Y}(t,\theta_1,\theta_2))$$

exists for all θ_1 and θ_2 in $[0,2\pi]$ outside an at most countable set $E_f \subset (0,\pi)$. Here, as above, $\overline{Y}(t,\theta_1,\theta_2) = \{z\in\mathbb{C}^+\colon |z| \le t,\ \theta_1 \le \arg \le \theta_2\}$.

The measure τ_f is positive on $\mathbb{C}^+$ and it follows from the definition of $\mu_{f,\partial}$ that the inequality $d\tau_f \ge -c_f|x|^\rho dx$ is valid on $\partial\mathbb{C}^+$, where c_f is a positive constant. Correspondingly, the measures $\tau_f^{[t]}$, $t > 1$, defined by $d\tau_f^{[t]} = t^{-\rho}\tau_f(tE)$ satisfy the conditions $\tau_f^{[t]} \geq 0$ in $\mathbb{C}^+$ and $d\tau_f^{[t]}(x) \geq -c_f dx$ on $\partial\mathbb{C}^+$, $\forall t > 1$. Therefore the statement of **Lemma 2.1.1** is valid for the family of measures $\{\tau_f^{[t]}\}_{t\geq 1}$. This implies that convergence of the measures $\tau_f^{[t]}$ as $t \to \infty$, regarded as functionals on $C(\bar{U}_r^+)$, is sufficient for the existence of the argument-boundary density. It is also clear that, as in the case of the ordinary angular density (see §2.4 of **Chapter 4**), existence of the argument-boundary density implies convergence of the measures $\tau_f^{[t]}$, $t \to \infty$, for any $r > 0$. Thus, existence of the argument-boundary density for $f \in H(\mathbb{C}^+,\rho]$ is equivalent to weak convergence of the measures $\tau_f^{[t]}$, i.e. it is equivalent to the existence of a measure $\hat{\tau}_f$ such that

$$\int_{\bar{U}_r^+} \varphi(z)d\hat{\tau}_f(z) = \lim_{t\to\infty} \int_{U_r^+} \varphi(z)d\tau_f^{[t]}(z), \quad \varphi \in C(\bar{U}_r^+) . \tag{4.2.3}$$

It follows from **Lemma 2.1.1** that the exceptional set E_f in the definition of argument-boundary density can be characterized as the set of those $\theta \in (0,\pi)$ such that $\hat{\tau}_f(l_\theta) \neq 0$, where l_θ is the ray $\{z = re^{i\theta}: r > 0\}$.

Note also that the measure $\hat{\tau}_f$ is positively homogeneous, i.e. $\hat{\tau}_f(tE) = t^\rho \hat{\tau}_f(tE)$, $\forall t > 0$, $E \subset \bar{\mathbb{C}}^+$, and that $a_f(\theta_1,\theta_2) = \hat{\tau}_f(Y(1,\theta_1,\theta_2))$, $\theta_1,\theta_2 \notin E_f$.

Theorem 4.2.5. If $f \in H(\mathbb{C}^+,\rho]$ is a function of c.r.g. in $\mathbb{C}^+$ then it has an argument-boundary density and the equality

$$\hat{\tau}_f(E) = \int_{E\cap\mathbb{C}^+} \sin(\arg z)d\hat{\mu}_f - \frac{1}{2\pi}h_f(+0)\int_{E\cap\mathbb{R}^+} x^{\rho-1}dx + \\ - \frac{1}{2\pi}h_f(\pi-0)\int_{E\cap(-\mathbb{R}^+)} |x|^{\rho-1}dx , \ E \subset \bar{\mathbb{C}}^+, \tag{4.2.4}$$

holds, where $\hat{\mu}_f$ is the Riesz associated measure of the indicator $\mathfrak{L}_f(z)$.

Proof. Let $\varphi \in C^2(U_\delta^+)$, $\delta > 1$, satisfy the condition $supp\varphi \subset U_\delta$. Assuming that $1 \notin \tilde{\Lambda}_f$ where the set $\tilde{\Lambda}_f$ is the same as in **Lemma 4.1.4**, we apply Green's formula (4.1.24) to the function $\varphi(z/R)$ in the semi-annulus $U_{1,R\delta}^+$. Taking into account the definition of τ_f, we obtain

$$\int_{\overline{\mathbb{C}}^+} \varphi(z/R)d\tau_f(z) = \int_{U^+_{1,\delta R}} \varphi(z/R)d\tau_f(z) =$$

$$= \frac{1}{2\pi} \int_{U^+_{1,\delta R}} ln|f(z)|\Delta(\varphi(z/R)sin(argz))d\omega_z + Q_1 + Q_2, \qquad (4.2.5)$$

where

$$Q_1 = \frac{1}{2\pi} Q^+_f\left(\varphi\left(\frac{1}{R}e^{i\theta}\right),1\right),$$

$$Q_2 = -\frac{1}{2\pi R}\int_0^{\pi} ln|f(e^{i\theta})|sin\theta \left.\frac{\partial\varphi}{\partial r}\right|_{r=1/R} d\theta.$$

Hence

$$\int_{\overline{\mathbb{C}}^+} \varphi(z)d\tau_f^{[R]}(z) = \frac{1}{2\pi R^\rho}\int_{U^+_{1/R,\delta}} ln|f(Rz)|\Delta(\varphi(z)sin(argz))d\omega_z +$$

$$+ \frac{1}{R^\rho}Q_1 + \frac{1}{R^\rho}Q_2 =$$

$$= \frac{1}{2\pi R^\rho}\int_{U^+_{1/R,\delta}} |z|^{-2}\Psi(z)ln|f(Rz)|d\omega_z + \frac{1}{R^\rho}Q_1 + \frac{1}{R^\rho}Q_2, \qquad (4.2.6)$$

where

$$\Psi(z) = |z|^2 sin(argz)\Delta\varphi(z) - \varphi(z)sin(argz) +$$

$$- 2|z|\frac{\partial\varphi}{\partial x}cos(argz)sin(argz) + 2|z|\frac{\partial\varphi}{\partial y}cos^2(argz).$$

It is readily seen that $\Psi(z)\in C(\overline{\mathbb{C}}^+)$. Therefore **Theorem 4.2.1** can be applied. We have

$$\lim_{R\to\infty}\frac{1}{2\pi R^\rho}\int_{U^+_{1/R,\delta}} |z|^{-2}\Psi(z)ln|f(Rz)|d\omega_z =$$

$$= \int_{\mathbb{C}^+}\varphi(z)sin(argz)d\hat{\mu}_{f,\partial}(z) - \frac{1}{2\pi}h_f(+0)\int_0^\infty \varphi(x)x^{\rho-1}dx +$$

$$-\frac{1}{2\pi} h_f(\pi-0) \int_0^\infty \varphi(x)|x|^{\rho-1}dx \overset{def}{=} \int_{C^+} \varphi(z)d\hat{\tau}_1(z). \quad (4.2.7)$$

Now note that (4.1.25) implies

$$\lim_{R\to\infty} \frac{1}{R^\rho} Q_1 = 0.$$

It is also clear that

$$\lim_{R\to\infty} \frac{1}{R^\rho} Q_2 = 0.$$

From this and (4.2.5), (4.2.6) we conclude that

$$\exists \lim_{R\to\infty} \int_{\overline{C}^+} \varphi(z)d\tau_f^{[R]}(z) \overset{def}{=}$$

$$\overset{def}{=} \int_{\overline{C}^+} \varphi(z)d\hat{\tau}_f(z), \ \forall\varphi\in C(\overline{C}^+), \ supp\varphi \subset U_\delta. \quad (4.2.8)$$

It is obvious that the measure $\hat{\tau}_1$ in this equality is positively homogeneous. Therefore $|\hat{\tau}_1|(S_r) = 0$, $\forall r > 0$, and by **Lemma 2.1.1** the equality (4.2.8) is true for every function $\varphi\in C(\overline{U}_1^+)$, $\varphi = 0$ on $C\setminus U_1$. Thus, the measure $\hat{\tau}_f$ sought for exists and coincides with $\hat{\tau}_1$. Note also that the equality defining $\hat{\tau}_1$ is equivalent to (4.2.4) if $\hat{\tau}_f$ is replaced by $\hat{\tau}_1$.

The proof is finished.

The converse theorem is true for $\rho\notin\mathbb{Z}$ only.

Theorem 4.2.6. Let $f\in H(\mathbb{C}^+,\rho]$, and let the measures $\tau_f^{[t]}(z)$ converge weakly on $\overline{U}_R^+$ to some measure $\hat{\tau}_f$ as $t\to\infty$. Then f is a function of c.r.g. in $\mathbb{C}^+$.

Proof. In accordance with **Theorem 4.1.1** we can represent f(z) in the form (4.1.37). It is obvious that the factors

$$e^{P(z)}, \quad \prod_{|z_k|<1} \frac{z-z_k}{z-\bar{z}_k} \quad \text{and } exp\left\{\frac{1}{\pi i}\int_{-1}^{1} \frac{d\mu_{f,\partial}(t)}{t-z}\right\}$$

figuring in the product (4.1.37) are functions of c.r.g. in $\mathbb{C}^+$.

Consider the remaining part of this product, i.e. the function

$$\Phi_f(z) = \prod_{|z_k|\geq 1} D_q(z,z_k)\cdot exp\left\{ \frac{1}{\pi i} \int_{|t|\geq 1} \frac{z^{q+1} d\mu_{f,\partial}(t)}{t^{q+1}(t-z)} \right\}. \qquad (4.2.9)$$

This function belongs to the class $H(\mathbb{C}^+,\rho]$, as does the original function $f(z)$. We set

$$K_q(z,\zeta) = \begin{cases} K_q^{(1)}(z,\zeta) = \dfrac{1}{sin\theta}\, ln|D_q(z,\zeta)| \text{ when } \zeta = re^{i\theta}\in\overline{\mathbb{C}}^+\setminus\{0\},\ z\in\mathbb{C}^+, \\ K_q^{(2)}(z,\zeta) = 2Im\, \dfrac{z^{q+1} sign\zeta}{\zeta^q(\zeta-z)} \text{ when } \zeta\in\partial\mathbb{C}^+\setminus\{0\},\ z\in\mathbb{C}^+. \end{cases} \qquad (4.2.10)$$

It can be directly verified that $K_q(z,\zeta)$ is continuous on $\{\mathbb{C}^+\times(\overline{\mathbb{C}}^+\setminus\{0\})\}\setminus\{\zeta=z\}$ and that

$$ln|\Phi_f(z)| = J_{\tau_f}(z),$$

where

$$J_{\tau_f} = \int_{\overline{\mathbb{C}}^+} K_q(z,\zeta)d\tau_f(\zeta).$$

Note also the following evident inequality:

$$K_q(z,\zeta) \leq const\cdot\frac{|z/\zeta|^{q+1}}{1+|z/\zeta|},\ \forall z\in\mathbb{C}^+, \zeta\in\overline{\mathbb{C}}^+\setminus\{0\}. \qquad (4.2.11)$$

Now we show that $\exists\mathcal{D}'\text{-}\lim_{R\to\infty} R^{-\rho} ln|f(Rz)|$. Let $\varphi\in D(\mathbb{C}^+)$. Then

$$\frac{1}{R^\rho}\int_{\mathbb{C}^+}\varphi(z)ln|\Phi_f(Rz)|d\omega_z = \frac{1}{R^\rho}\int_{\overline{\mathbb{C}}^+} d\tau_f(\zeta)\int_{\mathbb{C}^+}\varphi(z)K_q(Rz,\zeta)d\omega_z =$$

$$= \frac{1}{R^\rho}\left\{ \int_{\mathbb{C}^+}\int_{U^+_{R/T}\cup\left(\left(-\frac{R}{T},\frac{R}{T}\right)\right)} + \int_{\mathbb{C}^+}\int_{\overline{U}^+_{R/T,RT}} \right.$$

$$\left. + \int_{\mathbb{C}^+}\int_{\overline{U}^+_{RT,\infty}\setminus\overline{S}^+_{RT}} \varphi(z)K_q(Rz,\zeta)d\tau_f(\zeta)d\omega_z \right\} =$$

$$= \frac{1}{R^{\rho}} \int_{\mathbb{C}^+} \varphi(z)d\omega_z \int_{U^+_{R/T}\setminus \overline{S}^+_R} K_q(Rz,\zeta)d\tau_f(\zeta) +$$

$$+ \frac{1}{R^{\rho}} \int_{\mathbb{C}^+} \varphi(z)d\omega_z \int_{\overline{U}^+_{RT,\infty}\setminus \overline{S}^+_{RT}} K_q(Rz,\zeta)d\tau_f(\zeta) +$$

$$+ \int_{\overline{U}^+_{1/T,T}} d\tau_f^{[R]}(w) \int_{\mathbb{C}^+} \varphi(z)K_q(z,w)d\omega_z = I_1 + I_2 + I_3. \qquad (4.2.12)$$

The quantities I_1 and I_2 can be estimated in a standard manner, using some calculations analogous to those in the proofs of **Theorem 2.2.5** and **Lemmas 4.1.8, 4.1.10**). We obtain

$$\left.\begin{aligned} |I_1| &\le c_1\left(\frac{1}{T}\right)^{q-\rho}, \ \forall T > 1, R > 1. \\ |I_2| &\le c_2 T^{\rho-q-1}, \ \forall T > 1, R > 1. \end{aligned}\right\} \qquad (4.2.13)$$

Now it is easy to see that the function

$$\psi(z) = \int_{\mathbb{C}^+} \varphi(z)K_q(z.w)d\omega_z$$

is continuous on $\overline{\mathbb{C}}^+\setminus\{0\}$, and since the condition of the theorem implies weak convergence of the measures $\tau_f^{[R]}$ to $\hat{\tau}_f$, we find

$$\exists \mathit{lim} I_3 = \lim_{R\to\infty} \int_{\overline{U}^+_{1/T,T}} \psi(w)d\tau_f^{[R]}(w) =$$

$$= \int_{\overline{U}^+_{1/T,T}} \psi(w)d\hat{\tau}_f(w) = \int_{\mathbb{C}^+} \varphi(z)d\omega_z \int_{\overline{U}^+_{1/T,T}} K_q(z,w)d\hat{\tau}_f(w).$$

From this and estimates (4.2.13) it follows that

$$\exists \lim_{R\to\infty} \frac{1}{R^{\rho}} \int_{\mathbb{C}^+} \varphi(z) \ln|\Phi_f(Rz)| d\omega_z = \int_{\mathbb{C}^+} \varphi(z)\mathfrak{L}(z)d\omega_z ,$$

where

$$\mathfrak{L}(z) = \int_{\mathbb{C}^+} K_q(z,w)d\hat{\tau}_f(w).$$

As in the case of entire or subharmonic functions in $\mathbb{C}$ and $\mathbb{R}^n$, convergence of the functions $R^{-\rho}ln|f(Rz)|$ in $\mathcal{D}'(\mathbb{C}^+)$ to the subharmonic function $\mathfrak{L}(z)$ implies the equality $\mathfrak{L}(z) = \mathfrak{L}_f(z)$. Thus by **Theorem 4.2.4** we conclude that $\Phi_f(z)$ is a function of c.r.g. in $\mathbb{C}^+$. Therefore the original function $f(z)$ is also of c.r.g. in $\mathbb{C}^+$.

This finishes the proof of the theorem.

Remark. Since $\mathfrak{L}(z) = \mathfrak{L}_f(z)$, we also obtain the following representation of the indicator of a function of c.r.g. in $\mathbb{C}^+$ with non-integral order:

$$\mathfrak{L}_f(z) = \int_{\mathbb{C}^+} K_q(z,\zeta)d\hat{\tau}_f(\zeta).$$

3. Regular distribution of the measure τ_f ($\rho\in\mathbb{N}$). The case of integral order differs from that of non-integral order for functions of c.r.g. in $\mathbb{C}^+$, similar to the distinction in $\mathbb{C}$ or $\mathbb{R}^n$. When ρ is an integer, then the condition of existence of the argument-boundary density is not sufficient for the function to be of c.r.g. An additional condition has to be satisfied. Before introducing this condition, which we shall call the argument-boundary balance[20] condition, we state theorem concerning weak convergence on S_r^+.

Theorem 4.2.7. Let $f\in H(\mathbb{C}^+,\rho]$ be a function of c.r.g. and let the set $\tilde{\Lambda}_f$ and the quantity $Q_f^+(\eta,r)$ be defined as in **Lemma 4.1.4**. Then

a)

$$\exists \lim_{R\to\infty,\ R\notin\tilde{\Lambda}_f} \frac{1}{R^\rho}\int_0^\pi \eta(\theta)\sin\theta \ln|f(Re^{i\theta})|d\theta =$$

$$= \int_0^\pi \eta(\theta)\sin\theta h_f(\theta)d\theta,\quad \forall\eta\in C([0,\pi]).$$

b)

---

[20] N.Govorov [2] used the term "argument-boundary symmetry".

$$\exists \lim_{R \to \infty,\ R \notin \tilde{\Lambda}_f} \frac{1}{R^{\rho-1}} Q_f^+(\eta(\theta),R) = \rho \int_0^{\pi} \eta(\theta) \sin\theta h_f(\theta) d\theta, \quad \forall \eta \in C^2([0,\pi]).$$

Proof. Let $\eta(\theta) \in C^2([0,\pi])$, and set $\varphi(re^{i\theta}) = \eta(\theta)\gamma(r)$, where $\gamma(r) \in C^{\infty}([0,1])$ satisfies the conditions $supp\gamma \subset (0,1]$, $\gamma(1) = 1$, $\gamma'(1) = 0$. Then applying Green's formula (4.1.26) we have the following equality for $R > 0$ sufficient large:

$$\int_{U_1^+} \varphi(z) d\tau_f^{[R]}(z) = \frac{1}{2\pi R^{\rho}} \int_{U_1^+} \ln|f(Rz)| \Delta(\varphi(z) \sin(\arg z)) d\omega_z + \frac{1}{2\pi R^{\rho-1}} Q_f^+(\eta,R).$$

Then, using **Theorems 4.2.1** and **4.2.5**, we obtain

$$\exists \lim_{R \to \infty} \frac{1}{R^{\rho-1}} Q_f^+(\eta,R) = 2\pi \int_{U_1^+} \varphi(z) d\hat{\tau}_f(z) - \int_{U_1^+} \mathfrak{L}_f(z) \Delta(\varphi(z) \sin(\arg z)) d\omega_z.$$

Now to prove the statement b) of this theorem it is sufficient to transform the right-hand part of this equality using **Lemma 4.1.6** and the equality (4.2.7) defining the measure $\hat{\tau}_1 = \hat{\tau}_f$.

Let $\gamma(r)$ be a function satisfying the conditions: $\gamma'(1) \neq 0$, $\gamma(1) = 0$. Reasoning (as above) with the function $\gamma(z)$ and using the statement b) proved above, we find that the statement a) is valid for any function $\eta \in C^2([0,\pi])$. Since $C^2([0,\pi])$ is a dense subset of $C([0,\pi])$ statement a) is also true for any function $\eta \in C([0,\pi])$.

This finishes the proof.

Now we have everything required to obtain the above-mentioned property of argument-boundary balancing.

Theorem 4.2.8. Let $f \in H(\mathbb{C}^+,\rho]$, where $\rho = p \in \mathbb{N}$, be a function of c.r.g. in $\mathbb{C}^+$. Then

$$\exists \lim_{R \to \infty} \left\{ \frac{1}{p} \sum_{z_k \in U_{1,R}^+} \frac{\sin p\theta_k}{r_k^p} - \frac{1}{2\pi} \int_1^R \frac{d\mu_{f,\partial}(x)}{x^{p+1}} + \right.$$

$$\left. - \frac{1}{2\pi} \int_{-R}^{-1} \frac{d\mu_{f,\partial}(x)}{x^{p+1}} \right\} \stackrel{def}{=} \kappa_f^+ \qquad (4.2.14)$$

and the equality

$$\kappa_f^+ = \frac{1}{2\pi}\int^{\pi} ln|f(e^{i\theta})|sinp\theta d\theta +$$

$$- \frac{1}{2\pi p} Q_f^+(\frac{sinp\theta}{sin\theta},1) + \frac{1}{\pi}\int_0^{\pi} h_f(\theta)sinp\theta d\theta. \qquad (4.2.15)$$

holds.

Here, as above $z_k = r_k e^{i\theta_k}$ denote the zeros of $f(z)$ counted with multiplicities.

Proof. It is clearly sufficient to prove equality (4.2.15) under the assumption that $R \to \infty$ outside the set $\tilde{\Lambda}_f$ defined in **Lemma 4.1.5**. Under this assumption, i.e. for $R \notin \tilde{\Lambda}_f$, we set in formula (4.1.24)

$$\varphi(z) = \varphi(re^{i\theta}) = \begin{cases} r^{-p}\dfrac{sinp\theta}{sin\theta} & \text{when } 0 < \theta < \pi \\ r^{-p}p & \text{when } \theta = 0 \\ (-1)^{p+1}r^{-p}p & \text{when } \theta = \pi \end{cases}$$

Taking into account that $\Delta(r^{-p}sinp\theta) = 0$ we obtain

$$\sum_{k:z_k\in U^+_{1,R}} \frac{sinp\theta_k}{r_k^p} - \frac{p}{2\pi}\int_{1<|x|<R} \frac{d\mu_{f,\partial}(x)}{x^{p+1}} =$$

$$= \frac{1}{2\pi R^{p-1}} Q_f^+\left(\frac{sinp\theta}{sin\theta},R\right) + \frac{p}{2\pi R^p}\int_0^{\pi} ln|f(Re^{i\theta})|sinp\theta d\theta +$$

$$- \frac{p}{2\pi}\int_0^{\pi} ln|f(e^{i\theta})|sinp\theta d\theta - \frac{1}{2\pi} Q_f^+\left(\frac{sinp\theta}{sin\theta},1\right). \qquad (4.2.16)$$

Note that by **Theorem 4.2.7**,

$$\lim_{R\to\infty}\left(\frac{1}{2\pi R^{p-1}} Q_f^+\left(\frac{sinp\theta}{sin\theta},R\right) + \frac{p}{2\pi R^p}\int_0^{\pi} ln|f(Re^{i\theta})|sinp\theta d\theta\right) =$$

$$= \frac{p}{2\pi}\int_0^{\pi} h_f(\theta)\,\frac{sinp\theta}{sin\theta}\,d\theta + \frac{p}{2\pi}\int_0^{\pi} h_f(\theta)sinp\theta d\theta = \frac{p}{\pi}\int_0^{\pi} h_f(\theta)sinp\theta d\theta.$$

This and (4.2.16) imply the required statement on the existence of the argument-boundary balance and equality (4.2.15).

The proof is finished.

Remark. If the conditions of **Theorem 4.2.8** are satisfied, then the angular component $\hat{\nu}$ of the positively homogeneous Riesz associated measure[21)] $\hat{\mu}$ of the indicator $\mathfrak{L}_f(z)$ is such that

$$\int_0^{\pi} \sin p\theta d\hat{\nu}(\theta) = \frac{1}{2\pi}(h_f(+0) + (-1)^{p+1}h_f(\pi-0)). \tag{4.2.17}$$

Indeed it follows from (4.2.14) that

$$\lim_{R\to\infty}\left\{\frac{1}{p}\sum_{k:R<|z_k|<2R}\frac{\sin p\theta_k}{r_k^p} + \right.$$

$$\left. - \frac{1}{2\pi}\int_{[-2R,2R]\setminus[-R,R]}\frac{d\mu_{f,\partial}(x)}{x^{p+1}}\right\} = 0. \tag{4.2.18}$$

On the other hand, if $\varphi(z)$ is as in the proof of equality (4.2.16), then, taking into account the weak convergence of the measures $\tau_f^{[R]}$ to $\hat{\tau}_f$, we have

$$\lim_{R\to\infty}\left\{\frac{1}{p}\sum_{k:R<|z_k|<2R}\frac{\sin p\theta_k}{r_k^p} - \frac{1}{2\pi}\int_{[-2R,2R]\setminus[-R,R]}\frac{d\mu_{f,\partial}(x)}{x^{p+1}}\right\} =$$

$$= \lim_{R\to\infty}\int_{U^+_{R,2R}}\varphi(z)d\tau_f(z) = \lim_{R\to\infty}\int_{U^+_{1,2}}\varphi(w)d\tau_f^{[R]}(w) = \int_{U^+_{1,2}}\varphi(w)d\hat{\tau}_f(w) =$$

$$= \frac{1}{p}\int_1^2\frac{d(r^p)}{r^p}\int_0^{\pi}\sin p\theta d\hat{\nu}(\theta) - \frac{1}{2\pi}h_f(+0)\int_1^2\frac{x^{p-1}}{x^p}dx +$$

$$- \frac{(-1)^{p+1}}{2\pi}h_f(\pi-0)\int_{-2}^{-1}\frac{|x|^{p-1}}{|x|^p}dx =$$

---

21) In other words, $d\hat{\mu} = d(r^p)\otimes d\hat{\nu}(\theta)$.

$$= ln2\cdot\left(\int_0^\pi sinp\theta d\hat{\nu}(\theta) - \frac{1}{2\pi}(h_f(+0) + (-1)^{p+1}h_f(\pi-0))\right).$$

This and (4.2.18) imply the required equality (4.2.17).

The function f(z) only indirectly plays a role in the above definitions of argument-boundary density and argument-boundary balance: we use its zeros and boundary measure $\mu_{f,\partial}$, the information of which is contained in the measure τ_f. Therefore we can relate these notions to the measure τ_f directly and correspondingly introduce the notion of regular distribution of it.

We shall say that the associated measure τ_f of a function $f\in H(\mathbb{C}^+,\rho]$ is regularly distributed if it has an argument-boundary density and the argument-boundary balance condition (4.2.14) is satisfied for $\rho = p\in\mathbb{N}$.

Now **Theorems 4.2.8** and **4.2.6** can be combined to yield:

Theorem 4.2.9. If $f\in H(\mathbb{C}^+,\rho]$ is a function of c.r.g. in $\mathbb{C}^+$, then the corresponding measure τ_f is regularly distributed.

The converse theorem is also true.

Theorem 4.2.10. If $f\in H(\mathbb{C}^+,\rho]$ is a function such that the corresponding measure τ_f is regularly distributed, then f(z) is a function of c.r.g. in $\mathbb{C}^+$.

Proof. If $\rho\notin\mathbb{Z}$, then **Theorem 4.2.10** coincides with **Theorem 4.2.6**. Therefore only the case $\rho = p\in\mathbb{N}$ needs special consideration. Note also that in this situation the proof is a synthesis of the proofs of **Theorems 4.2.6** and **2.2.6**. Therefore we give it concise from.

By **Theorem 4.2.3**, f(z) is a function of c.r.g. if $\exists\mathcal{D}'\text{-}lim R^{-\rho} ln|f(Rz)|$. Here, as in **Theorem 4.2.6**, we have to consider functions of the form (4.2.9) only. The equality

$$ln|f(z)| = \int_{\bar{\mathbb{C}}^+} K_p(z,\zeta)d\tau_f(\zeta)$$

is valid for the kernel $K_p(z,\zeta)$ (see (4.2.10)). Taking into account the construction of $K_p(z,\zeta)$, we conclude that this representation implies:

$$R^{-p} ln|f(Rz)| = R^{-p}\int_{\bar{U}^+_{R/T}} K_{p-1}(Rz,\zeta)d\tau_f(\zeta) +$$

$$+ R^{-p} \int_{\overline{U}^+_{R/T,RT}} K_p(Rz,\zeta)d\tau_f(\zeta) + R^{-p} \int_{\overline{U}^+_{RT,\infty}} K_p(Rz,\zeta)d\tau_f(\zeta) +$$

$$+ Re \left\{ -2iz^p \left(\frac{1}{2} \sum_{|z_k|<R/T} \frac{sinp\theta_k}{r_k^p} - \frac{1}{2\pi} \int_{\left[-\frac{R}{T},\frac{R}{T}\right]\setminus[-1,1]} x^{-p-1} d\mu_{f,\partial}(x) \right) \right\} =$$

$$= I_1 + I_2 + I_3 + I_4.$$

Correspondingly, we then have

$$R^{-p} \int_{\mathbb{C}^+} ln|f(Rz)|\varphi(z)d\omega_z = \sum_{k=1}^{4} \int_{\mathbb{C}^+} \varphi I_k d\omega_k = \sum_{k=1}^{4} J_k(T,R), \ \forall\varphi\in C_0^\infty(\mathbb{C}^+).$$

As in Theorem 4.2.6, the estimates

$$|J_k(T,R)| \le cT^{-1}, \ \forall T > 1, \ R > 1, \ k = 1, \ 2, \tag{4.2.19}$$

are valid, and

$$\exists \lim_{R\to\infty} J_3(T,R) = \int_{\overline{U}^+_{1/T,T}} d\hat{\tau}_f(w) \int_{\mathbb{C}^+} \varphi(z)K_p(z,w)d\omega_z. \tag{4.2.20}$$

Finally, it follows from the argument-boundary balance condition that

$$\exists \lim_{R\to\infty} J_4(T,R) = 2\left\{ \frac{1}{2\pi} \int_0^\pi ln|f(e^{i\theta})|sinp\theta d\theta + \right.$$

$$\left. - \frac{1}{2\pi p} Q_f^+\left(\frac{sinp\theta}{sin\theta}, 1 \right) \right\} \int_{\mathbb{C}^+} \varphi(z)Imz^p d\omega_z. \tag{4.2.21}$$

Now, using (4.2.19) - (4.2.21) and taking into account (4.2.18), we obtain that

$$\exists \mathcal{D}' - \lim_{R\to\infty} \frac{ln|f(Rz)|}{R^p} = \int_{\overline{U}^+_\delta} K_{p-1}(z,\zeta)d\hat{\tau}_f(\zeta) + \int_{\overline{U}^+_{\delta,\infty}} K_p(z,\zeta)d\hat{\tau}_f(\zeta) - 2\kappa_f^+ Imz^p,$$

where δ is an arbitrary[22] positive number and κ_f^+ is defined as above by equality (4.2.14).

Consequently $f(z)$ is a function of c.r.g. in $\mathbb{C}^+$.

The proof is finished.

Remark. In proving the theorem we simultaneously obtain the following representation of the indicator $\mathfrak{L}_f(z)$ of a function $f\in H(\mathbb{C}^+,p]$, $p\in\mathbb{N}$, of c.r.g. in $\mathbb{C}^+$:

$$\mathfrak{L}_f(z) = \int_{\overline{U}_\delta^+} K_{p-1}(z,\zeta)d\hat{\tau}_f(\zeta) + \int_{\overline{U}_{\delta,\infty}^+} K_p(z,\zeta)d\hat{\tau}_f(\zeta) - 2\kappa_f^+ Im z^p. \quad (4.2.22)$$

Letting $\delta \to 0$ in this equality, we have

$$\mathfrak{L}_f(z) = \int_{\overline{\mathbb{C}}^+} K_p(z,\zeta)d\hat{\tau}_f(\zeta) - 2\kappa_f^+ Im z^p.$$

§3 Functions of completely regular growth in $\overline{\mathbb{C}}^+$

1. Criteria for c.r.g. in $\overline{\mathbb{C}}^+$. It follows from the definition of c.r.g. in $\mathbb{C}^+$ and $\overline{\mathbb{C}}^+$ given at the beginning of the previous section that, c.r.g. in $\mathbb{C}^+$ is a necessary condition fur c.r.g. in $\overline{\mathbb{C}}^+$. But this condition is not sufficient. To formulate a condition that together with the condition of c.r.g. in $\mathbb{C}^+$ gives a criterion for c.r.g. in $\overline{\mathbb{C}}^+$, we introduce the notion of boundary measure $\hat{\mu}_{f,\partial}$ of the indicator $\mathfrak{L}_f(z)$, $z = x+iy$. Namely, we set

$$\hat{\mu}_{f,\partial} = h_f(+0) \int_{E \cap \mathbb{R}_+} x^\rho dx + h_f(\pi-0) \int_{E \cap (-\mathbb{R}_+)} |x|^\rho dx \;, \; E \subset \mathbb{R}.$$

Also, we denote by $\hat{\mu}_{f,\partial}^{[R]}$ the measure defined on $\mathbb{R}$ by the equality

$$\hat{\mu}_{f,\partial}^{[R]} = R^{-\rho-1}\mu_{f,\partial}(RE), \; E \subset \mathbb{R}.$$

Note that

$$d\tau_f = sin(arg z)d\tilde{n}_f + \frac{1}{2\pi}\frac{d\mu_{f,\partial}}{|x|} \overset{def}{=} d\mu_{f,1} + d\mu_{f,2}$$

---

[22] By (4.2.17) the right-hand side is independent of δ.

on $\overline{\mathbb{C}}^+\setminus U_1$[23] and, correspondingly,

$$\tau_f^{[R]} = \mu_{f,1}^{[R]} + \mu_{f,2}^{[R]},$$

where, as above, the measure $\tau_f^{[R]}$ is defined by the equality $\tau_f^{[R]} = R^{-\rho}\tau_f(RE)$ and the measures $\mu_{f,1}^{[R]}$ and $\mu_{f,2}^{[R]}$ are defined similarly. It is natural to define $\mu_{f,1} = 0$ and $\mu_{f,2} = 0$ on $\overline{U}_1$.

The following theorem is the main result of this section.

Theorem 4.3.1. A function $f\in H(\mathbb{C}^+,\rho]$ is of c.r.g. in $\overline{\mathbb{C}}^+$ if and only if it is of c.r.g. in $\mathbb{C}^+$ and the measures $\mu_{f,\partial}^{[R]}$ converge in variation to the measure $\hat{\mu}_{f,\partial}$ as $R \to \infty$.

Before proving this theorem we note the following. If the conditions figuring in it are satisfied, then the measures $\mu_{f,\partial}^{[R]}$ converge as functionals on $C([-r,r])$ to the measure $\hat{\mu}_{f,2}$ defined by the equality

$$d\hat{\mu}_{f,2}(x) = \frac{1}{2\pi}\frac{d\hat{\mu}_{f,\partial}(x)}{|x|},$$

and the measures $\mu_{f,1}^{[R]}$ converge as functionals on $C(\overline{U}_r^+)$ to the measure $\hat{\mu}_{f,1}$ defined by the equality

$$d\hat{\mu}_{f,1}(z) = \sin(\arg z)d\hat{\mu}_f(z)$$

as $R \to \infty$.

The first statement follows in a simple manner from the convergence of $\mu_{f,\partial}^{[R]}$ to $\hat{\mu}_{f,\partial}$ and (4.4.16). The latter follows from the first and the fact that by **Theorem 4.2.5** the measures $\tau_f^{[R]}$ converge weakly on $\overline{U}_r^+$ to $\hat{\tau}_f = \hat{\mu}_{f,1} + \hat{\mu}_{f,2}$ as $R \to \infty$.

2.Necessity. Let $f\in H(\mathbb{C}^+,\rho]$ be a function of c.r.g. in $\overline{\mathbb{C}}^+$. Hence there exist a set E_f of relative measure zero such that the function

$$\gamma_f(r) \overset{\text{def}}{=} \sup_{0<\theta<\pi} |r^{-\rho}\ln|f(re^{i\theta})| - h_f(\theta)|$$

satisfies the condition

---

[23] Recall (see **Chapter 1**) that we denote by $\tilde{n}_f$ the measure concentrated at the zeros of the function $f(z)$ and equal to the multiplicity at each zero.

$$\lim_{r \to \infty,\ r \notin E_f} \gamma_f(r) = 0. \tag{4.3.1}$$

Without loss of generality we may assume that E_f is an open set, and hence

$$E_f = \bigcup_k I_k,$$

where $I_k = (r_k, r'_k)$, $k = 1, 2, \ldots$. We associate to each interval I_k the semi-annulus $T_j = \{z \in \mathbb{C}^+ : r_k < |z| < r'_k\}$ and the interval $I_k(h) = (r_k(h), r'_k(h)) = \{x: (x+ih) \in T_k, \ x > 0\}$ for any given $h \in (0,1/2)$. Set

$$I(h) = \bigcup_k I_k(h),$$

$$\delta_k(h) = r'_k(h) - r_k(h),$$

$$U^{(k,h)} = \{z \in (\mathbb{C}^+ + ih): |z - r_k(h) - ih| < 2\delta_k(h)\}.$$

We denote by $\tilde{z} = \tilde{z}_k(h)$ the upper intersection point of the circles S_{r_k} and $\{z: |z - r_k(h) - ih| = \delta_k(h)\}$. Since E_f is a set of relative measure zero, without loss of generality we may assume that $r_k > 1$, $\forall k$ and $\frac{\delta_k(h)}{r_k} < \frac{1}{4}$. It easily follows from this that $\tilde{y}_k = Im\tilde{z}_k \geq h + \frac{1}{2}\delta_k(h)$, $\tilde{x}_k = Re\tilde{z}_k \geq r_k(h) - \frac{1}{2}\delta_k(h)$.

We now estimate the integral

$$\int_{\pm I_k(h)} |ln|f(t+ih)||dt.$$

For this we use the standard representation of a subharmonic function using Green's function; namely,

$$ln|f(\tilde{z}_k)| = -\int_{U^{(k,h)}} G^{(k,h)}(\tilde{z}_k, \zeta)d\tilde{n}_f(\zeta) + $$

$$+ \int_{\partial U^{(k,h)}} ln|f(\zeta)| \frac{\partial}{\partial n_\zeta} G^{(k,h)}(\tilde{z}_k, \zeta)dl_\zeta ,$$

where $\zeta = t + i\xi$, $G^{(k,h)}(z,\zeta)$ is the Green's function of the semidisc $U^{(k,h)}$, $\frac{\partial}{\partial n_\zeta}$ is the interior normal derivative, and dl_ζ is the area element of $\partial U^{(k,h)}$.

This equality, and the positivity of the Green's function and its normal derivative, imply that

$$\ln|f(\tilde{z})| \le \int_{\partial U^{(k,h)}} \ln|f(\zeta)| \frac{\partial}{\partial n_\zeta} G^{(k,h)}(\tilde{z}_k,\zeta)dl_\zeta =$$

$$= \int_{\partial U^{(k,h)}} \left(\ln^+|f(\zeta)| - \ln^+\frac{1}{|f(\zeta)|}\right) \frac{\partial}{\partial n_\zeta} G^{(k,h)}(\tilde{z}_k,\zeta)dl_\zeta \le$$

$$\le - \int_{I_k(h)} \ln^+\frac{1}{|f(t+ih)|} \left(\frac{\partial}{\partial t} G^{(k,h)}(\tilde{z}_k,\zeta)\right)\Big|_{\xi=h} dt +$$

$$+ \int_{\partial U^{(k,h)}} \ln^+|f(\zeta)| \frac{\partial}{\partial n_\zeta} G^{(k,h)}(\tilde{z}_k,\zeta)dl_\zeta \le$$

$$\le \sup\{\ln^+|f(\zeta)|: \zeta\in U^{(k,h)}\} +$$

$$- \inf\left\{\left(\frac{\partial}{\partial \xi} G^{(k,h)}(\tilde{z}_k,\zeta)\right)\Big|_{\xi=h} : t\in I_k(h)\right\} \int_{I_k(h)} \ln^+\frac{1}{|f(t+ih)|} dt. \quad (4.3.2)$$

Note that

$$G^{(k,h)}(z,\zeta) = G\left(\frac{1}{2\delta_k(h)} (z-r_k(h) - ih), \frac{1}{2\delta_k(h)} (\zeta-r_k(h) - ih)\right),$$

where $G(z,\zeta)$ is the Green's function of the semidisc U_1^+. Therefore

$$\inf\left\{\left(\frac{\partial}{\partial \xi} G^{(k,h)}(\tilde{z}_k,\zeta)\right)\Big|_{\xi=h} : t\in I_k(h)\right\} \ge$$

$$\ge \frac{1}{2\delta_k(h)} \inf\left\{\left(\frac{\partial}{\partial \xi} G(\tilde{z}_k,\zeta)\right)\Big|_{\xi=0} :\right.$$

$$\left.\frac{1}{2} > Imz > \frac{1}{4},\ -\frac{1}{2} < Rez < 0,\ 0 < t < \frac{1}{2}\right\} = \frac{c_0}{2\delta_k(h)} > 0. \quad (4.3.3)$$

Further, since $f\in H(\mathbb{C}^+,\rho]$ we have

$$\ln|f(z)| \le c_1|z|^\rho,\ |z| > 1, \quad (4.3.4)$$

where $c_1 < \infty$, and hence

$$\sup_{0<\theta<\pi} h_f(\theta) \le c_1 . \tag{4.3.5}$$

In view of the trigonometrical ρ-convexity of the indicator $h_f(\theta)$ it follows that

$$\inf_{0<\theta<\pi} h_f(\theta) = -c_2 > -\infty, \tag{4.3.6}$$

and since $|\tilde{z}_k| = r_k \notin E_f$,

$$ln|f(\tilde{z}_k)| \ge r_k^{\rho}(-c_2 - \gamma_f(r_k)). \tag{4.3.7}$$

Substituting the estimates (4.3.3), (4.3.4) and (4.3.7) in inequality (4.3.2.), we find that

$$\int_{I_k(h)} ln^+ \frac{1}{|f(t+ih)|} dt \le \frac{2\delta_k(h)}{c_0} \{(c_2+\gamma(r_k)r_k^{\rho} + c_1(r_k^*+2\delta_k(h))^{\rho}\} \le$$

$$\le c_3\delta_k(h)r_k^{\rho}, \ \forall k. \tag{4.3.8}$$

Here c_3 is a positive constant independent of k and h. It follows also from (4.3.4) that

$$\int_{I_k(h)} ln^+|f(t+ih)|dt \le \delta_k(h)(r_k+\delta_k(h))^{\rho} \le 2\delta_k(h)r_k^{\rho} .$$

This and (4.3.6) imply the required estimate; namely,

$$\int_{I_k(h)} |ln|f(t+ih)||dt \le c_4\delta_k(h)r_k^{\rho}. \tag{4.3.9}$$

In view of an obvious symmetry the same estimate holds for the integral over the interval $-I_k(h)$.

Now we pass on directly to the proof of necessity in **Theorem 4.3.1**; more precisely, to the proof of convergence in variation of the measures $\mu^{[R]}_{f,\partial}$ to $\hat{\mu}_{f,\partial}$. We introduce the notation $\sigma = \mu_{f,\partial} - \hat{\mu}_{f,\partial}$, and note that convergence in variation holds if and only if

$$\lim_{R\to\infty} R^{-\rho-1}\|\sigma\|(R) = 0, \text{ where } \|\sigma\|(R) = |\sigma|([-R,R]). \tag{4.3.10}$$

It follows from the definitions of the measures $\mu_{f,\partial}$, $\hat{\mu}_{f,\partial}$ and σ that

$$\|\sigma\|(R-0) \le \overline{\lim_{h\to 0}} \int_{-R}^{R} |\mathcal{L}_f(x+ih) - \ln|f(x+ih)||dx \tag{4.3.11}$$

We estimate the integral on the right-hand side of this equality. We have:

$$\int_{-R}^{R} |\mathcal{L}_f(x+ih) - \ln|f(x+ih)||dx \le$$

$$\le \int_{[-R,R]\cap I(h)} |\mathcal{L}_f(x+ih) - \ln|f(x+ih)||dx +$$

$$+ \int_{[-R,R]\setminus I(h)} |\ln|f(x+ih)||dx + \int_{[-R,R]\cap I(h)} |\mathcal{L}_f(x+ih)|dx =$$

$$= A_1 + A_2 + A_3. \tag{4.3.12}$$

It follows from the definition of the set I(h) that

$$A_1 \le 2\int_0^R \gamma_f(\sqrt{x^2+h^2})\cdot\left(\sqrt{x^2+h^2}\right)^{\rho}dx \le 4\int_0^R \gamma(x)x^{\rho}dx. \tag{4.3.13}$$

Then, using the estimate (4.3.9), we obtain

$$A_2 \le \sum_{k:I_k(h)\subset[0,2R]}\left(\int_{-I_k(h)} + \int_{I_k(h)} |\ln|f(x+ih)||dx\right) \le$$

$$\le 2c_4(2R)^{\rho} \sum_{k:I_k(h)\subset[0,2R]} \delta_k(h) \le$$

$$\le 2c_4(2R)^{\rho}\cdot mes(I(h)\cap[-2R,2R]),\quad mes(\cdot) = mes_1(\cdot).$$

This implies the inequality

$$\overline{\lim_{h\to 0}}\ A_2 \le 2^{\rho+2}c_4R^{\rho}mes(E_f\cap[0,2R]). \tag{4.3.14}$$

Now, by (4.3.5) and (4.3.6), the inequality $|\mathcal{L}_f(z)| \le c_5|z|^\rho$ holds in $\mathbb{C}^+$ with $c_5 = \max(c_1, c_2)$. Hence

$$A_3 \le c_5R^\rho \int_{[-R,R] \cap I(h)} dx = c_5R^\rho mes(I(h) \cap [-R,R]).$$

In turn, this implies

$$\overline{\lim_{h \to 0}} A_3 \le 2c_5R^\rho mes(E_f \cap [0,R]). \tag{4.3.15}$$

Comparing (4.3.11) and (4.3.12) with (4.3.13)-(4.3.15) we find that

$\|\sigma\|(R) \le$

$$\le 4 \int_0^R \gamma_f(x)x^\rho dx + 2^{\rho+2}c_4R^\rho mes(E_f \cap [0,2R]) + 2c_5R^\rho mes(E_f \cap [0,R]).$$

Since $\gamma_f(x) \to 0$ as $x \to \infty$ (see (4.3.1)) and E_f is a set of relative measure zero, it follows that

$$\lim_{R \to \infty} \frac{1}{R^{\rho+1}} \|\sigma\|(R) = 0.$$

We noted above that this is equivalent to the convergence in variation of the measures $\mu_{f,\partial}^{[R]}$ to the measure $\hat{\mu}_{f,\partial}$. This finishes the proof of the necessity part in **Theorem 4.3.1**.

In conclusion of this section we note that the above estimates imply the following statement.

Let $f \in H(\mathbb{C}^+, \rho]$ be a function of c.r.g. and define the measures ν and ν_t, $t > 0$, on $\mathbb{R}^+$ by the equalities

$$d\nu(x) = t^{-\rho}h_f(\theta)dx,$$

$$d\nu_t(x) = t^{-\rho}\ln|f(txe^{i\theta})|dx.$$

Then the measures ν_t converge in variation to the measure ν on any interval $[0,r]$, $r > 0$, as $t \to \infty$.

3. Sufficiency ($\rho \notin \mathbb{Z}$). The proof of c.r.g. in $\overline{\mathbb{C}}^+$ of a function $f \in H(\mathbb{C}^+, \rho]$ under the conditions of **Theorem 4.3.1** be carried out using a modification of the ε-normal points method. A number of lemmas will be necessary. In the first lemma we shall obtain estimates of the kernel $K_q(z,\zeta)$ when ζ belongs to certain domains which depend on z. Before formulating this lemma, we agree to denote, as above, by $c_1 = c_1(\ldots)$, $c_2 = c_2(\ldots)$, ... positive constants depend on the indicated parameters only. Note also that each lemma has its own numbering of constants.

Lemma 4.3.1. The following estimates are valid for the kernel $K_q(z,\zeta)$ defined by equality (4.2.11)

a) if $z\in\mathbb{C}^+$, $\zeta\in\overline{\mathbb{C}}^+\backslash U_1$, $|z/\zeta| < p < 1$ then

$$|K_q(z,\zeta)| \le c_1|z/\zeta|^{q+1}, \quad c_1 = c_1(p,q), \tag{4.3.16}$$

b) if $z\in\mathbb{C}^+$, $\zeta\in\overline{\mathbb{C}}^+\backslash\{0\}$, $|z/\zeta| > p > 0$ then

$$K_q(z,\zeta) \ge \frac{2|z||\zeta|\sin(\arg z)}{|z-\zeta|^2} - c_2\sin(\arg z)|z/\zeta|^q, \quad c_2 = c_2(p,q), \tag{4.3.17}$$

c) if $z\in\mathbb{C}^+$, $\zeta\in\overline{\mathbb{C}}^+\backslash\{0\}$, $|z/\zeta| > p > 0$ then

$$K_q(z,\zeta) \le c_3\sin(\arg z)\left(\frac{|z|}{|\zeta|}\right)^q, \quad c_3 = c_3(p,q), \tag{4.3.18}$$

d) if $z\in\mathbb{C}^+$, $\zeta\in\overline{\mathbb{C}}^+\backslash\{0\}$, $0 < p_1 < |z/\zeta| < p_2 < \infty$, $|z-\zeta| < \frac{1}{2}\,Im z$ then

$$K_q(z,\zeta) \ge -c_4\sin(\arg z) - \frac{1}{\sin(\arg\zeta)}\,\frac{\ln\left(\frac{5}{2}\,Im z\right)}{|z-\zeta|}, \quad c_4 = c_4(p_1,p_2,q), \tag{4.3.19}$$

e) if $z\in\mathbb{C}^+$, $\zeta\in\overline{\mathbb{C}}^+\backslash\{0\}$, $0 < p_1 < |z/\zeta| < p_2 < \infty$, $|\sin(\arg\frac{z}{\zeta})| \ge p_3 > 0$, then

$$K_q(z,\zeta) \ge -c_5\sin(\arg z), \quad c_5 = c_5(p_1,p_2,p_3,q) \ . \tag{4.3.20}$$

Proof. First of all we note that since the kernel $K_q(z,\zeta)$ is continuous with respect to ζ in $\overline{\mathbb{C}}^+\backslash(\{0\}\cup\{z\})$, it is sufficient to prove the above estimates for $\zeta\in\mathbb{C}^+$ only. In this case

$$K_q(z,\zeta) = \frac{1}{\sin(\arg\zeta)}\ln|D_q(z,\zeta)| =$$

$$= \frac{1}{\sin(\arg\zeta)}\left\{\ln\left|\frac{1-z/\zeta}{1-z/\bar\zeta}\right| + Re\sum_{k=1}^{q}\frac{z^k}{k}\left(\frac{1}{\zeta^k} - \frac{1}{\bar\zeta^k}\right)\right\}.$$

Now note that the estimate a) has already been proved (see (4.1.30) or (4.2.12)).

We give a lower bound for of $K_q(z,\zeta)$ for $|z/\zeta| \ge p > 0$. Setting $z = re^{i\varphi}$, $\zeta = te^{i\theta}$, we have

$$K_q(z,\zeta) = \frac{1}{\sin\theta}\left\{ -\frac{1}{2}\ln\frac{r^2+t^2-2rt\cos(\theta+\varphi)}{r^2+t^2-2rt\cos(\theta-\varphi)} + 2\sum_{k=1}^{q}\frac{r^k}{kt^k}\sin k\varphi\sin k\theta \right\} =$$

$$= \frac{1}{\sin\theta}\left\{ -\frac{1}{2}\ln\left(1+\frac{4rt\sin\theta\sin\varphi}{|z-\zeta|^2}\right) + 2\sum_{k=1}^{q}\frac{r^k}{kt^k}\sin k\varphi\sin k\theta \right\} \ge$$

$$\ge \frac{1}{\sin\theta}\left\{ -\frac{2rt\sin\theta\sin\varphi}{|z-\zeta|^2} - 2\left(\frac{r}{t}\right)^q\sin\varphi\sin\theta\sum_{k=1}^{q}k\left(\frac{1}{p}\right)^{q-k} \right\} =$$

$$= -\frac{2rt\sin\varphi}{|z-\zeta|^2} - c_2\sin\varphi\left(\frac{r}{t}\right)^q.$$

Thus we have proved the estimate b).

Note that in proving b) we have also shown that for $|z/\zeta| > p$ the inequality

$$\left| Re\sum_{k=1}^{q}\frac{z^k}{k}\left(\frac{1}{\zeta^k}-\frac{1}{\bar\zeta^k}\right)\right| \le c_2\sin\theta\sin\varphi\left(\frac{r}{t}\right)^q \tag{4.3.21}$$

holds. From this and the fact that $\ln\left|\frac{z-\zeta}{z-\bar\zeta}\right| < 0$, $\forall\zeta\in\mathbb{C}^+$, $z\in\mathbb{C}^+$, the estimate c) follows.

To obtain the estimate d) we use the following statement : if $|z-\zeta| \le \frac{1}{2}Imz$, then $|z-\bar\zeta| \le \frac{5}{2}Imz$ and, therefore,

$$\ln\left|\frac{z-\zeta}{z-\bar\zeta}\right| \ge -\ln\frac{\frac{5}{2}Imz}{|z-\zeta|} \tag{4.3.22}$$

At the same time (4.3.21) implies that for $0 < p_1 < |z/\zeta| < p_2 < \infty$ the inequality

$$\left| Re\sum_{k=1}^{q}\frac{z^k}{k}\left(\frac{1}{\zeta^k}-\frac{1}{\bar\zeta^k}\right)\right| \le c_3\sin\varphi,\ c_3 = c_2p_2^q,$$

holds. From this and (4.3.22) we conclude that the estimate d) is true.

Finally, to obtain the estimate e) we use the already proved estimate b). Noting that if $|\sin(\arg(z/\zeta))| \ge p_3 > 0$, then $|z-\zeta| >$

$p_3|z|$, we obtain

$$K_q(z,\zeta) \geq -\frac{2rt\sin\varphi}{|z-\zeta|^2} - c_2\left(\frac{r}{t}\right)^q \sin\varphi \geq -\frac{2rt\sin\varphi}{p_3^2 r^2} - c_2 p_2^2 \sin\varphi \geq$$

$$\geq -\left(\frac{2}{p_3^2 p_1} + c_2 p_2^q\right)\sin\varphi = -c_4 \sin\varphi.$$

The proof of the lemma is finished.

In the following lemma we give as estimate of the potential (with kernel $K_q(z,\zeta)$) of a measure at those points which are not points of accumulation (in some sense) of this measure. It is natural to call these points ε-normal, because the conditions defining them are close to those which previously defined the concepts of normal points (see **Chapter 1** and 2).

Lemma 4.3.2. Let a real-valued measure τ, defined on $\mathbb{C}$, be such that $supp\tau \subset \bar{\mathbb{C}}^+ \setminus U_1$ and

$$\|\tau\|(t) \leq (\sigma+\gamma(t))t^\rho, \tag{4.3.23}$$

where $\|\tau\|(t) = |\tau|(U_t)$, $\sigma \geq 0$, $\gamma(t)\downarrow 0$ as $t \to \infty$, and $\rho \notin \mathbb{Z}$.

If at a point $z \in \mathbb{C}^+$ for prescribed $\varepsilon > 0$ the condition

$$\int_{U(z,\zeta)} \frac{d|\tau|(\zeta)}{|\zeta|^\rho} \leq \varepsilon \frac{s}{|z|}, \quad \forall s < \frac{1}{2}|z|, \tag{4.3.24}$$

is satisfied, then

$$\left| \int_{\bar{\mathbb{C}}^+} K_q(z,\zeta)d\tau(\zeta) \right| \leq c(\sigma+\varepsilon+\gamma_1(|z|))|z|^\rho,$$

where $q = [\rho]$, $c = c(q)$, $\gamma_1(|z|)$ is a function depending only on $\gamma(t)$ and q, and tending to zero as $|z| \to \infty$.

Proof. Let p_1, p_2, p_3 be numbers such that $0 < p_2 < 1 < p_1$, $0 < p_3 < 1$. Then we relate to each $z \in \mathbb{C}^+$, $|z| > 1/p_2$, the partition of the half-plane $\bar{\mathbb{C}}^+$ into the sets

$\Omega_0 = \bar{U}_1^+$,

$\Omega_1 = \Omega_1(z) = \{\zeta \in \bar{\mathbb{C}}^+ : |\zeta| \geq p_1|z|\}$,

$\Omega_2 = \Omega_2(z) = \{\zeta \in \bar{\mathbb{C}}^+ : 1 \leq |\zeta| \leq p_2|z|\}$,

$\Omega_3 = \Omega_3(z) = \{\zeta \in \bar{\mathbb{C}}^+ : p_2|z| < |\zeta| < p_1|z|,\ |\sin(\arg(z/\zeta))| \geq p_3\}$,

$$\Omega_4 = \Omega_4(z) = \{\zeta\in\overline{\mathbb{C}}^+ : p_2|z| < |\zeta| < p_1|z|, \ |sin(arg(z/\zeta))| < p_3\}.$$

In accordance with this partition

$$\int_{\overline{\mathbb{C}}^+} K_q(z,\zeta)d\tau(\zeta) = \sum_{j=1}^{4} \int_{\Omega_j} K_q(z,\zeta)d\tau(\zeta).$$

We estimate the integrals $\int_{\Omega_j}$. Unless otherwise stipulated, $\gamma_2(\zeta)$, $\gamma_3(\zeta)$, ... denote positive functions tending to zero as $|\zeta| \to \infty$.

Using (4.3.16) and (4.3.23), we obtain

$$\int_{\Omega_1} |K_q(z,\zeta)|d|\tau|(\zeta) \le c_1|z|^{q+1} \int_{\Omega_1} \frac{d|\tau|(\zeta)}{|\zeta|^{q+1}} \le$$

$$\le c_1(q+1)|z|^{q+1} \int_{p_1|z|}^{\infty} s^{-q-2}\|\tau\|(s)ds \le$$

$$\le c_1(q+1)|z|^{q+1} \int_{p_1|z|}^{\infty} s^{-q-2}(\sigma+\gamma(s))ds \le$$

$$\le c_2(\sigma+\gamma_2(|z|))|z|^{\rho}. \qquad (4.3.25)$$

From (4.3.17) and (4.3.23) it follows that

$$\int_{\Omega_2} |K_q(z,\zeta)|d|\tau|(\zeta) \le \int_{\Omega_2} (c_3|\frac{z}{\zeta}|^q + \frac{2|z||\zeta|}{|\zeta-z|^2})d|\tau|(\zeta) \le$$

$$\le c_3|z|^q \int_1^{p_1|z|} s^{-q}d\|\tau\|(s) + 2|z| \int_{\Omega_2} \frac{d|\tau|(\zeta)}{|\zeta|(|z/\zeta|-1)} \le$$

$$\le c_3p_2^{\rho-q}|z|^q(\sigma+\gamma(p_2|z|)) + c_3q|z|^q \int_1^{p_1|z|} (\sigma+\gamma(s)s^{\rho-q-1}ds +$$

$$+\frac{2|z|p_2^2}{(p_2-1)^2}\int_1^{p_1|z|}\frac{d\|\tau\|(s)}{s}\le$$

$$\le c_3|z|^q(\sigma+\gamma(p_2|z|))+c_3q|z|^q\int_1^{p_1|z|}(\sigma+\gamma(s))s^{-q-1+\rho}ds+$$

$$+\frac{2|z|p_2^2}{(p_2-1)^2}\left(\frac{\|\tau\|(p_2|z|)}{p_2|z|}+\int_1^{p_1|z|}\frac{(\sigma+\gamma(s))s^{\rho}}{s^2}ds\right)=$$

$$=c_4|z|^{\rho}(\sigma+\gamma_3(|z|)),\ c_4=c_4(p_2,q). \tag{4.3.26}$$

From the estimates (4.3.20) and (4.3.18) it is obvious that also

$$\int_{\Omega_3}|K_q(z,\zeta)|d|\tau|(\zeta)\le(c_5+c_6p_1^q)\int_{\Omega_3}d|\tau|(\zeta)\le$$

$$\le(c_5+c_6p_1^q)\|\tau\|(p_2|z|)\le(c_5+c_6p_1^q)(\sigma+\gamma(p_2|z|))|z|^{\rho}. \tag{4.3.27}$$

Finally we estimate

$$\int_{\Omega_4}K_q(z,\zeta)d|\tau|(\zeta).$$

First we represent Ω_4 in the form $\Omega_4=\Omega'+\Omega''$, where $\Omega'=\Omega_4\cap U\left(z,\frac{1}{2}y\right)$, $y=Imz$. Then we represent Ω' and Ω'' in the form

$$\Omega'=\bigcup_{k=1}^{\infty}\Omega'_k,$$

where

$$\Omega'_k=\{\zeta:\ \zeta\in\Omega',\ 2^{-k-1}y\le|z-\zeta|\le2^{-k}y\},$$

and

$$\Omega''=\bigcup_{k=1}^{\omega}\Omega''_k,$$

where

$$\Omega''_k=\{\zeta:\ \zeta\in\Omega'',\ 2^{k-2}y\le|z-\zeta|\le2^{k-1}y\},\ \omega<\infty.$$

Now, using the estimates (4.3.17) - (4.3.19) we have

$$\int_{\Omega_4} |K_q(z,\zeta)|d|\tau|(\zeta) \le \int_{\Omega'} \left(\frac{1}{\sin(\arg\zeta)} \ln \frac{\frac{5}{2}y}{|z-\zeta|} + c_7 \right) d|\tau|(\zeta) +$$

$$+ \int_{\Omega''} \left(\frac{2|z||\zeta|\sin(\arg z)}{|z-\zeta|^2} + c_8 \right) d|\tau|(\zeta). \qquad (4.3.28)$$

It is obvious that

$$\int_{\Omega'} c_7 d|\tau|(\zeta) + \int_{\Omega''} c_8 d|\tau|(\zeta) \le c_9(\sigma+\gamma(p_2|z|))|z|^\rho. \qquad (4.3.29)$$

Further, noting that if $\zeta = te^{i\theta} \in \Omega'$, then

$$\frac{1}{\sin\theta} = \frac{1}{Im\zeta} \le \frac{2t}{y} \le \frac{4|z|}{y} = \frac{4}{\sin(\arg z)},$$

and using (4.3.24), we obtain

$$\int_{\Omega'} \frac{1}{\sin\theta} \ln \frac{\frac{5}{2}y}{|z-\zeta|} d|\tau|(\zeta) = \sum_{k=1}^{\infty} \int_{\Omega'_k} \frac{1}{\sin\theta} \ln \frac{\frac{5}{2}y}{|z-\zeta|} d|\tau|(\zeta) \le$$

$$\le \frac{4|z|}{y} \sum_{k=1}^{\infty} \int_{\Omega'_k} |\zeta|^\rho \ln \frac{\frac{5}{2}y}{2^{-k-1}y} \frac{d|\tau|(\zeta)}{|\zeta|^\rho} \le$$

$$\le \frac{2^{\rho+2}|z|^{\rho+1}}{y} \sum_{k=1}^{\infty} \ln(5\cdot 2^k) \int_{\Omega'_k} \frac{d|\tau|(\zeta)}{|\zeta|^\rho} \le$$

$$\le \varepsilon 2^{\rho+2} \frac{|z|^{\rho+1}}{y} \sum_{k=1}^{\infty} \frac{y}{|z|} \frac{\ln(5\cdot 2^k)}{2^k} = c_{10}\varepsilon|z|^\rho,\ c_{10} = c_{10}(\rho). \qquad (4.3.30)$$

Similarly,

$$\int_{\Omega''} \frac{|z||\zeta|\sin(\arg z)}{|z-\zeta|^2} d|\tau|(\zeta) = \sum_{k=1}^{\omega} \int_{\Omega''_k} \frac{y|\zeta|^{\rho+1}}{|z-\zeta|^2} \frac{d|\tau|(\zeta)}{|\zeta|^\rho} \le$$

$$\le 2^{\rho+1}|z|^{\rho+1} \sum_{k=1}^{\omega} \frac{y}{(2^{k-2}y)^2} \int_{\Omega''_k} \frac{d|\tau|(\zeta)}{|\zeta|^\rho} \le$$

$$\le 2^{\rho+1} \frac{|z|^{\rho+1}}{y} \sum_{k=1}^{\omega} \frac{1}{2^{2k-4}} \frac{\varepsilon 2^{k-1} y}{|z|} \le$$

$$\le \varepsilon 2^{\rho+4} |z|^{\rho} \sum_{k=1}^{\infty} \frac{1}{2^k} = c_{11} \varepsilon |z|^{\rho} . \tag{4.3.31}$$

Combining these estimates, i.e. the estimates (4.3.25) - (4.3.31), we conclude that

$$\int_{\bar{\mathbb{C}}^+} |K_q(z,\zeta)| d|\tau|(\zeta) \le c_{12} |z|^{\rho} (\sigma+\varepsilon+\gamma_4(|z|)),$$

where the constant c_{12} depends on ρ only, the function $\gamma_4(t) \to 0$ as $t \to \infty$ and depends on the number $q = [\rho]$ and the function $\gamma(t)$ only.

This finishes the proof.

In the case $\rho = p \in \mathbb{N}$ **Lemma 4.3.2** is false, because instead of the integral $\int s^{\alpha} ds$, $\alpha \ne 1$, which plays a role in certain estimates, we have to deal with the integral $\int s^{-1} ds = \ln s$. The essential additional assumption $\sigma = 0$ or, equivalently, the condition

$$\lim_{R \to \infty} \frac{\|\tau\|(R)}{R^{\rho}} = 0, \tag{4.3.32}$$

does not change the situation. However, if we consider the stronger condition

$$\int_1^{\infty} \frac{\|\tau\|(t)}{t^{\rho+1}} dt < \infty,$$

then we can obtain a statement similar to **Lemma 4.3.2**. This statement (**Lemma 4.3.3.**) is not used in the proof of **Theorem 4.3.1** (nor in the case $\rho = p \in \mathbb{N}$), but we give it here in view of its connection with **Lemma 4.3.2.**

Lemma 4.3.3. Let a real-valued measure τ, defined on $\mathbb{C}$, be such that *supp*$\tau \subset \bar{\mathbb{C}}^+$ and

$$\int_{\mathbb{C}} |\zeta|^{-p} d|\tau|(\zeta) < \infty. \tag{4.3.33}$$

If condition (4.3.24) is satisfied at a point $z \in \mathbb{C}^+$, then

$$\left| \int_{\mathbb{C}^+} K_{p-1}(z,\zeta)d\tau(\zeta) \right| \le c|z|^p(\varepsilon+\gamma(|z|)), \tag{4.3.34}$$

where c = const and $\gamma(t) \to 0$ as $t \to \infty$.

Proof. The condition of this lemma differs from the one of **Lemma 4.3.2** by the replacement of $\rho \notin \mathbb{Z}$ by $\rho = p \in \mathbb{N}$, of $q = [\rho]$ by $p-1$, and of inequality (4.3.23) by (4.3.33). We indicate the changes in the proof of **Lemma 4.3.2** which are implied by these replacements. We take into account that (4.3.33) implies (4.3.32).

Estimate (4.3.25) will be replaced by the estimate

$$\int_{\Omega_1} |K_{p-1}(z,\zeta)|d|\tau|(\zeta) \le c_1|z|^p \int_{p_1|z|}^{\infty} \frac{d\|\tau\|(s)}{s^p} = c_1|z|^p\gamma_1(p_1|z|).$$ [24]

In the case $p > 1$ all calculations in (4.3.26) remain essentially the same, and the estimate can be written in the form

$$\int_{\Omega_2} |K_{p-1}(z,\zeta)|d|\tau|(\zeta) \le c_2|z|^p.$$

In the case $p = 1$ we have

$$\int_{\Omega_2} |K_0(z,\zeta)|d|\tau|(\zeta) \le c_3|\tau|(p_2|z|) + \frac{2|z|p_2^2}{(p_2-1)^2}\int_1^{p_2|z|} \frac{d\|\tau\|(\zeta)}{s} \le$$

$$\le |z| \left(\gamma_3(p_2|z|) + \int_1^{\infty} \frac{d\|\tau\|(s)}{s} \right),$$

where $c_3 = 2p_2^2(p_2-1)^{-2} \to 0$ as $p_2 \to 0$. Hence, without loss of generality we may assume that $c_3 < \varepsilon$.

It is obvious that the estimates (4.3.27) - (4.3.31) do not change when q is replaced by $q-1 = p-1$. Thus,

$$\int_{\Omega_3 \cup \Omega_4} |K_{p-1}(z,\zeta)|d|\tau|(\zeta) \le c_4(\varepsilon+\gamma(p_2|z|)).$$

Now it remains to note that the required estimate of the integral

---

[24] Recall that the numbering of the constants c_j and the functions $\gamma_j(t) = o(1)$ is done in each lemma separately.

$$\int_{\overline{\mathbb{C}}^+} |K_{p-1}(z,\zeta)|d|\tau|(\zeta)$$

follows from the estimates above of the integrals

$$\int_{\Omega_j} |K_{p-1}(z,\zeta)|d|\tau|(\zeta).$$

The proof is finished.

Now we estimate how "large" the set of points which are not ε-normal can be, i.e. the set of points at which condition (4.3.24) is not satisfied.

Lemma 4.3.4. Let a measure τ satisfy the conditions of **Lemma 4.3.2** for an arbitrary $\rho > 0$. Then the set of points which do not satisfy condition (4.3.4) can be covered by a family of discs $U^{(j)} = U(z_j,r_j)$ such that

$$\frac{1}{R} \sum_{j:|z_j|<R} r_j \le c\,\frac{\sigma+\tilde{\gamma}(R)}{\varepsilon}, \quad \forall R > 1,$$

where c is an absolute constant and $\tilde{\gamma}(R)$ is some function tending to zero as $R \to \infty$.

Proof. In accordance with the definition of the set $\Delta(\varepsilon)$ for every point $z\in\Delta(\varepsilon)$ there is a disc $U(z,\lambda_z)$, $\lambda_z < \frac{1}{2}|z|$, such that

$$\int_{U(z,\lambda_z)} \frac{d|\tau|(\zeta)}{|\zeta|^\rho} > \varepsilon\,\frac{\lambda_z}{|z|}.$$

Consider the set of all discs $U(z,\lambda_z)$ such that $z\in\Delta^k(\varepsilon) = \Delta(\varepsilon) \cap U_{4^k,4^{k+1}}$[25], $k = 1,2,\ \ldots$. Now, using **Ahlfors' Theorem** on covers of finite multiplicity, we extract from of this set a system of discs $\mathfrak{A}_k$ forming on at most countable cover of the set Δ^k of multiplicity $d \le 6$. Since $\lambda_z < \frac{1}{2}|z|$, the disc belonging to $\mathfrak{A}_k$ do not intersect with the ones belonging to $\mathfrak{A}_{k+2}$. Therefore the union of all systems $\mathfrak{A}_k$ is an at most countable cover of $\Delta(\varepsilon)$ of multiplicity less than 3d.

We denote by $\tilde{U}^{(j)} = U(\tilde{z}_j,\tilde{r}_j)$ the elements of this cover, i.e. the

---

25) Recall that $U_{r,R} = \{z\in\mathbb{C}: r < |z| < R\}$.

discs belonging to $\bigcup_k \mathfrak{A}_k$. Then let us estimate the value $\sum_{j:\tilde{z}_j\in\Delta^k(\varepsilon)} \tilde{r}_j$.

We have

$$\sum_{j:\tilde{z}_j\in\Delta^k(\varepsilon)} \tilde{r}_j \le 4^{k+1} \sum_{j:\tilde{z}_j\in\Delta^k(\varepsilon)} \frac{\tilde{r}_j}{|\tilde{z}_j|} \le \frac{4^{k+1}}{\varepsilon} \sum_{j:\tilde{z}_j\in\Delta^k(\varepsilon)} \int_{\tilde{U}^{(j)}} \frac{d|\tau|(\zeta)}{|\zeta|^{\rho}} \le$$

$$\le \frac{4^{k+1}\cdot 3d}{\varepsilon} \int_{\frac{1}{2}4^k \le|\zeta|\le 5\cdot 4^k} \frac{d|\tau|(\zeta)}{|\zeta|^{\rho}} \le$$

$$\le \frac{3d\cdot 4^{k+1}}{\varepsilon}\left\{ \sigma + \gamma(5\cdot 4^k) + \rho \int_{\frac{1}{2}4^k}^{5\cdot 4^k} \frac{\sigma+\gamma(s)}{s}\, ds \right\} =$$

$$= \frac{3d\cdot 4^{k+1}}{\varepsilon}(1+\rho \ln 10)(\sigma+\delta_k) = \frac{4^k}{\varepsilon} c_1(\sigma+\delta_k),$$

where $\delta_k \to 0$ as $k\to\infty$.

Assuming that $4^{N-1} \le R \le 4^N$, we obtain

$$\frac{1}{R}\sum_{j:1\le|\tilde{z}_j|<R} \tilde{r}_j \le \frac{1}{4^{N-1}} \sum_{k=0}^{N-1} \sum_{j:\tilde{z}_j\in\Delta^k(\varepsilon)} \tilde{r}_j \le \frac{c_1}{4^{N-1}\varepsilon}\sum_{k=0}^{N-1} 4^k(\sigma+\delta_k) \le$$

$$\le \frac{4}{3}\frac{c_1}{\varepsilon}\left(\sigma + \sum_{k=0}^{N-1}\frac{\delta_k}{4^{N-1-k}}\right) = \frac{c_2}{\varepsilon}(\sigma+\tilde{\gamma}(R)).$$

It is obvious that $\tilde{\gamma}(R) = \sum_{k=0}^{N-1} \delta_k 4^{1+k-N} \to 0$ as $N\to\infty$ or, equivalently, as $R\to\infty$.

This finishes the proof.

Remark. In the proof of **Lemma 4.3.4** we have obtained the estimate

$$\sum_{j:\tilde{z}_j\in\Delta^k(\varepsilon)} \frac{\tilde{r}_j}{|\tilde{z}_j|} \le \frac{3d}{\varepsilon} \int_{\frac{1}{2}4^k \le|\zeta|\le 5\cdot 4^k} \frac{d|\tau|(\zeta)}{|\zeta|^{\rho}},$$

which evidently implies the following characterization of the system $\tilde{U}^{(j)} = U(\tilde{z}_j, \tilde{r}_j)$ covering $\Delta(\varepsilon)$:

$$\sum_{j:|\tilde{z}_j|<R} \frac{\tilde{r}_j}{|\tilde{z}_j|} \le \frac{c_2}{\varepsilon} \int_{1\le|\zeta|\le 10R} \frac{d|\tau|(\zeta)}{|\zeta|^\rho} = \frac{c_2}{\varepsilon} \int_1^{10R} \frac{d\|\tau\|(\zeta)}{s^\rho} .$$

Now we consider one particular case of sufficient conditions for c.r.g. in $\overline{\mathbb{C}}^+$; namely, the case of functions that have zero argument-boundary density. This case is interesting in itself and will be used in a general situation.

Lemma 4.3.5. Let $f\in H(\mathbb{C}^+,\rho]$, $\rho\notin\mathbb{Z}$, be a function such that

$$\lim_{r\to\infty} \frac{\|\tau_f\|(r)}{r^\rho} = 0. \tag{4.3.35}$$

Then $h_f(\arg z) \equiv 0$, $f(z)$ is of c.r.g. in $\overline{\mathbb{C}}^+$, and, moreover,

$$|\ |z|^{-\rho}\ln|f(z)|\ | \to 0$$

as $z\to\infty$, $z\notin E$, where E is a C_0-set.

Proof. To prove the lemma it suffices to show that the function

$$\Phi_f(z) = \left\{ \prod_{|z_k|\ge 1} D_q(z,z_k) \right\}\cdot \exp\left\{ \frac{1}{\pi i} \int_{|t|>1} \frac{z^{q+1}d\mu_{f,\partial}(t)}{t^{q+1}(t-z)} \right\} \tag{4.3.36}$$

(the z_k are the zeros of $f(z)$) or, equivalently, the potential

$$J_{\tau_f,q} = \int_{\overline{\mathbb{C}}^+} K_q(z,\zeta)d\tau_f(\zeta), \tag{4.3.37}$$

has the required properties. This follows from the properties of the factors figuring in the representation (4.1.33).

From **Lemma 4.3.2**, taking into account (4.3.4), we find that the estimate

$$\frac{1}{|z|^\rho}\ |\ J_{\tau_f,q}(z)\ |\ \le c(\varepsilon+\gamma_1(|z|))$$

holds on the set $\mathbb{C}^+\setminus\Delta(\varepsilon)$, where $\Delta(\varepsilon)$ is the same as in **Lemma 4.3.4**. Using now**Lemma 4.3.4** we find that $\Delta(\varepsilon)$ is a C_0-set. Therefore, since $\gamma_1(|z|)\to\infty$ the set $\{z\in\mathbb{C}^+: |z|^{-\rho}|J_{\tau_f,q}(z)| > \varepsilon\}$ is a C_0-set for every $\varepsilon > 0$. From this it follows that the indicator of the function $J_{\tau_f,q}$ or, equivalently, the indicator of the function $\Phi_f(z)$, vanishes identically. Moreover, these functions $J_{\tau_f,q}$ and Φ_f are of c.r.g. in $\overline{\mathbb{C}}^+$ and $||z|^{-\rho}\ln|\Phi_f(z)| - \mathfrak{L}_{\Phi_f}(z/|z|)| \to 0$ as $z \to \infty$, $z \notin E$, where E is a C_0-set.

The proof is finished.

Recall that the product of two functions of c.r.g. is a function of c.r.g. Therefore to prove that of the function f(z) in **Theorem 4.3.1** is of c.r.g. it suffices to show that the functions

$$\Psi_f(z) = \prod_{|z_k|<1} D_q(z,z_k) \tag{4.3.38}$$

and

$$F_f(z) = \exp\left\{ \frac{1}{\pi i} \int_{|t|>1} \frac{z^{q+1}}{t^{q+1}(t-z)} d\mu_{f,\partial}(t) \right\} \tag{4.3.39}$$

in the factorization of f(z) (**Theorem 4.1.1**) are of c.r.g. We first consider the function $F_f(z)$.

Lemma 4.3.6. Let $f \in H(\mathbb{C}^+,\rho]$ be a function of c.r.g. in $\mathbb{C}^+$ and let the measures $\mu_{f,\partial}^{[R]}$ converge in variation to the measure $\hat{\mu}_{f,\partial}$ as $R \to \infty$. Then the function $F_f(z)$ defined by (4.3.39) is of c.r.g. in $\overline{\mathbb{C}}^+$.

Proof. Since the measure $\hat{\mu}_{f,\partial}$ is clearly positively homogeneous of degree $\rho+1$, then the potential

$$J^{(1)}(z) = \frac{1}{\pi} \int_{-\infty}^{\infty} Im \frac{z^{q+1}}{t^{q+1}(t-z)} d\hat{\mu}_{f,\partial}(t)$$

is positively homogeneous of degree ρ. It is also clear that the function $J^{(1)}(z)$ is harmonic in $\mathbb{C}^+$ and continuous on $\overline{\mathbb{C}}^+\setminus\{0\}$. Moreover, $J^{(1)} = \lim_{y\to 0} \mathfrak{L}_f(x+iy)$, $\forall x \in \mathbb{R}$. From these properties of the function $J^{(1)}(z)$ it follows that the functions

$$F_\pm = \exp\left\{ \pm \frac{1}{\pi i} \int_{-\infty}^{\infty} \frac{z^{q+1}}{t^{q+1}(t-z)} d\hat{\mu}_{f,\partial}(t) \right\}$$

belong to the class $H(\mathbb{C}^+,\rho]$ and satisfy the condition $\mu_{F_\pm,\partial} = \pm \hat{\mu}_{f,\partial}$. Now we represent $F_f(z)$ in the form $F_f(z) = F_+(z)\varphi(z)$, where $\varphi(z) = F_-\cdot F_f$. From the above-mentioned properties of the functions $F_\pm$ and from the properties of the function F_f given in **Lemma 4.1.10**, it follows that $\varphi \in H(\mathbb{C}^+,\rho]$ and $\mu_{\varphi,\partial} = \mu_{f,\partial} - \hat{\mu}_{f,\partial}$. Note that since the measures $\mu^{[R]}_{f,\partial}$ converge in variation to $\hat{\mu}_{f,\partial}$ as $R \to \infty$, the measure $\mu_{\varphi,\partial}$ satisfies the condition

$$\lim_{R\to\infty} \frac{\|\mu_{\varphi,\partial}\|(R)}{R^{\rho+1}} = 0$$

or, equivalently, the condition

$$\lim_{R\to\infty} \frac{\|\tau\|(R)}{R^{\rho}} = 0.$$

In accordance with **Lemma 4.3.5** we find that the function $\varphi(z)$ is of c.r.g. in $\overline{\mathbb{C}}^+$. Regularity of growth of the function F_+ is evident. Hence $F_f(z)$ is of c.r.g. in $\overline{\mathbb{C}}^+$, as the product of two functions of c.r.g.

The proof is finished.

Finally we consider the function $\Psi_f(z)$.

Lemma 4.3.7. Let $f\in H(\mathbb{C}^+,\rho]$ be a function of c.r.g. in $\mathbb{C}^+$ and let the measures $\mu^{[R]}_{f,\partial}$ converge in variation to the measure $\hat{\mu}_{f,\partial}$ as $R \to \infty$. Then the function $\Psi_f(z)$ defined from the zeros z_k of the function $f(z)$ using the equality (4.3.38) is of c.r.g. in $\overline{\mathbb{C}}^+$.

Proof. First of all we recall that, as noted after the formulation of **Theorem 4.3.1**, convergence of the measures $\mu^{[R]}_{f,\partial}$ in variation to $\hat{\mu}_{f,\partial}$ implies convergence of the measures[26] $\mu^{[R]}_{f,1}$ to $\hat{\mu}_{f,1}$ as functionals on the space $C(\overline{U}^+_r)$, $\forall r$. Hence if $\hat{\mu}_{f,1}(l_\theta) = 0$, where $l_\theta = \{z = te^{i\theta}: t > 0\}$, then

$$\lim_{R\to\infty} \mu^{[R]}_{f,1}(Y(1;0,\theta)) = \hat{\mu}_{f,1}(Y(1;0,\theta)),$$

where, as before, $Y(t;\theta_1,\theta_2) = \{z\in\mathbb{C}: 0 < |z| < t, \theta_1 < \mathit{argz} < \theta_2\}$. Since $\hat{\mu}_{f,1}(U^+) < \infty$, we have

---

[26] Recall that $d\mu_{f,1}(z) = \sin(\arg z)d\tilde{n}_f(z)$ on $\mathbb{C}^+\backslash U_1$, $\mu^{[R]}_{f,1}(E) = R^{-\rho}\mu_{f,1}(RE)$, $\forall E \subset \mathbb{C}^+$, $d\hat{\mu}_{f,1} = \sin(\arg z)d\hat{\mu}_f$, $d\hat{\mu}^{[R]}_f = R^{-\rho}\hat{\mu}_f(RE)$, and $\hat{\mu}_f$ is the Riesz associated measure of the indicator $\mathfrak{L}_f(z)$.

$$\lim_{\theta \to +0} \hat{\mu}_{f,1}(Y(1;0,\theta)) = 0,$$

$$\lim_{\theta \to \pi-0} \hat{\mu}_{f,1}(Y(1;\theta,\pi)) = 0.$$

Therefore we conclude that the inequality

$$\mu_{f,1}(Y(R;0,\delta)) + \mu_{f,1}(Y(R;\pi-\delta,\pi)) \le \varepsilon R^{\rho}$$

holds for any $\varepsilon > 0$, for all sufficiently small $\delta > 0$ and for $R \ge R_0 = R_0(\varepsilon)$. If ε and δ are as above, then we represent the measure $\mu_{f,1}$ in the form $\mu_{f,1} = \mu' + \mu''$, where μ' is the restriction of $\mu_{f,1}$ to the sector $\overline{Y}(\delta,\pi-\delta)$ and μ'' is the restriction of $\mu_{f,1}$ to the domain $Y'' = \mathbb{C}^+\backslash Y'$. In accordance with this partition the potential

$$I_{\mu_f} = \int_{\mathbb{C}^+} K_q(z,\zeta)d\mu_{f,1}(\zeta) = \int_{\mathbb{C}^+\backslash U_1} ln|D_q(z,\zeta)|d\tilde{n}_f$$

can be represented in the form $I_{\mu_f} = I_{\mu'} + I_{\mu''}$. Similarly $\Psi_f = \Psi_1 \cdot \Psi_2$, where

$$\Psi_1 = \prod_{z_k \in Y'\backslash U_1} D_q(z,z_k),$$

$$\Psi_2 = \prod_{z_k \in Y''\backslash U_1} D_q(z,z_k).$$

Consider the potential $I_{\mu''}$. By **Lemmas 4.3.2 and 4.3.4**, for each $\varepsilon_1 > 0$ the set of the points $z \in \mathbb{C}^+$ satisfying the condition

$$|I_{\mu''}(z)| \ge \varepsilon_1 |z|^{\rho}$$

can be covered by a system of discs $U(\tilde{z}_j,\tilde{r}_j)$ such that for all sufficiently largeR the inequality

$$\frac{1}{R} \sum_{j:|\tilde{z}_j|<R} \tilde{r}_j \le c_1 \frac{\varepsilon}{\varepsilon_1} \tag{4.3.40}$$

holds with a constant c_1 independent of ε and ε_1.

Now we consider the potential $I_{\mu'}$. We show that the function Ψ_1 connected with $I_{\mu'}$ by the equality $ln|\Psi_1(z)| = I_{\mu'}(z)$ is of c.r.g. in

$\overline{\mathbb{C}}^+$. First of all we note that since $\sin(\arg z) \geq \sin\delta$, $\forall z \in Y'$, the condition

$$\varlimsup_{R\to\infty} R^{-\rho}\|\mu_{f,1}\|(R) < \infty$$

implies

$$\varlimsup_{R\to\infty} R^{-\rho}\|n_{f,1}\|(R) < \infty ,$$

where $\|n_{f,1}\|(R) = \tilde{n}_f((Y' \cap U_R))\setminus U_1$. In turn, this implies the convergence of the products[27)]

$$\Psi_1^+ = \prod_{z_k\in Y'\setminus U_1} G_q(z/z_k) ,$$

$$\Psi_1^- = \prod_{z_k\in Y'\setminus U_1} G_q(z/\bar{z}_k)$$

and the fact that the functions $\Psi_1^{\pm}$ belong to the class $H(\mathbb{C},\rho]$ of entire functions of at most normal type with respect to the order ρ. Further, as has been already noted, the measures $\mu_{f,1}^{[R]}$ converge weakly to the measure $\hat{\mu}_{f,1}$ as $R \to \infty$. This clearly implies that

$$\exists\mathcal{D}' - \lim_{R\to\infty} \mu'^{[R]},$$

and hence

$$\exists\mathcal{D}' - \lim_{R\to\infty} \tilde{n}_1^{[R]},$$

where $\tilde{n}_1$ is the restriction of the measure $\tilde{n}_f$ to the sector Y'.

Thus the set $\{z_k\colon z_k\in Y'\setminus U_1\}$ has an angular density, and since $\rho\notin\mathbb{Z}$ the $\Psi^{\pm}$ are entire functions of c.r.g.. The function Ψ_1^- has not have zeros in the domain $\{z\in\mathbb{C}\colon \bar{z}\notin Y'\}$. Therefore, the fact that Ψ_1^- is of c.r.g. in $\mathbb{C}$ implies that the function $(\Psi_1^-)^{-1}$ is of c.r.g. in $\overline{\mathbb{C}}^+$ and that its indicator in $\overline{\mathbb{C}}^+$ equals $-\mathfrak{L}_{\Psi_1^-}(z)$. Taking into account the equality $\Psi_1 = \Psi_1^+/\Psi_1^-$, we conclude that Ψ_1 is a function of c.r.g. in $\overline{\mathbb{C}}^+$. Moreover, in accordance with the properties of entire functions of c.r.g. that were mentioned in **Chapter 1**, we can state that there is a

---

[27)] Recall that $G_q(z)$ is a Weierstrass' primary factor.

C_0-set $E \subset \mathbb{C}^+$ such that

$$\lim_{\substack{z \to_+ \infty \\ z \in \mathbb{C}^+ \setminus E}} \left| |z|^{-\rho} \ln|\Psi_1(z)| - \mathfrak{L}_{\Psi_1}(z/|z|) \right| = 0.$$

Consider now the indicator of the function $\Psi_f(z)$. In accordance with the **Remark** following **Theorem 4.2.7**, we have

$$\mathfrak{L}_{\Psi_f}(z) = \int_{\mathbb{C}^+} K_q(z,\zeta) d\hat{\mu}_{f,1}(\zeta).$$

It is also clear that for the indicators of the functions Ψ_1 and Ψ_2 the following representations hold:

$$\left.\begin{aligned} \mathfrak{L}_{\Psi_1}(z) &= \int_{Y'} K_q(z,\zeta) d\hat{\mu}_{f,1}(\zeta), \\ \mathfrak{L}_{\Psi_2}(z) &= \int_{Y''} K_q(z,\zeta) d\hat{\mu}_{f,1}(\zeta). \end{aligned}\right\} \tag{4.3.41}$$

Consider the set $E_{\varepsilon_1} = \{z \in \mathbb{C}^+ : \left| \ln|\Psi_f(z)| - \mathfrak{L}_{\Psi_f}(z) \right| \geq 4\varepsilon_1 |z|^\rho\}$. This set is clearly contained in the union of the sets

$$E' = \{z \in \mathbb{C}^+ : \left| \ln|\Psi_1(z)| - \mathfrak{L}_{\Psi_2}(z) \right| \geq 2\varepsilon_1 |z|^\rho\}$$

and

$$E'' = \{z \in \mathbb{C}^+ : \left| \ln|\Psi_2(z)| - \mathfrak{L}_{\Psi_2}(z) \right| \geq 2\varepsilon_1 |z|^\rho\}.$$

The the positive homogeneity of the measure $\hat{\mu}_{f,1}$ and the representation (4.3.41) easily imply that the inequality

$$|\mathfrak{L}_{\Psi_2}(z)| \leq \varepsilon_1 |z|^\rho, \ \forall z \in \mathbb{C}^+$$

holds for sufficiently small $\delta > 0$. Hence, in this situation,

$$E'' \subset \{z \in \mathbb{C}^+ : \left| \ln|\Psi_2(z)| \right| \geq \varepsilon_1 |z|^\rho\} = \{z \in \mathbb{C}^+ : |I_{\mu''}(z)| \geq \varepsilon_1 |z|^\rho\},$$

and, as mentioned before, the set E'' can be covered by a system of discs $U(\tilde{z}_j, \tilde{r}_j)$ satisfying the condition (4.3.40). At the same time, by the above-mentioned properties of the function $\Psi_1(z)$, the set E' is a C_0-set. Therefore the union of the sets E' and E'', and hence the set

E_{ε_1}, can be covered by a system of discs $U(\zeta_j, r'_j)$ such that

$$\frac{1}{R} \sum_{j: |\zeta_j| < R} r'_j \le c_1 \left(\frac{\varepsilon}{\varepsilon_1} + \varepsilon \right), \ \forall R > R_0.$$

Since $\varepsilon > 0$ is arbitrary, the last inequality implies that the set E_{ε_1} can be covered by a system of discs $U(\zeta'_j, r''_j)$ such that

$$\lim_{R \to \infty} \frac{1}{R} \sum_{j: |\zeta'_j| < R} r''_j = 0.$$

Thus E_{ε_1} is a C_0-set. Since this is true for any $\varepsilon_1 > 0$, there is a C_0-set $E \subset \mathbb{C}^+$ such that

$$\lim_{\substack{z \to \infty \\ z \in \mathbb{C}^+ \setminus E}} \frac{1}{|z|^\rho} \left| ln|\Psi_f(z)| - \mathfrak{L}_{\Psi_f}(z) \right| = 0.$$

Hence there is a set $E^* \subset \mathbb{R}_+$ of relative measure zero such that

$$\frac{1}{r^\rho} ln|\Psi_f(re^{i\theta})| \rightrightarrows h_{\Psi_f}(\theta)$$

as $r \to \infty$, $r \notin E^*$, $0 < \theta < \pi$. Thus $\Psi_f(z)$ is a function of c.r.g. in $\overline{\mathbb{C}}^+$.

This finishes the proof.

Lemmas 4.3.6 and **4.3.7** state that the functions $\Psi_f(z)$ and $F_f(z)$ are of c.r.g. All the other factors figuring in the representation (4.3.38) of the function $f(z)$ are obviously of c.r.g. Hence, as mentioned above, $f(z)$ is of c.r.g. Thus we have proved the sufficiency part of **Theorem 4.3.1** in the case $\rho \notin \mathbb{Z}$.

4. Sufficiency ($\rho \in \mathbb{N}$). The case $\rho = p \in \mathbb{N}$ can be rather simply reduced to the case of non-integral ρ. Indeed, let $f \in H(\mathbb{C}^+, p]$, $p \in \mathbb{N}$, be a function of c.r.g. in $\mathbb{C}^+$ and let the measures $\hat{\mu}^{[R]}_{f,\partial}$ converge in variation to $\hat{\mu}_{f,\partial}$. Comparing the definitions of regularity of growth in $\mathbb{C}^+$, $\overline{\mathbb{C}}^+$, and $\overline{Y}(\theta_1, \theta_2)$, we conclude that for $f(z)$ being of c.r.g. in $\overline{\mathbb{C}}^+$ it is sufficient that $f(z)$ be of c.r.g. in the closed sectors $\overline{Y}(0, \delta)$ and $\overline{Y}(\pi - \delta, \pi)$ for some $\delta > 0$. Let $\delta = \frac{1}{2p}$. Then note that, as has been proved at the beginning of the previous section a function $f(z)$ is of c.r.g. in the sector $\overline{Y}(0, \frac{1}{2p})$ if and only if the function $\varphi(w) = f(w^{\frac{1}{2p}})$, which belongs to $H\left[\mathbb{C}^+, \frac{1}{2}\right]$, is of c.r.g. in $\overline{\mathbb{C}}^+$. It is obvious that $\varphi(w)$

is a function of c.r.g. in $\mathbb{C}^+$. It is also easy to see that from convergence in variation of the measures $\mu^{[R]}_{f,\partial}$ to the measure $\hat{\mu}_{f,\partial}$ and the same statement concerning the measures induced by the function $ln|f(te^{i\theta})|$ (see the remark at the end of §4.2) it follows that the $\mu^{[R]}_{\varphi,\partial}$ converge in the variation to $\hat{\mu}_{\varphi,\partial}$ as $R \to \infty$. Thus, the conditions of the theorem are satisfied for the function $\varphi(w)$ with $\rho = \frac{1}{2}$ and, therefore, according to the statement proved in the previous section, $\varphi(w)$ is of c.r.g. in $\overline{\mathbb{C}}^+$. Hence $f(z)$ is of c.r.g. in $\overline{Y}\left(0,\frac{1}{2p}\right)$. The fact that $f(z)$ is of c.r.g. in $\overline{Y}\left(\pi-\frac{1}{2p},\pi\right)$ can be proved in a similar way. Thus $f(z)$ is of c.r.g. in $\overline{\mathbb{C}}^+$, which finishes the proof of the theorem.

5. Bounded functions. Functions bounded in $\mathbb{C}^+$ are often met in problems of complex analysis and its applications. For example, Fourier transforms of functions in $L^1(\mathbb{R})$ which have supports in $\mathbb{R}_+$ are bounded functions in $\mathbb{C}^+$. To use these functions we often need to know their asymptotic behaviour as $z \to \infty$. Here we shall show that such functions, as well as functions belonging to some certain wider classes, are of c.r.g. in $\overline{\mathbb{C}}^+$. Moreover, the corresponding "exceptional" sets are thinner than it is dictated by the definition and properties of general functions of c.r.g.

A set $E \subset \mathbb{C}^+$ is said to be a set of finite field of view there is an most countable family of discs $U(z_j,r_j)$, $j = 1, 2, \dots$, such that $E \subset \bigcup_j U(z_j,r_j)$ and $\sum_j (r_j/|z_j|) < \infty$. It is clear that a set of finite field view is a C_0-set, but the converse is false.

The following lemma is the main element in the proof the above-mentioned facts.

Lemma 4.3.8. Let a real-valued measure τ, defined on $\overline{\mathbb{C}}^+$, satisfy the conditions: *supp*$\tau \subset \overline{\mathbb{C}}^+\backslash U_1$ and, for some $p\in\mathbb{N}$, the inequality

$$\int_0^\infty \frac{d\|\tau\|(t)}{t^p} < \infty. \tag{4.3.42}$$

Then there is a set of finite field of view E such that

$$\lim_{\substack{z \to \infty \\ z\in\mathbb{C}^+\backslash E}} \frac{1}{|z|^p} \int_{\overline{\mathbb{C}}^+} |K_{p-1}(z,\zeta)|d|\tau|(\zeta) = 0.$$

Proof. To prove the lemma it clearly suffices to show that for each $\varepsilon > 0$ the set

$$E_\varepsilon = \left\{ z\in\mathbb{C}^+ : \int_{\overline{\mathbb{C}}^+} |K_{p-1}(z,\zeta)|d|\tau|(\zeta) > \varepsilon |z|^p \right\}$$

is a set of finite field of view. Note that in accordance with **Lemma 4.3.4** for $R > 0$ sufficiently large the inclusion $E_\varepsilon \setminus U_R \subset \Delta(c\varepsilon)$ holds, where c is a constant and, as above, $\Delta(\varepsilon)$ is the set of ε-normal points of the measure τ.

At the same time, in view of the **Remark** following **Lemma 4.3.4**, from (4.3.42) it follows that $\Delta(\varepsilon)$ can be covered by a countable system of discs $U(z_j, r_j)$, $j = 1, 2, \ldots$, such that $\sum_j (r_j/|z_j|) < \infty$. Therefore, in this situation $\Delta(\varepsilon)$ is a set of finite field of view. Hence E_ε is also a set of finite field of view. This implies the existence of the set E.

The proof is finished.

Using **Lemma 4.4.8** we can easily establish the following statement, which can be considered as the analogue of **Lemma 4.3.5** for integral order.

Theorem 4.3.2. Let a function $f\in H(\mathbb{C}^+, p]$ satisfy the condition

$$\int_{\overline{\mathbb{C}}^+} \frac{d|\tau_f|(\zeta)}{|\zeta|^p} < \infty . \tag{4.3.43}$$

Then f is a function of c.r.g. in $\overline{\mathbb{C}}^+$ and, moreover,

$$\frac{1}{|z|^p} \, ln|f(z)| - \mathfrak{L}_f\left(\frac{z}{|z|}\right) \to 0$$

as $z \to \infty$ outside a set of finite field of view. Moreover, $\mathfrak{L}_f(z) = \mathrm{const}\cdot Im z^p$.

Proof. As in the proof of **Lemma 4.3.5**, here it suffices to consider the case

$$f(z) = \left\{ \prod_{|z_k|<1} D_p(z, z_k) \right\} \cdot exp\left\{ \frac{1}{\pi i} \int_{|t|>1} \frac{z^{p+1}}{t^{p+1}(t-z)} \, d\mu_{f,\partial}(t) \right\}$$

and, correspondingly,

$$\ln|f(z)| = \int_{\mathbb{C}^+} K_p(z,\zeta)d\tau_f(\zeta).$$

Note that since condition (4.3.42) is satisfied for any $z \neq z_k$, $k = 1, 2, \dots$, the integral

$$\int_{\mathbb{C}^+} K_{p-1}(z,\zeta)d\tau_f(\zeta)$$

exists and converges absolutely. Hence

$$\ln|f(z)| = \int_{\mathbb{C}^+} K_p(z,\zeta)d\tau_f(\zeta) = \int_{\mathbb{C}^+} K_{p-1}(z,\zeta)d\tau_f(\zeta) +$$

$$+ \int_{\mathbb{C}^+} [K_p(z,\zeta)-K_{p-1}(z,\zeta)]d\tau_f(\zeta) = \int_{\mathbb{C}^+} K_{p-1}(z,\zeta)d\tau_f(\zeta) + 2\kappa_f^+ Im z^p,$$

where κ_f^+ is as above[28] (see **Theorem 4.2.8**).

From this equality and **Lemma 4.3.8** it follows that

$$|z|^{-p}\ln|f(z)| - 2\kappa_f^+ sin(p\arg z) \to 0$$

as $z \to \infty$ outside a set of finite field of view. Hence $h_f(\theta) = \kappa_f^+ sin p\theta$, $\mathfrak{L}_f(z) = 2\kappa_f^+ Im z^p$, and $f(z)$ is of c.r.g. in $\overline{\mathbb{C}}^+$.

The proof of the theorem is finished.

The case $p = 1$ is of most interest, because we can replace condition (4.3.43) by one in which only the measure $\mu_{f,\partial}$ figures. More exactly, in this condition the positive component $\mu_{f,\partial}^+$ of the standard representation $\mu_{f,\partial} = \mu_{f,\partial}^+ - \mu_{f,\partial}^-$ figures.

Theorem 4.3.3. Let $f \in H(\mathbb{C}^+,1]$ be a function such that

$$\int_{|t|>1} t^{-2}d\mu_{f,\partial}^+(t) < \infty . \qquad (4.3.44)$$

Then f is of c.r.g. in $\overline{\mathbb{C}}^+$, its indicator is proportional to $sin\theta$ and there is a set of finite field of view $E \subset \mathbb{C}^+$ such that

---

[28] The existence of κ_f^+ clearly follows from condition (4.3.42).

$$\lim_{\substack{z\to\infty \\ z\in\mathbb{C}^+\setminus E}} \left|\, |z|^{-1} \ln|f(z)| - \mathcal{L}_f(z/|z|) \,\right| = 0.$$

Moreover, the zeros $z_k = r_k e^{i\theta_k}$ of $f(z)$ satisfy the condition

$$\sum_k \frac{\sin\theta_k}{r_k} < \infty.$$

Proof. It follows from **Theorem 4.3.2** that the theorem will be proved if we show that the function $f(z)$ satisfies the condition

$$\int_{\overline{\mathbb{C}}^+} \frac{d|\tau_f|(\zeta)}{|\zeta|} < \infty. \tag{4.3.45}$$

First we prove that

$$\int_{|t|>1} t^{-2} d|\mu_{f,\partial}|(t) < \infty.$$

To show this we use the generalized Carleman's formula once more. We have

$$\int_{1<|t|<R/2} t^{-2} d\mu^-_{f,\partial}(t) \le \frac{4}{3} \int_{1<|t|<R} \left(\frac{1}{t^2} - \frac{1}{R^2} \right) d\mu^-_{f,\partial}(t) \le$$

$$\le \frac{4}{3} \int_{1<|t|<R} \left(\frac{1}{t^2} - \frac{1}{R^2} \right) d\mu^+_{f,\partial}(t) + \frac{4}{3}\,\frac{1}{\pi R} \int_0^\pi \ln|f(Re^{i\theta})|\sin\theta d\theta + c(R),$$

where $c(R) = O(1)$, $R \to \infty$. Taking into account that $f(z)\in H(\mathbb{C},1]$ and that condition (4.3.44) is satisfied, we conclude that

$$\int_{|t|>1} t^{-2} d\mu^-_{f,\partial}(t) < \infty.$$

Comparing this inequality with (4.3.42) we obtain

$$\int_{|t|>1} t^{-2} d|\mu_{f,\partial}|(t) < \infty. \tag{4.3.46}$$

Recall that $\tau_f = \mu_{f,1} + \mu_{f,2}$, where the measure $\mu_{f,2}$, which is concentrated on $\mathbb{C}^+\setminus U_1$, is defined by the equality $d\mu_{f,1} = \sin(\arg z) d\tilde{n}_f(z)$ and the measure $\mu_{f,2}$, which is concentrated on $\mathbb{R}\setminus[-1,1]$, is defined by the equality $d\mu_{f,2}(x) = |x|^{-1} d\mu_{f,\partial}(x)$. Thus, condition (4.3.45) is equivalent to

$$\int_{|t|\geq 1} \frac{d|\mu_{f,2}|(t)}{|t|} < \infty.$$

Therefore to prove inequality (4.3.44) it is sufficient to show that

$$\int_{\mathbb{C}^+} \frac{d|\mu_{f,1}|(\zeta)}{|\zeta|} < \infty,$$

or, equivalently,

$$\sum_{1<r_k<\infty} \frac{\sin\theta_k}{r_k} < \infty,$$

where the $z_k = r_k e^{i\theta_k}$ are the zeros of $f(z)$.

Using Carleman's formula once more we obtain

$$\sum_{1<r_k<R/2} \frac{\sin\theta_k}{r_k} \leq \frac{4}{3} \sum_{1<r_k<R} \left(\frac{1}{r_k} - \frac{r_k}{R^2} \right) \sin\theta_k \leq$$

$$\leq \frac{4}{3}\frac{1}{\pi R} \int_0^{\pi} \ln|f(Re^{i\theta})|\sin\theta d\theta + \frac{4}{3} \int_{1<|t|<R} \frac{1}{t^2}\, d|\mu_{f,\partial}|(t) + O(1),\ R \to \infty.$$

Taking into account (4.3.46) we immediately conclude that

$$\sum_{1<r_k<\infty} \frac{\sin\theta_k}{r_k} < \infty. \tag{4.3.47}$$

The proof is finished.

Corollary. If an entire function $f \in H(\mathbb{C},1]$ satisfies the condition

$$\int_{-\infty}^{\infty} \frac{\ln^+|f(x)|}{1+x^2}\, dx < \infty$$

(in particular if $\sup|f(x)| < \infty$) then it is of c.r.g. and $h_f(0) = h_f(\pi) = 0$.

In conclusion we consider the case of functions $f(z)$ which are analytic in the closed half-plane $\overline{\mathbb{C}}^+$. For such functions, as for entire functions, $d\mu_{f,\partial}(x) = \ln|f(x)|dx$, while if $|f(z)| \leq c_1 \exp\{c_2|z|\}$, $\forall z \in \mathbb{C}^+$, then the indicator $h_f(\theta)$ is also defined at the points $\theta = 0$ and $\theta = \pi$. Besides, if we additionally assume that $f(z)$ is of c.r.g. in $\overline{\mathbb{C}}^+$, then, as follows from **Theorem 4.3.1**, the equalities $h_f(+0) = h_f(0)$,

$h_f(\pi-0) = h_f(\pi)$ are valid.

Theorem 4.3.4. If a function $f\in H(\overline{\mathbb{C}}^+)$ satisfies the conditions

$$h_f(0) + h_f(\pi) = 0, \tag{4.3.48}$$

$$|f(z)| \le c_1 e^{c_2|z|}, \quad \forall z\in\mathbb{C}^+,$$

and

$$\exists \lim_{R\to\infty} \int_{1<|x|<R} \frac{ln|f(x)|}{x^2}\,dx,$$

then $f\in H(\mathbb{C}^+,1]$, and $f(z)$ is of c.r.g. in $\overline{\mathbb{C}}^+$.

Proof. First of all we note that

$$\lim_{R\to\infty} \frac{1}{R^2} \int_{1<|x|<R} ln|f(x)|dx = 0. \tag{4.3.49}$$

Indeed, setting

$$A = \lim_{R\to\infty} \int_{1<|x|<R} \frac{ln|f(x)|}{x^2}dx, \tag{4.3.50}$$

we have

$$\lim_{R\to\infty} \frac{1}{R^2} \int_{1<|x|<R} ln|f(x)|dx = \lim_{R\to\infty} \frac{1}{R^2}\int_1^R ln|f(x)f(-x)|dx =$$

$$= \lim_{R\to\infty} \frac{1}{R^2}\left\{ R^2 \int_1^R \frac{ln|f(x)f(-x)|}{x^2}\,dx - 2\int_1^R tdt \int_1^t \frac{ln|f(x)f(-x)|}{x^2}\,dx \right\} =$$

$$= A - 2\lim_{R\to\infty} \frac{1}{R^2}\int_1^R t(A+o(1))dt = \lim_{R\to\infty} \frac{1}{R^2}\int_1^R o(t)dt = 0.$$

By (4.3.49) and (4.3.50) it follows that

$$\lim_{R\to\infty} \int_{1\le|x|\le R} \left(\frac{1}{x^2} - \frac{1}{R^2}\right) ln|f(x)|dx = A\ . \tag{4.3.51}$$

From this using Carleman's formula, we find that

$$\lim_{R\to\infty} \frac{1}{R} \int_0^{\pi} |ln|f(Re^{i\theta})||sin\theta d\theta < \infty. \qquad (4.3.52)$$

Hence $f \in H(\mathbb{C}^+, 1]$.

We now consider the function $\varphi(z) = e^{az}$, $a \in \mathbb{C}$. It can be immediately verified that $\varphi(z)$ satisfies the conditions of the theorem. Therefore the function $F(z) = f(z)e^{az}$ also satisfies these conditions, and hence, according to the statement proved above, $F \in H(\mathbb{C}^+, 1]$. It is obvious that $f(z)$ is a function of c.r.g. in $\overline{\mathbb{C}}^+$ if $F(z)$ of c.r.g. in $\overline{\mathbb{C}}^+$ for even one $a \in \mathbb{C}$. We choose a value "a" in such a way that $h_{\varphi}(0) = -h_f(0)$. Then $h_F(0) = 0$, $h_F(\pi) = 0$. Moreover, since the indicator $h_F(\theta)$ is trigonometrically convex on $[0, \pi]$, we have $h_F(+0) + h_f(\pi - 0) \geq 0$, $h_F(+0) \leq h_F(0)$ and $h_F(\pi - 0) \leq h_F(\pi)$. Hence $h_f(+0) = 0$, $h_F(\pi - 0) = 0$ or, equivalently, $\hat{\mu}_{F,\partial} = 0$.

Now, according to **Theorem 4.3.1**, in order that $F(z)$ be a function of c.r.g. in $\overline{\mathbb{C}}^+$ it is sufficient that

$$\lim_{R\to\infty} \frac{1}{R^2} \int_{1\leq|x|\leq R} |ln|F(x)||dx = 0. \qquad (4.3.53)$$

Since $h_f(0) = h_F(\pi) = 0$, we have

$$ln|F(x)| \leq \varepsilon|x|, \ \forall x: \ |x| > x_0(\varepsilon).$$

Therefore

$$\lim_{R\to\infty} \frac{1}{R^2} \int_{1\leq|x|\leq R} ln^+|F(x)|dx = 0,$$

and hence in this situation condition (4.3.53) is equivalent to the equality

$$\lim_{R\to\infty} \frac{1}{R^2} \int_{1\leq|x|\leq R} ln|F(x)|dx = 0,$$

which is valid for any function that satisfies the conditions of the theorem (see (4.3.49)).

This fact implies that the function $F(z)$, and therefore the original function $f(z)$, are of c.r.g. in $\overline{\mathbb{C}}^+$.

The proof is finished.

The following theorem is the converse of **Theorem 4.3.4**.

Theorem 4.3.5. If $f \in H(\mathbb{C}^+, 1] \cap H(\overline{\mathbb{C}}^+)$ is a function of c.r.g. in $\overline{\mathbb{C}}^+$ and its zeros $r_k = r_k e^{i\theta_k}$, k=1,2,... , satisfy condition (4.3.47), then

$$h_f(0) + h_f(\pi) = 0$$

and

$$\exists \lim_{R\to\infty} \int_{1<|x|<R} \frac{ln|f(x)f(-x)|}{x^2}\,dx. \tag{4.3.54}$$

Proof. Since f is of c.r.g. in $\overline{\mathbb{C}}^+$, the limit

$$\lim_{\substack{R\to\infty\\ R\notin\Lambda_f}} \frac{1}{R}\int_0^\pi ln|f(Re^{i\theta})|sin\theta d\theta = \int_0^\pi h_f(\theta)sin\theta d\theta \tag{4.3.55}$$

exists (see **Theorem 4.2.7**) and (see (4.2.17)) the equality

$$\int_0^\pi sin\theta d\hat{\nu}_f(\theta) = \frac{1}{2\pi}\,(h_f(+0) + h_f(\pi-0))$$

holds. Taking into account that f(z) is of c.r.g. in $\overline{\mathbb{C}}^+$, we obtain

$$\int_0^\pi sin\theta d\hat{\nu}_f(\theta) = \frac{1}{2\pi}(h_f(0) + h_f(\pi)). \tag{4.3.56}$$

By (4.3.47) and (4.3.55) according to Carleman's formula it follows that

$$\lim_{R\to\infty} \int_{1\le|x|\le R} \left(\frac{1}{x^2} - \frac{1}{R^2}\right) ln|f(x)|dx$$

exists. In turn, this fact easily implies the existence of the limit (4.3.54).

Further, by (4.3.47) it follows that

$$\lim_{R\to\infty} \frac{\|\mu_{f,1}\|(R)}{R} = 0.$$

From this taking into account that f(z) is of c.r.g. in $\overline{\mathbb{C}}^+$ and therefore $\mu_{f,1}^{[R]} \to \hat{\mu}_{f,1}$ as $R \to \infty$, we find that

$$\|\hat{\mu}_{f,1}\|(1) = \lim_{R\to\infty} \|\hat{\mu}_{f,1}^{[R]}\|(1) = \lim_{R\to\infty} \frac{\|\hat{\mu}_{f,1}\|(R)}{R} = 0.$$

Thus $\hat{\nu}_f = 0$ on $(0,\pi)$,

$$\int_0^\pi \sin\theta d\hat{\nu}_f(\theta) = 0,$$

and therefore (4.3.56) implies (4.3.48).

The proof of the theorem is finished.

§4 Functions of completely regular growth in a cone

The method used in §1-§3 of this chapter for the study of the asymptotic behaviour of functions in a cone can also be applied to construct of the theory of subharmonic functions of finite order, which includes functions of c.r.g. in a cone of $\mathbb{R}^n$. The results obtained in this way will be presented without proofs, since with regard to the basic ideas these proofs are close to those of the corresponding facts for functions $f \in H(\mathbb{C}, \rho]$. However, sometimes the technique differences are essential. Note also that the mathematical tools used for studying functions in a cone is less elementary than in §1-§3.

1. Boundary measure. We consider spherical coordinates (r, θ) in $\mathbb{R}^n$, i.e. we set

$$r = |x|, \ x = (x_1, \dots, x_n) \in \mathbb{R}^n,$$

$$x_1 = r\cos\theta_1, \ x_2 = r\cos\theta_2 \sin\theta_1, \ \dots,$$

$$x_{n-1} = r\cos\theta_{n-1}\sin\theta_{n-2}\dots\sin\theta_1,$$

$$x_n = r\sin\theta_{n-1}\dots\sin\theta_1.$$

In these coordinates the Laplace Operator takes the form

$$\Delta = \frac{\partial^2}{\partial r^2} + \frac{n-1}{r}\frac{\partial}{\partial r} + \frac{1}{r^2}\Delta^*,$$

where Δ^* is the so-called spherical part of the Laplace Operator[29].

---

[29] $$\Delta^* = \left\{\frac{\partial^2}{\partial\theta_1^2} + (n-2)ctg\theta_1 \frac{\partial}{\partial\theta_1}\right\} + \frac{1}{\sin^2\theta_1}\left\{\frac{\partial^2}{\partial\theta_2^2} + (n-3)ctg\theta_2 \frac{\partial}{\partial\theta_2}\right\} + \dots$$

Let Γ be a domain on the sphere $S_1 \subset \mathbb{R}^n$ whose boundary is twice continuously differentiable. Then we consider the following boundary value problem on Γ:

$$\Delta^* \varphi + \lambda\varphi = 0, \ \varphi\big|_{\partial\Gamma} = 0. \tag{4.4.1}$$

We denote by $\lambda_j = \lambda_j^\Gamma$, $0 < \lambda_1 < \lambda_2 \le \lambda_3 \ldots$, the eigenvalues of this problem and by $\varphi_j = \varphi_j^\Gamma$ the corresponding eigenfunctions. We may assume that these functions are normalized by the condition $\int_{S_1} |\varphi_j|^2 dS_1 = 1$, where dS_1 is the (n-1)-dimensional volume element of the sphere S_1 induced by the Euclidean metric. Further, the domain Γ is supposed to be such that $\varphi_j \in C^2(\bar\Gamma)$, $j = 1, 2, \ldots$. Besides, for the sake of being specific we assume that $\frac{\partial\varphi_1}{\partial n} > 0$. Here and below, $\frac{\partial}{\partial n}$ denotes the interior normal derivative.

We set $K = K^\Gamma = \{x \in \mathbb{R}^n: \ x = ty, \ y \in \Gamma, \ t > 0\}$. Then we extend each function φ_j as a positively homogeneous function of degree 0 from $\bar\Gamma$ to the closure $\bar K$ of K. We reserve the notation φ_j for these functions. We set

$$\kappa_j^\pm = \frac{1}{2}\left(-n+2 \pm \sqrt{(n-2)^2 + 4\lambda_j}\right), \quad j = 1, 2, \ldots . \tag{4.4.2}$$

Then each function $|x|^{\kappa_j^\pm}\varphi_j(x)$ is harmonic in K, belongs to $C^2(K\setminus\{0\})$ and vanishes on $\partial K\setminus\{0\}$. For brevity we write φ instead of φ_1, and $\kappa^\pm$ instead of $\kappa_1^\pm$ [30].

Now we denote by $d\sigma$ and dS_R the (n-1)-dimensional volume elements induced by the Euclidean metric on ∂K and S_R, respectively. We also set

$$\Gamma_R = S_R \cap K = R\Gamma, \ K_R = K \cap B_R,$$

$$B_{r,R} = \{x \in \mathbb{R}^n: \ r < |x| < R\},$$

$$+ \left(\prod_{j=1}^{n-3} \sin^{-2}\theta_j\right) \cdot \left\{\frac{\partial^2}{\partial\theta_{n-1}^2} + ctg\theta_{n-2}\frac{\partial}{\partial\theta_{n-2}}\right\} + \prod_{j=1}^{n-2} \sin^{-2}\theta_j \frac{\partial^2}{\partial\theta_{n-1}^2} .$$

[30] Note that if n = 2, $K = \mathbb{C}^+$, then $\varphi(re^{i\theta}) = \sin\theta$ and $\kappa^\pm = \pm 1$.

$$K_{r,R} = K\cap B_{r,R}\ ,\ \Gamma_{r,R} = \partial K\cap B_{r,R}\ .$$

We supplement the information about derivatives of subharmonic functions given in §1 of chapter 2 by the following lemma, which will be proved by applying Green's formula[31] with a special choice of the functions figuring in it.

Lemma 4.4.1. Let $G\subset\mathbb{R}^n$ be a domain containing the sets $\bar{\Gamma}_r$ for all $r\in(a,b)$. Let $u_j\in SH(G)\cap C^{\infty}(G)$ be a monotonically decreasing sequence of functions converging to u as $j\to\infty$. Then for any $r\in(a,b)$ such that $\mu_u(\bar{\Gamma}_r) = 0$ the following limit exists:

$$\lim_{j\to\infty}\int_{\Gamma_r}\frac{\partial u_j}{\partial r}\,\eta dS_r \overset{\text{def}}{=} \int_{\Gamma_r}\frac{\partial u}{\partial r}\,\eta dS_r,\ \forall\eta\in C^1(S_r),\ \operatorname{supp}\eta\subset\bar{\Gamma}_r.$$

This limit is independent of the choice of the sequence $\{u_j\}$. Besides, if $\eta\in D(K_{a,b})$, then for almost all $r\in(a,b)$ the equality

$$\lim_{j\to\infty}\int_{\Gamma_r}\frac{\partial u_j}{\partial r}\,\eta dS_r = \int_{\Gamma_r}\Phi\eta dS_r$$

holds where $\Phi(x)$ is a function which is locally summable in K and which, considered as a distribution, coincides with the distribution

$$\sum_{i=1}^{n}\frac{\partial u}{\partial x_i}\,\frac{x_i}{|x|}\,.$$

Similarly to the case of holomorphic functions in $\mathbb{C}^+$ Carleman's formula, more exactly its analogue for the case under consideration, plays an important role in the construction of the theory of subharmonic functions in a cone.

First we give an the analogue of the usual (i.e. not generalized) Carleman's formula.

Lemma 4.4.2. Let G be a domain containing the set $\bar{K}_{r,R}$, $0 < r < R < \infty$, and set $\Psi_R(x) = (|x|^{\kappa^-} - R^{\kappa^- -\kappa^+}|x|^{\kappa^+})\varphi(x) \overset{\text{def}}{=} \gamma_R(x)\varphi(x)$. If $\mu_u(\bar{\Gamma}) =$

---

[31] I.e. the ordinary Green's formula, i.e. the equality

$$\int_G (u\Delta v - v\Delta u)d\omega = -\int_{\partial G}\left(u\,\frac{\partial v}{\partial n} - v\,\frac{\partial u}{\partial n}\right)dS \qquad (4.4.3)$$

0, then

$$\theta_n \int_{K_{r,R}} \Psi_R d\mu_u = \gamma_R(r)\int_{\Gamma_r} \varphi \frac{\partial u}{\partial r}\, dS_r + \gamma_R'(r)\int_{\Gamma_r} u\varphi dS_r +$$

$$- \gamma_R'(R) \int_{\Gamma_R} u\varphi dS_r + \int_{\Gamma_{r,R}} u\gamma_R(|x|)\, \frac{\partial\varphi}{\partial n}\, d\sigma \,. \tag{4.4.4}$$

This lemma can be proved for twice continuously differentiable functions u(x) with the aid of Green's formula (4.4.3), and the general case can be reduced to this one by approximation of the original function by infinitely differentiable functions.

We denote by $SH^*(K)$ the set of functions $u\in SH(K)$ that are bounded on bounded subsets of the cone K. We say that the cone $K = K^{\Gamma}$ satisfies the condition of "possibility of displacement into the cone $K' = K^{\Gamma'}$", where $\Gamma \subseteq \Gamma'$, or, simply, "the cone K can be displaced into K'", if there is $x^0\in K$ such that for any $t>0$ and any $y\in K$ the point $y+tx^0$ belongs to K'. In this situation we denote by $u_h(x)$ the function $u(x+hx^0)$.

Using **Lemma 4.4.2** in a manner similar to the use of Carleman's formula in the previous sections, it is easy to show that the following statement is true.

Lemma 4.4.3. Let $u\in SH^*(K')$, suppose the cone K can be displaced into K', $K \subseteq K'$, and set $\Delta^{\delta}_{r,R} = \{x\in K_{r,R} : \varphi(x) < \delta\}$. Then for any r and R, $0 < r < R < \infty$, there are constants $c_j = c_j(u,r,R)$, $j = 1, 2, 3$, and $\delta_1 = \delta_1(K)$ such that

$$\int_{K_{r,R}} |u(x)|\varphi(x)d\omega_x \le c_1,$$

$$\int_{K_{r,R}} \varphi(x)d\mu_u(x) \le c_2,$$

$$\int_{\Delta^{\delta}_{r,R}} |u(x)|d\omega_x \le \delta c_3 \,, \ \forall\delta\in(0,\delta_1).$$

These estimates give the possibility of using in the study of limit transition for $u_h(x)$ the fact that a bounded set in L_1 is a weakly compact. Relying on this and using, as before, Green's formula we can obtain the following lemma. which includes, in particular, a statement concerning the existence of a boundary measure.

Lemma 4.4.4 ("Lemma on limit transitions"). Suppose the cone K can be "displaced" into a cone K', $K \subseteq K'$, and let $u \in SH^*(K')$. Then the following statements are true:

a) $$\lim_{h\to 0} \int_{K_{r,R}} u_h(x)Q(x)d\omega_x = \int_{K_{r,R}} u(x)Q(x)d\omega_x, \quad \forall Q \in C(\overline{K}_{r,R});$$

b) for any $r > 0$ and $R > 0$ outside the set $\Lambda_u^1 = \{t: \mu_u(\Gamma_t) > 0\}$ the equality

$$\lim_{h\to 0} \int_{K_{r,R}} \varphi(x)Q(x)d\mu_h(x) = \int_{K_{r,R}} \varphi(x)Q(x)d\mu_u(x), \quad \forall Q \in C^1(\overline{K}_{r,R})$$

holds;

c) the functions $u_h(x)$, considered as functionals on the space of functions that belong to $C(\partial K)$ and have compact supports in $\partial K\setminus\{0\}$, weakly converge, as $h \to 0$, to a measure $\mu_{u,\partial}$, the boundary measure of $u(x)$;

d) outside of an most countable set $\Lambda_u^2 = \{r: |\mu_{f,\partial}|(\partial\Gamma_r) \neq 0\}$ the functions $u_h(x)\varphi(x)$, considered as functionals on the space $C(\overline{\Gamma}_r)$ for all r converge weakly, as $h \to 0$, to a measure $\mu_{f,r}$ on $\overline{\Gamma}_r$;

e) if $r \notin \Lambda_u^1 \cup \Lambda_u^2$ and $h \to 0$ outside the set of values such that $\mu_{u_h}(\overline{\Gamma}_r) \neq 0$, then for any function $\eta \in C^2(\overline{\Gamma}_r)$ the following limit exists:

$$\lim_{h\to 0} \int_{\Gamma_r} \varphi\eta \frac{\partial u_h}{\partial r} dS_r \overset{\text{def}}{=} Q(\eta;r,u,K) = Q(\eta;r,u).$$

Note that $d\mu_{u,\partial} \le u^* d\sigma$ where $u^* = u^*(x) = \lim\limits_{\substack{x'\to x \\ x'\in K}} u(x')$. If, in addition the function u is subharmonic in a neighbourhood of a set $E \subset \partial K$ that is open in the topology of ∂K then on E the equality $d\mu_{u,\partial} = ud\sigma$ holds. Moreover, if $r \notin (\Lambda_u^1 \cup \Lambda_u^2)$ then $d\mu_{u,\partial} \le u\varphi dS_r$ on $\overline{\Gamma}_r$ and there is a set $\Lambda_u^3 \subset \mathbb{R}_+$ of Lebesgue measure zero such that $d\mu_{u,\partial} = u\varphi dS_r$ on $\overline{\Gamma}_r$.

We set $\Lambda_u = \Lambda_u^1 \cup \Lambda_u^2 \cup \Lambda_u^3$. This set has Lebesgue measure zero. It depends on the choice of the function u and the cone K.

The concept of boundary measure $\mu_{u,\partial}$ and functional Q can also be introduced in case K = K' and K is "non-displacable" into itself. To do this we represent the domain Γ generating the cone K as a finite union of domains $\Gamma^{(j)} \subset S_1$, j = 1,...,N, such that $\Gamma^{(j)}$ satisfies the same conditions as Γ and such that the cone $K^{(j)} = K^{\Gamma^{(j)}}$ can be "displaced" into K for any j = 1, ..., N. It is easy to see that such a representation is possible. We set $\Gamma^{(j)} \cup (\partial\Gamma^{(j)} \cap \partial\Gamma) = \tilde{\Gamma}^{(j)}$ and consider the functions η_j, j = 1,..., N, satisfying the conditions: 1) $\eta_j \in C^2(\overline{\Gamma})$, 2) $supp\eta_j \subset \tilde{\Gamma}^{(j)}$, and 3) $\eta_1 + \ldots + \eta_N = 1$. We extend η_j from Γ to K as a positively homogeneous function of degree 0. We denote by η_j these extensions.

The conditions of **Lemma 4.4.4** are satisfied for a function $u \in SH^*(K)$ and the cones K and $K^{(j)}$. Therefore the boundary measure $\mu_{u_j,\partial}$ of the function $u_j = u|_{K^{(j)}}$ is defined on $\partial K^{(j)} \setminus \{0\}$. Besides, there are corresponding sets $\Lambda_{u_j}^1$, $\Lambda_{u_j}^2$, $\Lambda_{u_j}^3$ and hence $\Lambda_{u_j} = \Lambda_{u_j}^1 \cup \Lambda_{u_j}^2 \cup \Lambda_{u_j}^3$ is defined as well. Without loss of generality we may assume that $\mu_{u_j,\partial}(E) = 0$ for any set E belonging to the boundary (in the topology of ∂K) of the set $\partial K^{(j)} \cap \partial K$. We define the measure $\mu_{u,\partial}$ on $\partial K \setminus \{0\}$ by the equality

$$\mu_{u,\partial}(E) = \sum_{j=1}^{N} \int_E \eta_j d\mu_{u_j,\partial}, \quad E \subset\subset \partial K \setminus \{0\}$$

or, equivalently, by the equality

$$\mu_{u,\partial}(E) = \sum_{1 \le j \le N} \mu_{u_j,\partial}(E_j), \quad E \subset\subset \partial K \setminus \{0\},$$

where $\{E_j\}$ is a partition of the set E such that $E_j \subset \partial K^{(j)}$, j = 1,...,N.

It can be shown that the measure defined above is independent of the choice of the sets $\Gamma^{(j)}$ and, correspondingly, of the functions η_j. We $\Lambda_u = \bigcup_j \Lambda_{u_j}$, $\Lambda_u^{(i)} = \bigcup_j \Lambda_{u_j}^{(i)}$, i = 1, 2, 3. We denote by $x^{(j)}$ a vector in whose direction the "displacement" of the cone $K^{(j)}$ is realized. From the construction of the measure $\mu_{u,\partial}$ it follows that for any function $\psi\in C(\bar\Gamma_{r,R}\cap\bar\Gamma^{(j)}_{r,R})$, $1\le j\le N$, if $r\in\Lambda^2_u$, then

$$\lim_{h\to 0}\int_{\Gamma_{r,R}\cap\Gamma^{(j)}_{r,R}}\psi(x)u(x+hx^{(j)})d\sigma = \int_{\Gamma_{r,R}\cap\Gamma^{(j)}_{r,R}}\psi(x)d\mu_{u,\partial}(x).$$

In this sense the measure $\mu_{u,\partial}$ is locally the boundary value of the function $u\in SH^*(K)$.

Now we define the functional Q for a "non-displacable" cone K. We denote by $Q^{(j)}(\eta;r,u)$, $\eta\in C^2(\bar\Gamma_r)$, $r\notin\Lambda^2_{u_j}$, the corresponding functional for a function $u\in SH^*(K)$ and the cone $K^{(j)}$, i.e. the functional $Q(\eta:r,u,K^{(j)})$ (which exists by **Lemma 4.4.4 e)**). We denote by $\varphi^{(j)}$ the first eigenfunction of the problem (4.4.1) in the domain $\Gamma^{(j)}$ and its positively homogeneous extension on K of degree 0. For $\eta\in C^2(\bar\Gamma_r)$ and $r\notin\Lambda^2_u$ we set

$$Q(\eta) = Q(\eta;r,u,K) = \sum_{1\le j\le N} Q^{(j)}\left(\frac{\eta_j\varphi}{\varphi^{(j)}}\eta;r,u\right),$$

where the functions η_j and φ are as above. It can be shown that the value $Q(\eta)$ is independent of the choice of the sets $\Gamma^{(j)}$ and the functions η_j.

2. The Measure τ_u, Green's and Carleman's formulas. In the previous section we have associated to each function $u\in SH^*(K)$ its boundary measure $\mu_{u,\partial}$. Besides, to each function $u\in SH^*(K)$ corresponds its Riesz associated measure μ_u. Since these measures are defined on $\partial K\setminus\{0\}$ and K, respectively it is natural to combine them by introducing a measure on $\bar K\setminus\{0\}$ whose its restrictions to $\partial K\setminus\{0\}$ and K would be defined by (and in turn define) the measures $\mu_{u,\partial}$ and μ_u. We introduce such a measure τ_u, assuming that for Borel sets $E\subset\subset\bar K\setminus\{0\}$,

$$\tau_u(E) = \int_{E\cap K}\varphi(x)d\mu_u(x) - \int_{E\cap\partial K}\frac{1}{\theta_n}\frac{\partial\varphi}{\partial n}d\mu_{u,\delta}(x), \qquad (4.4.5)$$

where the constant θ_n is the same as in the definition of Riesz masses (see §1.1 of **Chapter 1**). The fact that $\tau_u(E)$ is finite follows from **Lemma 4.4.3**.

The equality given in the following lemma can be considered as a

Green's formula for subharmonic functions in a cone.

Lemma 4.4.5. Let $u\in SH^*(K)$. Then for any function $\psi\in C^2(\overline{K}_{r,R})$ and any numbers $\lambda\notin\Lambda_u$, $R\notin\Lambda_u$, $0 < \lambda < R$ the equality

$$\theta_n \int_{\overline{K}_{\lambda,R}} \psi d\tau_u - \int_{K_{\lambda,R}} u\Delta(\varphi\psi)d\omega_z =$$

$$= \int_{\Gamma_\lambda} u\varphi \frac{\partial\psi}{\partial r}\Big|_{r=\lambda} dS_\lambda - \int_{\Gamma_R} u\varphi \frac{\partial\psi}{\partial r}\Big|_{r=R} dS_R - Q(\psi;\lambda,u) + Q(\psi;R,u) \quad (4.4.6)$$

holds.

For a function $u(x)\in SH^*(K)\cap C^2(\overline{K}\setminus\{0\})$ the statement of the lemma, i.e. the last equality, is an obvious consequence of the ordinary Green's formula. In case the cone K can be "displaced" into itself, the statement of the lemma can be obtained for an arbitrary function u with the aid of standard approximation of the "displaced" function u_h by smooth subharmonic functions $u_{h,j}$. First we write equality (4.4.6) for the functions $u_{h,j}$, then we pass to the limit as $j \to \infty$ and finally, we pass to the limit as $h \to 0$. The possibility of these limit transitions is guaranteed by **Lemmas 4.4.1** and **4.4.4**. The general case is reduced to the case of a "displacable" cone by partition of the cone K into the cones $K^{(j)}$ figuring in the definition of the measure $\mu_{u,\partial}$ for a "non-displacable" cone.

Applying **Lemma 4.4.5** to the function $\psi(x) = \Psi_R(|x|)$, where the function $\Psi_R(|x|)$ is the same as in **Lemma 4.4.2** we obtain the following statement.

Lemma 4.4.6 (Generalized Carleman's formula for subharmonic functions in a cone). Let $u\in SH^*(K)$. Then for any $\lambda\notin\Lambda_u$ and $R\notin\Lambda_u$ the equality

$$\theta_n \int_{\overline{K}_{\lambda,R}} \Psi_R(|x|)d\tau_u = \Psi_R'(\lambda) \int_{\Gamma_\lambda} u\varphi dS_\lambda +$$

$$- \Psi_R'(R) \int_{\Gamma_R} u\varphi dS_R - \Psi_R(\lambda)Q(1;\lambda,u) \qquad (4.4.7)$$

holds.

The following lemma is useful for the study of functions $u\in SH^*(K)$ in a neighbourhood of the origin. It follows from **Lemma 4.4.5** when $\psi(x) = \tilde{\Psi}_\lambda(|x|)$, where $\tilde{\Psi}_\lambda(t) = t^{\kappa^+} - \lambda^{\kappa^+ - \kappa^-} t^{\kappa^-}$.

Lemma 4.4.7. Let $u \in SH^*(K)$ and let $\lambda \notin \Lambda_u$, $R \notin \Lambda_u$. Then

$$\theta_n \int_{\bar{K}_{\lambda,R}} \tilde{\Psi}_\lambda(|x|) d\tau_u = - \Psi_\lambda'(R) \int_{\Gamma_R} u\varphi dS_R + \\ + \Psi_\lambda'(\lambda) \int_{\Gamma_\lambda} u\varphi dS_\lambda - \tilde{\Psi}_\lambda(R) Q(1;R,u). \tag{4.4.8}$$

Another useful statement can be obtained if we make the additional assumption on that the function u figuring in **Lemma 4.4.5** is positively homogeneous.

Lemma 4.4.8. Let $L(x) \in SH^*(K)$ be a positively homogeneous function of degree $\rho > 0$ which is bounded above on Γ. Then for any function $\psi \in C^2(\bar{K}_1)$ the equality

$$\theta_n \int_{\bar{K}_1} \psi d\tau_L - \int_{K_1} L\Delta(\varphi\psi) d\omega = \rho \int_{\Gamma_1} L\psi\varphi dS_1 + \int_{\Gamma_1} L\varphi \frac{\partial\psi}{\partial n} dS_1 \tag{4.4.9}$$

holds.

Note that for $\rho = \kappa^+$ and $\psi = |x|^{\kappa^+}$ the equality (4.4.9) assumes the following form

$$\int_{K_1} |x|^{\kappa^+} \varphi(x) d\mu_L = \frac{1}{\theta_n} \int_{\Gamma_{0,1}} |x|^{\kappa^+} \frac{\partial\varphi}{\partial n} d\mu_{L,\partial} .$$

It follows that

$$\int_{\partial\Gamma} \frac{\partial\varphi}{\partial n} d\nu \ge 0,$$

where ν is the spherical component of the positively homogeneous measure $\mu_{L,\partial}$ [32]. In particular, if $n = 2$, and hence $L(z) = r^\rho h(\theta)$, the well-known inequality $h(\theta) + h\left(\theta + \frac{\pi}{\rho}\right) \ge 0$, which is valid for any ρ-trigonometrically convex , 2π-periodic function $h(\theta)$, is obtained.

---

[32] I.e. the measure ν is defined by the equality

$$d\mu_{L,\partial} = \frac{1}{\kappa^+ + n - 1} d(r^{\kappa^+ + n - 1}) \otimes d\nu.$$

3. Functions of finite order. Let $u \in SH^*(K)$. We set

$$\hat{M}_u(r) = \sup_{x \in \Gamma_r} u(x),$$

$$\tilde{M}_u(r) = r^{-n+1} \cdot \int_{\Gamma_r} |u| \varphi dS_r .$$

The order of a function u (the order of u in a cone K) is the value

$$\hat{\rho}_u = \max\left(\overline{\lim_{r \to \infty}} \frac{\ln^+ \hat{M}_u^+(r)}{\ln r} , \overline{\lim_{r \to \infty}} \frac{\ln \tilde{M}_u(r)}{\ln r} \right).$$

Correspondingly we define the class $SH(K,\rho]$. Namely, we say that a function $u(z) \in SH^*(K)$ belongs to the class $SH(K,\rho]$, $\rho > 0$, if

$$\overline{\lim_{r \to \infty}} \frac{\hat{M}_u(r)}{r^\rho} < \infty \tag{4.4.10}$$

and

$$\overline{\lim_{r \to \infty}} \frac{\tilde{M}_u(r)}{r^\rho} < \infty . \tag{4.4.11}$$

Using **Lemma 4.4.6** we can show that for $\rho > \kappa^+$ the estimate (4.4.11) is a consequence of the estimate (4.4.10). Thus, if

$$\rho_u \overset{\text{def}}{=} \overline{\lim_{r \to \infty}} \frac{\hat{M}_u^+(r)}{r^\rho} \geq \kappa^+,$$

then $\hat{\rho}_u \geq \rho_u$.

We shall obtain a number of estimates of measures related to a function $u \in SH(K,\rho]$, as well as estimates different from (4.4.10) and (4.4.11) for the function u itself. All these estimates are contained in the following lemma, which can be proved with the aid of the generalized Carleman's formula, similar to the estimates for functions $u \in H(\mathbb{C}^+,\rho]$.

Lemma 4.4.9. Let $u \in SH(K,\rho]$, $\rho > 0$. Then for any $\lambda > 0$ there are constants c_1, c_2, c_3, c_4 depending on λ and the function u only, such that for any $R > \lambda$ the following inequalities hold:

$$\int_{K_{\lambda,R}} |u(x)| |x|^{-2} \varphi(x) d\omega_x \leq c_1 R^{\rho+n-3} (R-\lambda) ; \tag{4.4.12}$$

$$\int_{\Delta^{\delta}_{\lambda,R}} |u(x)||x|^{-2}d\omega_x \le \delta c_2 R^{\rho+n-2}, \tag{4.4.13}$$

where $\Delta^{\delta}_{\lambda,R} = \{x\in K_{\lambda,R}: 0 < \varphi(x) < \delta\}$, $\delta > 0$;

$$\int_{K_{\lambda,R}} \varphi(x)d\mu_u(x) \le c_3 R^{\rho+n-2} ; \tag{4.4.14}$$

$$\int_{\Gamma_{\lambda,R}} \frac{\partial\varphi}{\partial n}\, d|\mu_{f,\partial}|(x) \le c_4 R^{\rho+n-2}. \tag{4.4.15}$$

Note that we can combine the estimates (4.4.14) and (4.4.15) as follows:

$$|\tau_u|(\bar{K}_{\lambda,R}) \le c_5 R^{\rho+n-2},\ c_5 = c_3 + c_4. \tag{4.4.16}$$

Using **Lemma 4.4.7** instead of **Lemma 4.4.6**, we obtain estimates for the function u(x) near the origin.

Lemma 4.4.10. Let a function $u\in SH(K_R)$, $R > 0$, satisfy the condition *sup*$\{u(x): x\in K_R\} < \infty$. Then there are constants c_1 and c_2 such that for $0 < \lambda < R' < R$ the following inequalities holds:

$$\int_{\Gamma_\lambda} |u|\varphi dS_\lambda \le c_1(\lambda^{1-\kappa^+}+1), \tag{4.4.17}$$

$$|\tau_u|(\bar{K}_{\lambda,R'}) \le c_2\lambda^{-\kappa^+}. \tag{4.4.18}$$

For functions $u\in SH^*(K)$ there exists a representation which coincides in the case $n = 2$, $K = \mathbb{C}^+$ and $u = ln|f|$, $f\in H(\mathbb{C}^+,\rho]$, with the representation of the function $ln|f|$ following from **Theorem 4.1.1** (the factorization theorem). To give a precise formulation of the corresponding statement we need some definitions and notations.

We denote by G(x,y) the Green's function of a cone K. The following representation holds (see J.Lelong-Ferrand [1]):

$$G(x,y) = \beta\sum_{j=1}^{\infty} |x|^{\kappa_j^+}|y|^{\kappa_j^-}\frac{\varphi_j(x)\varphi_j(y)}{\alpha_j},\ |x| < |y|,$$

where $\beta = \int_{S_1} dS_1$, $\alpha_j = \kappa_j^+ - \kappa_j^-$.

We define the canonical kernel $G_p(x,y)$, $p\in N$, by the equality

$$G_p(x,y) = -G(x,y) + \beta \sum_{j=1}^{p} |x|^{\kappa_j^+} |y|^{\kappa_j^-} \frac{\varphi_j(x)\varphi_j(y)}{\alpha_j}.$$

For $p = 0$ we set $G_p(x,y) = -G(x,y)$.

This kernel is subharmonic in each variable in the cone K, harmonic for $x \neq y$, continuous on $\{K\times K\}\setminus\{(x,y): x = y\}$ and such that $G_p(x,y) = 0$ on $\{\partial K\setminus\{0\}\}\times K \cup K\times\{\partial K\setminus\{0\}\}$. Besides $\Delta_x G_p(x,y) = \delta(x-y)$ ($\delta(x)$ is the Dirac function).

Now, using the estimates of the eigenvalues λ_j and eigenfunctions φ_j following from the condition of smoothness of the boundary of the domain Γ, we obtain the following estimates of the canonical kernel and its normal derivative (on ∂K):

$$|G_p(x,y)| \le c_1 |x|^{\kappa_{p+1}^+} |y|^{\kappa_{p+1}^-} \varphi(x)\varphi(y)$$

for $|x| \le t|y|$, $p \ge 0$, $c_1 = c_1(K,p,t)$, $0 < t < 1$;

$$G_p(x,y) \le c_2 \frac{|x|^{\kappa_{p+1}^+} |y|^{\kappa_{p+1}^-} \varphi(x)\varphi(y)}{|x|^{\kappa_{p+1}^+ - \kappa_p^+} + |y|^{\kappa_{p+1}^+ - \kappa_p^+}}, \quad \forall x,y\in K,\ p\ge 1,\ c_2 = c_2(K,p);$$

$$\left|\frac{\partial G_p(x,y)}{\partial n_y}\right| \le c_1 |x|^{\kappa_{p+1}^+} |y|^{\kappa_{p+1}^-} \varphi(x) \frac{\partial \varphi(y)}{\partial n}, \quad \forall x\in K, y\in\partial K\setminus\{0\},\ |x|\le t|y|,\ p\ge 0;$$

$$\frac{\partial G_p(x,y)}{\partial n_y} \le c_2 \frac{|x|^{\kappa_{p+1}^+} |y|^{\kappa_{p+1}^-} \varphi(x) \frac{\partial \varphi(y)}{\partial n}}{|x|^{\kappa_{p+1}^+ - \kappa_p^+} + |y|^{\kappa_{p+1}^+ - \kappa_p^+}}, \quad \forall x\in K, y\in\partial K\setminus\{0\},\ p\ge 1.$$

We consider a positive measure μ on K such that the function

$$\tilde{\mu}(R) = \int_{K_{\lambda,R}} \varphi(x)d\mu(x), \ R > 1,$$

satisfies the condition

$$\int_{\lambda}^{\infty} t^{\overline{\kappa}_{p+1}} d\tilde{\mu}(t) < \infty$$

for some $\lambda > 0$. Then it follows from the above-mentioned estimates of $G_p(x,y)$ that the integral

$$J'_{\mu,p}(x) = \int_{K_{\lambda,\infty}} G_p(x,y)d\mu(y)$$

converges for every $x \in K$. It is also easy to see that the function $v(x) = J'_{\mu,p}(x)$ is subharmonic in K, and that the corresponding measures μ_v and $\mu_{v,\partial}$ satisfy the conditions: $\mu_v = \mu|_{K_{\lambda,\infty}}$, $\mu_{v,\partial} = 0$. Besides,

$$\hat{\rho}_v = \overline{\lim_{R\to\infty}} \frac{ln\tilde{\mu}(R)}{lnR} - (n-2)$$

and the estimate

$$v(x) \le c\varphi(x)\left\{ |x|^{\overset{+}{\kappa}_p} \int_{\lambda}^{|x|} t^{\overline{\kappa}_p - 1} \tilde{\mu}(t)dt + |x|^{\overset{+}{\kappa}_{p+1}} \int_{|x|}^{\infty} t^{\overline{\kappa}_{p+1} - 1} \tilde{\mu}(t)dt \right\}$$

holds.

We call the function $J'_{\mu,p}(x)$ the cone potential of the measure μ, and denote it by $v(x)$.

Now we consider a real-valued measure μ_∂ on $\partial K \backslash \{0\}$ which for some $\lambda > 0$ satisfies the condition

$$\int_{\lambda}^{\infty} t^{\overline{\kappa}_{p+1} - 1} d\tilde{\mu}_\partial(t) < \infty,$$

where

$$\tilde{\mu}_\partial(t) = \int_{\Gamma_{\lambda,\infty}} \frac{\partial\varphi(x)}{\partial n} d|\mu_\partial|(x).$$

From the estimates of the function $\frac{\partial G_p}{\partial n}$ given above it follows that the integral

$$J''_{\mu_{\partial},p}(x) = -\frac{1}{\theta_n}\int_{\Gamma_{\lambda,\infty}} \frac{\partial G_p(x,y)}{\partial n_y}\, d\mu_{\partial}(y)$$

converges for every $x \in K$. Moreover, the function $w(x) = J''_{\mu_{\partial},p}(x)$ is harmonic in K, and its boundary measure $\mu_{w,\partial}$ coincides with the restriction of μ_{∂} to $\Gamma_{\lambda,\infty}$. The order $\hat{\rho}_w$ is equal to $\left(\overline{\lim_{R\to\infty}} \frac{\ln\tilde{\mu}_{\partial}(R)}{\ln R} - (n-2)\right)$, and the estimate

$$w(x) \le c\cdot\left\{ |x|^{\kappa_p^+} \int_{\lambda}^{|x|} t^{\kappa_p^- -1}\tilde{\mu}(t)dt + |x|^{\kappa_{p+1}^+} \int_{|x|}^{\infty} t^{\kappa_{p+1}^- -1}\tilde{\mu}(t)dt \right\}$$

holds.

We call this function $w(x) = J''_{\mu_{\partial},p}$ the boundary potential of μ_{∂}.

Now, using the above properties of the potentials $J'_{\mu,p}$ and $J''_{\mu_{\partial},p}$ it is easy to prove the following theorem on the canonical representation of a function $u \in SH^*(K)$ of order $\hat{\rho}_u \le \rho$.

Theorem 4.4.1. Let $u \in SH^*(K)$ be a function of order $\hat{\rho}_u \le \rho$, and let there be numbers p, λ such that $\kappa_p^+ \le \rho \le \kappa_{p+1}^+$, $\lambda \notin \Lambda_u$. Then the following representation holds:

$$u(x) = J'_{\mu,p}(x) + J''_{\mu_{\partial},p}(x) - \int_{K_{\lambda}} G(x,y)d\mu_u(y) +$$

$$+ H_{\lambda}(x) + \sum_{j=1}^{p} a_j |x|^{\kappa_j^+}\varphi_j(x), \qquad (4.4.19)$$

where $a_1, \ldots, a_p$ are constants and $H_{\lambda}(x)$ is an harmonic function in K such that

$$\overline{\lim_{x\to\infty}} H_{\lambda}(x)|x|^{-\kappa^-} < \infty$$

and

$$\mu_{H_{\lambda},\partial} = \mu_{u,\partial}\big|_{\Gamma_{0,\lambda}} .$$

If we use the measure τ_u instead of $\mu_{u,\partial}$ and μ_u, then the representation (4.4.19) can be written in another form. For this we need to introduce a kernel $W_p(x,y)$, as follows:

$$W_p(x,y) = \begin{cases} \dfrac{G_p(x,y)}{\varphi(x)} & \text{when } x,y \in K \\ \dfrac{G_p(x,y)}{\partial n_y} \left(\dfrac{\partial \varphi(y)}{\partial n} \right)^{-1} & \text{when } x \in K,\ y \in \partial K/\{0\} \end{cases}$$

The representation (4.4.19) then becomes

$$u(x) = \int_{\bar{K}_{\lambda,\infty}} W_p(x,y)d\tau_u(y) + \sum_{j=1}^{p} a_j|x|^{\kappa_j^+}\varphi_j(x) - \int_{K_\lambda} G(x,y)d\mu_u(y) + H_\lambda(x).$$

Note that the function

$$u_1(x) = -\int_{K_\lambda} G(x,y)d\mu_u(y) + H_\lambda(x)$$

is subharmonic in K, harmonic in $K_{\lambda,\infty}$ and satisfies the condition

$$|u_1(x)| \le \text{const}\cdot|x|^{-\kappa^+}$$

for $|x| > \lambda$.

We use the representation (4.4.19) for a function L(x) that is subharmonic in K, positively homogeneous of degree ρ and bounded on Γ. Then we have

$$L(x) = \int_{\bar{K}\setminus\{0\}} W_p(x,y)d\tau_L(y) \tag{4.4.20}$$

when $\kappa_p^+ < \rho < \kappa_{p+1}^+$ (this is analogous to the case of non-integral order for functions in $\mathbb{C}_+$), and

$$L(x) = \int_{\bar{K}\setminus U_\delta} W_p(x,y)d\tau_L(y) + \sum_{j=0}^{1} a_{p-j}\varphi_{p-j}(x) +$$

$$+ \int_{\bar{U}_\delta^+} W_{p-1}(x,y)d\tau_L(y), \tag{4.4.21}$$

where δ is an arbitrary positive number, when $\kappa_{p-l-1}^+ < \kappa_{p-l}^+ = \ldots = \kappa_p^+ = \rho < \kappa_{p+1}^+$ (this is analogous to the case of integral order).

Since the measure is positively homogeneous we can obtain from (4.4.20) and (4.4.21) a representation for L(x) in which integration over K is replaced by integration over Γ.

To formulate this result we need the measure $\hat{\nu}_L$ and the Green's function $g^*(x,y)$ of the operator $\Delta^*_\rho = \Delta^* + \rho(\rho+n-2)$ for the domain Γ. We define $\hat{\nu}_L$ on the Borel subsets $E \subset \bar{\Gamma}$ by the equality

$$\hat{\nu}_L(E) = \tau_L(K_1^E), \quad K_1^E = \{x = ty:\ y \in E,\ 0 < t < 1\}.$$

The function $g^*(x,y)$ is defined for $\kappa_p^+ < \rho < \kappa_{p+1}^+$ in the standard manner and for $\kappa_{p-l-1}^+ < \kappa_{p-l}^+ = \ldots = \kappa_p^+ = \rho < \kappa_{p+1}^+$ as the fundamental solution of the corresponding equation that vanishes on $\partial\Gamma$ and satisfies the conditions

$$\int_\Gamma g^*(x,y)\varphi_{p-j}(y)dS_1,\ 0 \le j \le l.$$

We set

$$W_p^*(x,y) = \frac{g^*(x,y)}{\varphi(x)},\ \forall x \in \bar{\Gamma},\ y \in \Gamma,$$

and define in addition, the function $W_p^*(x,y)$ at the points $y \in \partial K$ by the equality

$$W_p^*(x,y) = \lim_{\substack{y' \to y \\ y' \in \Gamma}} W_p^*(x,y') = \frac{\partial g^*(x,y)}{\partial n_y} \left(\frac{\partial\varphi(y)}{\partial n}\right)^{-1}.$$

Note that the function $W_p^*(x,y)$ is continuous on $\{\bar{\Gamma}\times\bar{\Gamma}\}\setminus\{(x,y):\ x=y\}$.

Theorem 4.4.2. Let $L(x) \in SH^*(K)$ be a positively homogeneous function of degree ρ. If $\kappa_p^+ < \rho < \kappa_{p+1}^+$, then

$$L(x) = -\frac{\beta}{\rho+n-2}\int_{\bar{\Gamma}} W_p^*(x,y)d\hat{\nu}_L(y)$$

and if $\kappa_{p-l-1}^+ < \kappa_{p-l}^+ = \ldots = \kappa_p^+ = \rho < \kappa_{p+1}^+$ then

$$L(x) = \sum_{j=0}^{l} a_{p-j}\varphi_{p-j}(x) - \frac{\beta}{\rho+n-2}\int_{\bar{\Gamma}} W_p^*(x,y)d\hat{\nu}_L(y).$$

In conclusion of this section we note that in the case $\kappa_{p-l-1}^+ < \kappa_{p-l}^+ = \ldots = \kappa_p^+ = \rho < \kappa_{p+1}^+$ the function $L(x)$ figuring in the statement of **Theorem 4.4.2** satisfies the condition

$$\int_{\bar{\Gamma}} \frac{\varphi_{p-j}(x)}{\varphi(x)}\, d\hat{\nu}_L(x) = 0,\ j = 0,\ 1,\ \ldots,\ l.$$

4. Functions of c.r.g. To define the concept of regular growth in a cone we shall use the notion of a C_0^α-set, just as in the case of the whole space $\mathbb{R}^n$. Similarly to the case $K = \mathbb{C}^+$ we shall distinguish c.r.g. in an open cone K and in the closed cone $\overline{K}$.

A function $u \in SH(K,\rho]$, $\rho > 0$, is said be of c.r.g. in $\overline{K}$ (with respect to the order ρ) if there is a C_0^{n-1}-set $E \subset K$ such that

$$\overline{\lim_{\substack{x \to \infty \\ x \in K \setminus E}}} \frac{|u(x) - \mathfrak{L}_u^*(x)|}{|x|^\rho} = 0.$$

A function $u \in SH(K,\rho]$ is said to be of c.r.g. in K if it is of c.r.g. in every closed cone $\overline{K}' = K^{\overline{\Gamma}'}$, where $\overline{\Gamma}' \subset\subset \Gamma$.

The following theorem gives a relationship between c.r.g. of a function u(x) and weak convergence of the functions $u^{[t]}(x) = t^{-\rho}u(tx)$.

Theorem 4.4.3. Let $u \in SH(K,\rho]$. Then the following conditions are equivalent

a) u(x) is of c.r.g. in K;

b) $\exists \lim_{t \to \infty} \int_K u^{[t]}(x)\psi(x)d\omega_x$, $\forall \psi \in D(K)$;

c) $\exists \lim_{t \to \infty} \int_{K_{1/t,1}} u^{[t]}(x)\psi(x)|x|^{-2}d\omega_x$, $\forall \psi \in C(\overline{K}_1)$.

Moreover, if ne of these conditions is satisfied, then the equalities

$$\lim_{t \to \infty} \int_K u^{[t]}(x)\psi(x)d\omega_x = \int_K \mathfrak{L}_u^*(x)\psi(x)d\omega_x, \ \forall \psi \in D(K),$$

$$\lim_{t \to \infty} \int_{K_{1/t,1}} u^{[t]}(x)\psi(x)|x|^{-2}d\omega_x = \int_{K_1} \mathfrak{L}_f^*(x)\psi(x)|x|^{-2}d\omega_x, \ \forall \psi \in C(\overline{K}_1)$$

hold.

The proof of this theorem is based on **Lemmas 2.1.6, 2.1.9** and the estimates of **Lemma 4.4.9**. In some parts it is close to the proofs of **Theorems 4.2.2** and **4.2.3**.

The weak convergence mentioned in **Theorem 4.4.3** implies the weak convergence of $u^{[t]}$ on Γ. Namely, the following theorem holds.

Theorem 4.4.4. Let $u \in SH(K,\rho]$ be of c.r.g. in a cone K. Then

a)
$$\varlimsup_{\substack{t \to \infty \\ t \notin \Lambda_u}} \int_{\Gamma} u^{[t]}(x)\varphi(x)\eta(x)dS_1 = \int_{\Gamma_1} \mathfrak{L}_u^*(x)\varphi(x)\eta(x)dS_1, \ \forall\eta\in C(\overline{\Gamma}_1);$$

b)
$$\varlimsup_{\substack{t \to \infty \\ t \notin \Lambda_u}} Q(\eta(x);1,u^{[t]}) = \rho\int_{\Gamma} \mathfrak{L}_u^*(x)\varphi(x)\eta(x)dS_1, \ \forall\eta\in C^2(\overline{\Gamma}_1).$$

To characterize the relationship between regularity of growth of a function $u(x)$ and the distribution of its associated measure τ_u, we introduce the concept of regular distribution of the measure τ_u. This concept is the analogue of the concept of regular distribution of the measures τ_f (for functions $f\in H(\mathbb{C}^+,\rho]$) and μ_u (for functions $u\in SH(\mathbb{R}^n,\rho]$). First of all we introduce the cone-boundary density, which is the analogue of the argument-boundary density.

Let τ be a real-valued measure on $\overline{K}\setminus\{0\}$ that satisfies the conditions

i) $\varlimsup_{R \to \infty} R^{-\rho-n+2}\|\tau\|(R) < \infty$, where $\|\tau\|(R) = |\tau|(\overline{K}_{1,R})$;

ii) $\tau \geq 0$ on K;

iii) $d\tau(x) \leq A|x|^{\rho-1}d\sigma$ on $\partial K\setminus\{0\}$.

Let $\tau^{[t]}$ be the measure on $\overline{K}\setminus\{0\}$ defined on Borel sets $E \subset \overline{K}\setminus\{0\}$ by $\tau^{[t]}(E) = t^{-\rho-n+2}\tau((tE) \cap \overline{K}_{1,\infty})$.

We say that the measure τ has a cone-boundary density if for any $\psi\in C(\overline{K}_1)$,

$$\exists \lim_{t \to \infty} \int_{\overline{K}_1} \psi d\tau^{[t]} \overset{\text{def}}{=} \int_{\overline{K}_1} \psi d\hat{\tau},$$

i.e. if the measures $\tau^{[t]}$ converge weakly on $\overline{K}_1$ to some measure $\hat{\tau}$ as $t \to \infty$.

This definition is equivalent to the following one.

A measure τ has a cone-boundary density if there is a measure $\hat{\nu}$ on $\overline{\Gamma}$ (the cone-boundary density) such that for every cone $K' = K^{\Gamma'}$, where the domain $\Gamma' \subset \Gamma$ satisfies the condition $|\hat{\nu}|(\overline{\partial\Gamma' \cap \Gamma}) = 0$, the equality

$$\lim_{R\to\infty} R^{-\rho-n+2}\tau(\overline{K}'_{1,R}) = \hat{\nu}(\Gamma')$$

holds.

It is obvious that the above-mentioned measures $\hat{\tau}$ and $\hat{\nu}$ are related by the equality $d\hat{\tau} = \frac{1}{\rho+n-2}\, d(r^{\rho+n-2})\otimes d\hat{\nu}$, equivalently, by $\hat{\nu}(\overline{\Gamma}') = \hat{\tau}(\overline{K}_1)$.

In the case $\rho \neq \kappa_j^+$, $\forall j = 0, 1, 2, \ldots$, a measure τ having a the cone-boundary density will also be called a regularly distributed measure. In case $\rho\in\{\kappa_j^+\}_{j=1}^{\infty}$ we call a measure τ regularly distributed if it has a cone-boundary density and if, in addition, the the cone-boundary balance condition holds, i.e.

$$\exists \lim_{R\to\infty} \int_{\overline{K}_{1,R}} \varphi_q\varphi^{-1}|x|^{-\rho-n+2}d\tau,\ \forall q\in\{j\in\mathbb{N}:\ \rho = \kappa_j^+\}.$$

Theorem 4.4.5. A function $u\in SH(K,\rho]$ is of c.r.g. in K if and only if its associated measure τ_u is regularly distributed.

In the proof of this theorem, which is in some the parts close to the proof of **Theorems 4.2.5 - 4.2.7**, the following facts are used in an essential manner: **Lemma 4.4.5** (Green's formula for the functions $u\in SH^*(K)$), **Theorem 4.4.1** (on the canonical representation of functions $u\in SH(K,\rho]$) and **Theorem 4.4.3**. Note also that when in the proof of **Theorem 4.4.5** we obtained the following representation for the limits in the cone-boundary balance condition:

$$\lim_{R\to\infty} \theta_n \int_{\overline{K}_{1,R}} \frac{\varphi_q(x)}{\varphi(x)}|x|^{-\rho-n+2}d\tau_u =$$

$$= \kappa_q^- \int_{\Gamma_1} u\varphi_q dS_1 - Q\left(\frac{\varphi_q}{\varphi};1,u\right) + (2\rho+n-2)\int_{\Gamma_1} \mathfrak{L}_u^*\varphi_q dS_1.$$

Note also that in all equalities and definitions related to **Theorem 4.4.5** we can take K_λ and $K_{\lambda,\infty}$, where λ is arbitrary positive, instead of K_1 and $K_{1,\infty}$.

Now we consider functions of c.r.g. in a closed cone. First of all we note that, as in the case of the half-plane, a function of c.r.g. in K is also of c.r.g. in $\overline{K}$ if its boundary measure is asymptotically close to the boundary measure of its indicator. More exactly, the following theorem holds.

Theorem 4.4.6. In order that $u \in SH(\mathbb{C},\rho]$ be a function of c.r.g. in $\bar{K}$ it is sufficient and, under the additional assumption $\sup_{x\in\Gamma} |\mathcal{L}_u(x)| < \infty$ also necessary, that u be of c.r.g. in K and

$$\lim_{R\to\infty} \frac{1}{R^{\rho+n-1}} |\mu_{u,\partial} - \mu_{\mathcal{L}_u^*,\partial}|(\Gamma_{1,R}) = 0. \tag{4.4.22}$$

We reformulate the latter equality. For this purpose we consider the family of measures $\mu_{u,\partial}^{[R]}$, defined on Borel sets $E \subset\subset \partial K \setminus \{0\}$ by the equalities

$$\mu_{u,\partial}^{[R]}(E) = \frac{1}{R^{\rho+n-1}} \mu_{u,\partial}((RE) \cap \Gamma_{1,\infty}).$$

Now note that (4.4.22) is none other than the condition of convergence in variation (as $R \to \infty$) of the measures $\mu_{u,\partial}^{[R]}$ to the measure $\mu_{\mathcal{L}_u^*,\partial}$.

The proof of **Theorem 4.4.6** is similar to that of **Theorem 4.3.1**, but it is essentially more complicated. We single out the following statement, which was obtained in the proof and is interesting in itself.

Theorem 4.4.7. If a function $u \in SH(K,\rho]$, where $\rho \notin \{\kappa_j^+\}_{j=1}^{\infty}$, satisfies the condition

$$\lim_{R\to\infty} \frac{\|\tau_u\|(R)}{R^{\rho+n-2}} = 0,$$

then it is of c.r.g. in $\bar{K}$ and $\mathcal{L}_u^*(x) \equiv 0$.

We think that in **Theorem 4.4.6** the assumption of boundedness of the indicator $\mathcal{L}_u^*(x)$ is unnecessary. In particular, this opinion is based on the following fact. If a function $u \in SH(K,\rho]$ can be extended to a cone $K' = K^{\Gamma'}$, $\Gamma' \supset \bar{\Gamma}$, as a function belonging to $H(K',\rho]$, then the necessity part of **Theorem 4.4.6** holds without this assumption.

We consider sets of finite field of view, i.e. sets that can be covered by a countable system of balls $B(x^{(j)},r_j)$ such that

$$\sum_j \frac{r_j}{|x^{(j)}|} < \infty.$$

These sets are more "thinner" than C_0^{n-1}-sets in the case $n > 2$.

The following theorem supplements **Theorem 4.4.7.**

Theorem 4.4.8. If a function $u \in SH(K,\rho]$, where $\rho \in \{\kappa_j^+\}_{j=1}^{\infty}$, is such that

$$\int_{\overline{K}_{1,\infty}} \frac{d|\tau_u|(x)}{|x|^{\rho+n-2}} < \infty,$$

then $\mathfrak{L}_u^*(x) = |x|^{\rho}\Big(\sum_{j:\kappa_j^+=\rho} a_j \varphi_j(x) \Big)$ and the following condition is satisfied:

A) There is a set E of finite field of view such that

$$\lim_{\substack{x \to \infty \\ x \in K \setminus E}} \frac{|\mathfrak{L}_u^*(x) - u(x)|}{|x|^{\rho}} = 0.$$

In particular, u(x) is a function of c.r.g. in $\overline{K}$.

Now we turn to the theorems concerning the case $\rho = \kappa_1^+ = \kappa^+$. As in the case of a half-plane, the characteristic feature of this situation is the possibility of defining regularity of growth of a function by the behaviour of its boundary measure.

Theorem 4.4.9. If a function $u \in SH^*(K)$ satisfies the conditions

$$u(x) \le c_1 |x|^{\kappa^+} + c_2, \ \forall x \in K$$

and

$$\int_{\Gamma_{1,\infty}} |x|^{-\kappa^+-n+1} d\mu^+_{u,\partial}(x) < \infty,$$

then :

1) its indicator $\mathfrak{L}_u^*$ is proportional to the function $|x|^{\kappa^+}\varphi(x)$;

2) u(x) satisfies condition A) in **Theorem 4.4.8** and the condition

$$\int_{K_{1,\infty}} \varphi(x) d\mu_u(x) < \infty.$$

Note that the functions bounded in K satisfy the condition of **Theorem 4.4.9**. Hence these functions are of c.r.g. in $\overline{K}$ and, moreover, the corresponding exceptional sets are of finite field of view.

The following theorem is similar to **Theorems 4.3.4** and **4.3.5**.

Theorem 4.4.10. Let $u \in SH(K,\kappa^{+}]$ be a function of c.r.g. in $\bar{K}$. Let also $\sup_{x\in\Gamma} |\mathfrak{L}^{*}_{u}(x)| < \infty$. Then

$$\int_{K_{1,\infty}} \varphi(x) d\mu_{u}(x) < \infty$$

if and only if the limit

$$\lim_{R\to\infty} \int_{\Gamma_{1,R}} \frac{\partial\varphi}{\partial n} |x|^{-\kappa^{+}-n+2} d\mu_{u,\partial}(x)$$

exists and the equality

$$\int_{\partial\Gamma} \frac{\partial\varphi}{\partial n} d\hat{\nu}_{\mathfrak{L}^{*}_{u}} = 0$$

is valid.

Notes

The existence of the boundary measure $\mu_{f,\partial}$ (in equivalent form) was established by N.Govorov [1]-[4]. The generalized Carleman's formula (**Lemma 4.1.3**) is contained at the same papers. The estimates of **Lemma 4.1.4** were established by L.Ronkin [3], [4], for $\rho > 1$ and were extended to the case $\rho \le 1$ by A.Rashkovskiĭ. Some of these estimates were obtained previously by A.Grishin [1], who defined the class $H(\mathbb{C}^{+},\rho]$ in a different way, and by N.Govorov [1], [4] for functions of c.r.g. The definition of the order $\hat{\rho}_{f}$ used here is equivalent to the one introduced by A.Rashkovskiĭ [1] (in [1] functions subharmonic in a cone of the space $\mathbb{R}^{n}$ are considered). If the functions are of c.r.g. then the definition is also equivalent to the definitions of order given by N.Govorov [1], [4] and E.Titchmarsh [1]. All the definitions are equivalent when $\hat{\rho}_{f} > 1$. If $\hat{\rho}_{f} \le 1$, then Govorov's and Titchmarsh's definitions are equivalent (see N.Govorov [1], [4]), and are not equivalent to the one adopted here (see A.Rashkovskiĭ [1]).

The Green's formula (**Lemma 4.1.5**) was obtained by L.Ronkin [3], [4]. The factorization representation in **Theorem 4.1.1** for functions of finite order in $\mathbb{C}^{n}$ was obtained by N.Govorov [1], [4]. He used the Govorov-Titchmarsh definition, and did not estimate the order of the factors participating in this representation. The basic facts of the theory of f.c.r.g. in $\mathbb{C}^+$ and $\overline{\mathbb{C}}^+$ not concerned with the relationship between such functions and weak convergence of the corresponding families were obtained by N.Govorov [1] - [4]. Thus, he introduced the notions of argument-boundary density and argument-boundary balance symmetry, proved **Theorems 4.2.5** (without the formula for $\hat{\tau}_f$), **4.2.6** (in an equivalent form), **4.2.8** (without the formula for κ_f), and **4.2.10.** L.Ronkin [3], [4] obtained Theorems **4.2.1 - 4.2.4** and **Theorem 4.2.7. Theorem 4.3.1** is equivalent to the corresponding statement proved previously by N.Govorov [1], [4]. For proving this theorem we applied some a technique of V.Azarin [6]. **Theorem 4.3.3** is a strengthening of a corresponding **Theorem of M.Cartwright** [1]. **Theorems 4.3.4** and **4.3.5** were obtained by A.Pfluger [4] in the case when f(z) is holomorphic in $\mathbb{C}$ and by N.Govorov [1], [4] in the general case. The results of §4.1 and 4.2, as well as **Lemma 4.4.10** and **Lemma 4.4.9** for $\rho > \kappa^+$, were obtained jointly by the author and A.Rashkovskiĭ [1]. A.Rashkovskiĭ [1] considered the case $\rho \le \kappa^+$. He also established [2] the representation of a subharmonic function of finite order in a cone K. The estimates in this potential representation were obtained by A.Rashkovskiĭ and L.Ronkin.

All the results of §4.4 related to with functions of c.r.g. in an open cone belong to A.Rashkovskiĭ and L.Ronkin [1], [2], and the ones related to f.c.r.g. in a closed cone belong to L.Ronkin [5]. Note that **Theorem 4.4.9** includes V.Azarin's theorem on bounded functions in a cone (see V.Azarin [6]).

CHAPTER 5

FUNCTIONS OF EXPONENTIAL TYPE AND BOUNDED ON THE REAL SPACE (FOURIER TRANSFORMS OF DISTRIBUTION OF COMPACT SUPPORT)

In this chapter we consider entire functions of at most normal type with respect to the order 1 (i.e. functions of exponential type) that are bounded for real values of the variables. The importance of this class of functions is lies in the fact that it contains the Fourier transforms of functions of compact support and belonging to $L^1(\mathbb{R}^n)$. Besides, since the Fourier transforms of distributions of compact support are entire functions of polynomial growth on $\mathbb{R}^n$, by "suppression of growth" their study can often be reduced to that of corresponding functions which are bounded on $\mathbb{R}^n$. Side by side with the problem of regularity of growth, which is the main topic of this book, we study some problems on the connection between the behaviour of the functions under consideration on the whole space and their behaviour on certain discrete sets. Such problems are of interest for some applications, in particular, for completeness problems of systems of exponentials or monomials.

§1 Regularity of growth of entire functions of exponential type and bounded on the real space

It was noted in §3 of **Chapter 4** (see the **Corollary** to **Theorem 4.3.3**) that if an entire function of one variable is of exponential type and bounded on the real axis, then it is of c.r.g. J.Vauthier [1] has shown (and we give an account of this in §1.3) that the analogue of this statement for functions of several variables is false. In connection with this, the problem of finding sufficient conditions for c.r.g. of the functions under consideration naturally appears; we consider it in §1.4.

In §1.3 we give a construction of a function that is not of c.r.g. This construction is based on a refinement of the Martineau-Kiselman theorem on the existence of an entire function with prescribed indicator. This refinement was given by W.Wirtinger [1] for indicators of special form. It is of independent interest, and we present it in §1.2. Before this we give an account of some general facts concerning the functions the class under consideration.

1. The Fourier transform of a function of compact support, the Polya-Plancherel indicator. As already mentioned, the Fourier transform of a distribution of compact support is an entire function of exponential type and of polynomial growth on $\mathbb{R}^n$. The converse statement is also true, and there is a close relation between the support and the asymptotic behaviour of the corresponding entire function (the Fourier transform). For the description of this relation in the multidimensional case we need the concept of P-indicator $P_f(\lambda)$ (Polya-Plancherel indicator), which in the case $n = 1$ coincides with the ordinary indicator. This concept is introduced in the following way.

Let $f \in H(\mathbb{C}^n,1]$. We set

$$P_f(\lambda,x) = \overline{\lim_{r \to \infty}} \frac{1}{r} \ln|f(x+ir\lambda)|, \quad \lambda \in \mathbb{R}^n,$$

and define the P-indicator by the equality

$$P_f(\lambda) = \sup_{x \in \mathbb{R}^n} P_f(\lambda,x).$$

It is obvious that $P_f(\lambda,0) = \mathfrak{L}_f(i\lambda)$, where $\mathfrak{L}_f$ is the ordinary (non-regularized) indicator of the function $f(z)$. Therefore

$$\mathfrak{L}_f(i\lambda) \le P_f(\lambda), \quad \forall \lambda \in \mathbb{R}^n.$$

Using **Hartogs' Lemma** it is not difficult to show also that

$$P_f(\lambda) \le \mathfrak{L}_f^*(i\lambda), \quad \forall \lambda \in \mathbb{R}^n.$$

For the Fourier transform of an ordinary function the following theorem is true. This theorem extends the well-known **Paley-Wiener Theorem** to the multidimensional case.

Theorem 5.1.1. In order that a function $f(z)$, $z\in\mathbb{C}^n$, belong to $H(\mathbb{C}^n,1]$ and $f|_{\mathbb{R}^n}\in L^2(\mathbb{R}^n)$, it is necessary and sufficient that it admit the representation

$$f(z) = \left(\frac{1}{\sqrt{2\pi}}\right)^n \int_{\mathbb{R}^n(t)} F(t)e^{-i<z,t>}dt,$$

where $F(t)$ is some function satisfying the conditions: $F\in L^2(\mathbb{R}^n)$, $suppF \subset\subset \mathbb{R}^n$. If this representation is valid then the support function[33] $K(\lambda)$ of the smallest convex domain outside which $F(t)$ vanishes almost everywhere, coincides with the P-indicator $P_f(\lambda)$ of the function $f(z)$, and

$$|f(x+iy)| \leq ce^{K(y)}, \ \forall y\in\mathbb{R}^n, \ x\in\mathbb{R}^n,$$

where

$$c = \left(\frac{1}{\sqrt{2\pi}}\right)^n \int_{\mathbb{R}^n} |F(t)|dt.$$

A proof of this theorem can be found in, for example, L.Ronkin [1].

The general case, i.e. the case of the Fourier transform of a distribution, is as follows.

Theorem 5.1.2. In order that a function $f(z)$, $z\in\mathbb{C}^n$, belong to $H(\mathbb{C}^n,1]$ and satisfy the condition $|f(x)| \leq |x|^a + c$, $\forall x\in\mathbb{R}^n$, for some $c > 0$ and $a > 0$, it is necessary and sufficient that $f(z)$ be the Fourier transform[34] of a distribution $F\in\mathcal{D}'(\mathbb{R}^n)$ of compact support. If

---

[33] Recall that the support function $K(\lambda)$ of a set $D \subset \mathbb{R}^n$ is defined by

$$K_D(\lambda) = \sup_{x\in D} <x,\lambda> , \ \lambda\in\mathbb{R}^n.$$

The support function is convex, and hence continuous in $\mathbb{R}^n$ if the set D is bounded.

[34] Recall that the Fourier transform or, as it is sometimes called, the Fourier-Laplace transform, of a distribution $F\in\mathcal{D}'(\mathbb{R}^n)$ of compact support is the result of applying this distribution (as a functional) to the test function $\left(\frac{1}{\sqrt{2\pi}}\right)^n e^{-i<z,t>}\kappa(t)$, where $\kappa(t)\in\mathcal{D}(\mathbb{C}^n)$ is some function which equals 1 on $suppF$. One can find more detailed

the above is true, then the support function $K(\lambda)$ of the convex hull of *supp*F (*convsupp*F) coincides with the P-indicator $P_f(\lambda)$, and

$$|f(x+iy)| \le c(1+|z|^N)e^{K(y)}, \quad \forall y\in\mathbb{R}^n,\ x\in\mathbb{R}^n,$$

where N is the order[35)] of the distribution F and $c = c(f)$ is a constant.

A proof of **Theorem 5.1.2** can be found in, for example, V.Napalkov [1], L.Hörmander [1].

Note that the estimate figuring in **Theorem 5.1.2** evidently implies the inequality

$$\mathfrak{L}(z) \le K(y),\ \forall z\in\mathbb{C}^n,\ z = x+iy,$$

and since $K(y)$ is a continuous function, the inequality

$$\mathfrak{L}^*(z) \le K(y),\ \forall z\in\mathbb{C}^n,$$

follows too. From this taking into account that $K(y) = P_f(y)$ and $P_f(y) \le \mathfrak{L}_f^*(iy)$, we conclude that for the functions figuring in **Theorems 5.1.1** and **5.1.2** the following equality

$$P_f(\lambda) = \mathfrak{L}_f^*(iy),\ \forall\lambda\in\mathbb{R}^n,$$

is valid.

2. On the existence of a function with prescribed indicator. In **Chapter 3** we obtained **Theorem 3.3.2**, which is a strengthening of the Martineau-Kiselman theorem on the existence of an entire function with prescribed indicator. This theorem was obtained as a consequence of Sigurdsson's theorem (**Theorem 3.3.1**) on the approximation of

---

information on Fourier transforms of distributions in, for example, V.Vladimirov [1], L.Hörmander [1].

35) The smallest integer m such that for some $c = c(m)$ the inequality

$$|F(\varphi)| \le c \sum_{k\in\mathbb{Z}_+^n,\ \|k\|\le m} \left| \frac{\partial^{\|k\|}\varphi}{\partial x^k} \right|,\ \forall\varphi\in D(\mathbb{R}^n),$$

is valid, is called to as the order of the distribution $F\in\mathcal{D}'(\mathbb{R}^n)$ of compact support. It is known that the order of a distribution of compact support is finite.

plurisubharmonic functions by functions of the form $ln|f|$, $f\in H(\mathbb{C}^n)$. Here we give a refinement of the Martineau-Kiselman theorem for indicators having some special properties.

Theorem 5.1.3. Let $h(z)$ be a positively homogeneous plurisubharmonic function of degree 1 that satisfies the conditions[36)]

1) $h(z) \le c_1|y|$, $\forall z = x+iy\in\mathbb{C}^n$;

2) $\exists\eta > 0$: $|h(z') - h(z'')| < c_2|z' - z''|^{\eta}, \forall z', z''\in S_1 \subset \mathbb{C}^n$;

where c_1 and c_2 are constants.

Then there is a function $f\in H(\mathbb{C}^n, 1]$ such that

$$\mathcal{L}^*_f(z) = h(z),\ \forall z\in\mathbb{C}^n,$$

and for certain constants c_3 and N, depending on F, the inequality

$$|f(z)| \le c_3(1+|z|^2)^N e^{h(z)} \tag{5.1.1}$$

is valid.

Proof. We follow of Martineau's proof of the general theorem on the existence of an entire function with prescribed indicator (see, for example, L.Ronkin [1], P.Lelong, L.Gruman [1]). First of all we prove a lemma on the possibility to of choosing a countable family of functions that separates an arbitrary function from a fixed majorant. This lemma differs from the one used in Martineau's proof by the replacement of the semicontinuous functions figuring there by continuous functions.

Lemma 5.1.1. Let $u(z)$ be a real-valued continuous function on a compact set $K \subset \mathbb{C}^n$ and let P_u be the family of all upper semicontinuous functions $v \not\equiv u$ which are majorized by $u(z)$. Then there is a sequence of continuous functions $u_k\in P_u$ satisfying the condition: for any $v\in P_u$ there exist $k_0\in\mathbb{Z}_+$ and $\delta > 0$ such that for every $z\in K$ the inequality $v(z) \le u_{k_0}(z)$ is valid, and at the points $z\in K$ at which $u_{k_0}(z) < u(z)$, the inequality $v(z) < u_{k_0}(z)-\delta$ is true.

Proof. We choose a sequence of numbers $\varepsilon_k\downarrow 0$, and for each $m\in\mathbb{N}$ a cover the set K by a finite number of balls $B(z^{(j,m)},\varepsilon_m)$, $j = 1,\dots,\omega(m)$. Then we construct functions $u_{j,m}\in C(K)$ in such a way that $u_{j,m} = u$

---

36) It is easy to see that condition 1) is equivalent to: $h(x)=0$, $\forall x\in\mathbb{R}^n$.

on $K\setminus B(z^{(j,m)},\varepsilon_m)$, $u_{j,m} \ge u-\varepsilon_m$ on K, and $u_{j,m}(z^{(j,m)}) = u(z^{(j,m)})-\varepsilon_m$. This is clearly possible.

We show that the sequence obtained by the enumeration of the functions belonging to the countable set $\bigcup_m \bigcup_j u_{j,m}$ as the required. First we consider the case of a continuous function $v\in P_u$. Then there exist $z^0\in K$, $\varepsilon > 0$, and $\delta > 0$ such that $v(z) \le u(z) -\varepsilon$ for $|z-z^0| < \varepsilon$. Now we choose m in such a way that $\varepsilon_m < min\{\delta/2,\varepsilon/2\}$. Further we find some j such that $z^0\in B(z^{(j,m)},\varepsilon_m)\cap K$. It is obvious that $B(z^{(j,m)},\varepsilon_m) \subset B(z^0,\delta)$. Hence $v(z) \le u(z) - \varepsilon \le u^{(j,m)}(z) - \frac{\varepsilon}{2}$, $\forall z\in B(z^{(j,m)},\varepsilon_m)\cap K$. At the same time, since $v\in P_u$ we have $v(z) \le u^{(j,m)} = u(z)$ on $K\setminus B(z^{(j,m)},\varepsilon_m)$. Thus we have proved the lemma in the case of continuous functions. To reduce the general case to this case it is clearly sufficient to show that $\forall v\in P_u$ there is a continuous function $\tilde{v}\in P_u$ such that $v \le \tilde{v} \le u$. To construct such a function we represent the function v in the form $v = lim v_j$, where $v_j\in C(K)$ and $v_j \ge v_{j+1}$, $\forall j$ (this is possible because of the semicontinuity of the function). Then we set $\tilde{v} = min\{u,v_{j_0}\}$, where j_0 is chosen so large that $\tilde{v} \not\equiv u$. It is obvious that $\tilde{v}\in P_u$ and $v \le \tilde{v} \le u$.

The proof is finished.

Next to the statement which is similar to the above lemma, an essential role in Martineau's proof of the Martineau-Kiselman theorem is played by **Hörmander's Theorem** on holomorphic continuation from a subspace (see, for example, L.Hörmander [2], and also L.Ronkin [1], Lelong-Gruman [1]). The following lemma is a variant of **Hörmander's Theorem**, adapted to our problem.

Lemma 5.1.2. Let a function $h(z)\in PSH(\mathbb{C}^n)$ satisfy the condition

$$|h(z')-h(z'')| \le A(|z'|+|z''|)^{\alpha}|z'-z''|^{\beta} + B, \ \forall z\in\mathbb{C}^n, \tag{5.1.2}$$

for certain constants $\alpha,\beta,A,B > 0$. Let also $\sum$ be a subspace of complex codimension k and $d\sigma$ be the 2(n-k)-dimensional volume element on $\sum$ which is induced by the metric of $\mathbb{C}^n$. Then there is constant c such that for every function $f\in H(\sum)$ satisfying the condition

$$\int_{\Sigma} |f|^2 e^{-h} d\sigma < \infty , \tag{5.1.3}$$

there is a function $F\in H(\mathbb{C}^n)$ coinciding with f on Σ and such that

$$\int_{\mathbb{C}^n} |F|^2 e^{-h}(1+|z|^2)^{-q} d\omega_z \le c\cdot \int_{\Sigma} |f|^2 e^{-h} d\sigma, \tag{5.1.4}$$

where $q = \left(3+ \frac{\alpha}{\beta}\right)k$.

Proof. Since the function $ln(1+|z|^2)$ clearly satisfies the condition (5.1.2), the proof of the lemma in the general case can readily be reduced to the case $k = 1$, $\Sigma = \{ z\in\mathbb{C}^n: z_n = 0\}$ and $f = f(z')\in H(\mathbb{C}^{n-1})$, $z' = (z_1,\dots,z_{n-1})$, $d\sigma = d\omega_{z'}$. Under this condition we look for the function F in the form

$$F(z) = \Phi(z)f(z') - z_n v(z), \tag{5.1.5}$$

where Φ and v are smooth functions and $\Phi(z',0) = 1$. It is clear that the function F defined in such a way satisfies the conditions $F|_{\Sigma} = f$, i.e. $F(z',0) = f(z')$, and $F\in H(\mathbb{C}^n)$ if

$$\bar{\partial}v = \frac{f\bar{\partial}\Phi}{z_n} . \tag{5.1.6}$$

We define an appropriate function $\Phi(z)$.

Let $\psi\in C^\infty(\mathbb{C})$ be a function such that $0 \le \psi \le 1$, $\psi(w) = 1$ for $|w| < \frac{1}{2}$, $\psi(w) = 0$ for $|w| > 1$. We set $\Phi(z',z_n) = \psi(|z'|^{\alpha/\beta}z_n)$ for $|z'| > 1$ and for $|z'| \le 1$, $|z_n| < \frac{1}{2}$, and $\Phi(z',z_n) = 0$ for $|z'| \le 1$, $|z_n| > 2$. Further we define $\Phi(z',z_n)$ for $|z'| < 1$, $\frac{1}{2} \le |z_n| \le 2$ such that the function Φ be smooth in $\mathbb{C}^n$ and $0 \le \Phi \le 1$. Note that by the above choice of the function Φ the inequality

$$|\bar{\partial}\Phi(z)| \le c_1|z'|^{\alpha/\beta}, \ \forall z\in\mathbb{C}^n, \tag{5.1.7}$$

is valid for certain constant c_1.

We estimate $|h(z',z_n)-h(z',0)|$ on *supp*Φ. If $z\in$*supp*Φ and $|z'| > 1$, then $|z_n| \le |z'|^{-\alpha/\beta}$. Therefore by (5.1.2) we have

$$|h(z',z_n) - h(z',0)| \le A(|z|+|z'|)^2|z_n|^\beta + B \le$$

$$\le A(3|z'|)^{\alpha}|z_n|^{\beta} + B \le A(3|z_n|^{-\alpha/\beta})^{\alpha}|z_n|^{\beta} + B = A3^{\alpha} + B.$$

From this and the boundedness of the set $\{ z \in supp\Phi\colon |z'| < 1\}$ we obtain that

$$sup\{|h(z',z_n) - h(z',0)|: (z',z_n) \in supp\Phi\} = c_2 < \infty. \qquad (5.1.8)$$

Since we intend to solve equation (5.1.6) using **Hörmander's Theorem** on the solvability of the $\bar{\partial}$-problem in weighted spaces, we estimate the integral

$$I = \int_{\mathbb{C}^n} \left|\frac{f(z')}{z_n}\bar{\partial}\Phi\right|^2 (1+|z|^2)^{-\alpha/\beta} e^{-h(z)} d\omega_z .$$

Using the estimates (5.1.8) and (5.1.7) we obtain

$$I \le c_1 \int_{supp\bar{\partial}\Phi} \left|\frac{f(z')}{z_n}\right|^2 (1+|z|^2)^{-\alpha/\beta} e^{-h(z',0)} e^{h(z',0)-h(z',z_n)} d\omega_z \le$$

$$\le e^{c_2} c_1 \int_{supp\bar{\partial}\Phi} e^{-h(z',0)} |z_n|^{-2}|f(z')|^2 d\omega_{z'} d\omega_{z_n} \le$$

$$\le e^{c_2}\cdot c_1 \int_{\mathbb{C}^{n-1}\setminus B_1} e^{-h(z',0)} |f(z')|^2 d\omega_{z'} \cdot \int_{\frac{1}{2}|z'|^{-\alpha/\beta}<|z_n|<|z'|^{-\alpha/\beta}} \frac{d\omega_{z_n}}{|z_n|^2} +$$

$$+ e^{c_2} c_1 \int_{|z'|<1} e^{-h(z',0)} |f(z')|^2 d\omega_{z'} \cdot \int_{\frac{1}{2}<|z_n|<2} \frac{d\omega_{z_n}}{|z_n|^2} \le$$

$$\le c_3 \int_{\mathbb{C}^{n-1}} e^{-h(z',0)} |f(z')|^2 d\omega_{z'} < \infty .$$

Referring now to the above-mentioned **Hörmander's Theorem**, we conclude that equation (5.1.6) has a solution $v(z)$ that satisfies the condition

$$\int_{\mathbb{C}^n} |v(z)|^2(1+|z|^2)^{-(2+\alpha/\beta)}e^{-h(z)}d\omega_z \le c_3 \int_{\mathbb{C}^{n-1}} e^{-h(z',0)}|f(z')|^2 d\omega_{z'} .$$

With this choice of v(z) the function $F = \Phi\cdot f - z_n\cdot v$ is entire, coincides with f on Σ, and satisfies the estimate:

$$\int_{\mathbb{C}^n} |F|^2(1+|z|^2)^{-(3+\alpha/\beta)}e^{-h}d\omega_z \le$$

$$\le 2 \int_{supp\Phi} |f|^2(1+|z|^2)^{-(3+\alpha/\beta)}e^{-h}d\omega_z +$$

$$+ 2 \int_{supp\bar\partial\Phi} |z_n|^2|v|^2(1+|z|^2)^{-(3+\alpha/\beta)}d\omega_z \le$$

$$\le 2e^{c_2}\int_{\mathbb{C}^{n-1}} e^{-h(z',0)}|f(z')|^2 d\omega_{z'}\cdot \int_{|z_n|<2} d\omega_{z_n} +$$

$$+ 2 \int_{|z_n|<2} |v|^2(1+|z|^2)^{-(2+\alpha/\beta)}e^{-h(z)}d\omega_z \le$$

$$\le (2e^{c_2}4\pi + 2c_3)\int_{\mathbb{C}^{n-1}} e^{-h(z',0)}|f(z')|^2 d\omega_{z'} = c\int_{\Sigma} |f|^2e^{-h}d\sigma.$$

This finishes the proof of the lemma.

Let $\varphi(z)$ be a positive continuous function on $\mathbb{C}^n$. We denote by E_φ the family of all entire functions satisfying the condition

$$\|f\|_\varphi \overset{def}{=} sup\{ |f(z)|\varphi^{-1}(z) : z\in\mathbb{C}^n\} < \infty.$$

This family is a Banach space with norm $\|f\|_\varphi$. If $\varphi_1 \le \varphi_2$, then $E_{\varphi_1} \subset E_{\varphi_2}$, and the operator I_{φ_1,φ_2} that identically maps E_{φ_1} into E_{φ_2} is bounded.

We set

$$H_N(z) = (1+|z|)^N e^{h(z)},$$

where the function $h(z)$ is the same as in the theorem to be prove. It is clear that if $f\in E_{H_N}$, then $\mathfrak{L}_f^* \le h(z)$. **Theorem 5.1.3** will be proved if we show the following. For some N there is at least one $z \ne 0$ such that the set of all functions $f\in E_{H_N}$ for which $\mathfrak{L}_f^* < h(z)$ does not coincide with all of space E_{H_N}. As a matter of fact, we prove somewhat more, namely, that the above set of functions is a set of the first category. For this purpose we choose by **Lemma 5.1.1** a countable set of functions $u_k\in C(S_1)$, $k = 1, 2, \dots$, approximating the function $u(z) = h(z)$ on the compact set $K = S_1$. We continue these functions to the whole space $\mathbb{C}^n$ as positive homogeneous functions of degree 1. We denote them continuations also by u_k. Then we set $V_{k,N} = (1+|z|^2)^N e^{u_k(z)}$, and consider the spaces $E_{V_{k,N}}$ and corresponding operators $I_{V_{k,N},H_N}$.

Lemma 5.1.3. For each $k = 1, 2, \dots$ and

$$N = n\cdot\left(3 + \frac{\eta}{1-\eta}\right) + \left(\frac{1}{\eta} - 1\right)^n,$$

where η is the same as in **Theorem 5.1.3**, the operator $I_{V_{k,N},H_N}$ is not surjective.

Proof. In accordance with the definition of the functions u_k there is for each k a point $z_0 = z_0(k)$ such that $u_k(z_0) < h(z_0)$. It is also clear that $\mathfrak{L}_f^* \le u_k$, $\forall f\in E_{u_k}$. Therefore, to prove the lemma it suffices to show that there is a function $F(z)$ in E_{H_N} such that $\mathfrak{L}_F^*(z_0) = h(z_0)$.

We set $\Sigma = \{ z\in\mathbb{C}^n\colon z = z_0 w,\ w\in\mathbb{C}\}$ and $\tilde{h}(w) = h(z_0 w)$. The function $\tilde{h}(w)$ is subharmonic and positively homogeneous of degree 1. Hence, it is convex and, therefore, is the support function of some convex set, namely of the set

$$\Gamma = \bigcap_{w\in\mathbb{C}} \{ \lambda\in\mathbb{C}\colon Re\lambda\bar{w} < \tilde{h}(w)\}.$$

Let a be a support point on the line $Re\lambda = \tilde{h}(1)$, i.e. $a\in\Gamma$. Then we define the function $f\in H(\Sigma)$ by the equality $f(z_0 w) = e^{\bar{a}w}$. From the definitions of the function $\tilde{h}(w)$ and the point a it follows that

$$|f(z_0w)| \le e^{\tilde{h}(w)} = e^{h(z_0w)}, \quad \forall w\in\mathbb{C},$$

and therefore

$$\int_{\Sigma} |f(z)|^2 e^{-2h(z)}(1+|z|^2)^{-2} d\sigma < \infty.$$

It is easy to see that if h(z) satisfies condition 2) of **Theorem 5.1.3**, then the conditions (5.1.2) are valid with $\alpha = 1-\eta$, $\beta = \eta$ and some constants A, B for this function as well as for the function $h(z) + N\ln(1+|z^2|)$. Thus, **Lemma 5.1.2** can be applied. In accordance with this lemma there is an entire function F(z) which coincides with f on Σ and such that

$$\int_{\mathbb{C}^n} |F|^2 e^{-2h}(1+|z|^2)^{-q} d\omega_z = C < \infty, \tag{5.1.9}$$

where $q = \left(3 + \frac{1-\eta}{\eta}\right)n$. We show that $F\in E_{H_N}$. Indeed,

$$|F(z)| \le \frac{1}{\omega_{2n}r^{2n}} \int_{B(z,r)} |F(\zeta)|^2 d\omega_\zeta \le$$

$$\le \frac{1}{\omega_{2n}r^{2n}} \mathit{sup}\{ e^{2h(\zeta)}(1+|\zeta|^2)^q\colon \zeta\in B(z,r)\}\times$$

$$\times \int_{B(z,r)} |F(\zeta)|^2 e^{-2h(\zeta)}(1+|\zeta|^2)^{-q} d\omega_\zeta \le$$

$$\le \frac{1}{\omega_{2n}r^{2n}} \sup_{B(z,r)} \{ e^{2h(\zeta)}(1+|\zeta|^2)^q\}.$$

Here we set $r = |z|^{1-1/\eta}$. Again taking into account that the function $h(z) + q\ln(1+|z|^2)$ satisfies the condition (5.1.2) with $\alpha = 1-\eta$, $\beta = \eta$, we obtain

$$|F(z)| \le \sqrt{\frac{c}{\omega_{2n}}}\ \frac{(1+|z|^2)^q e^{h(z)}}{|z|^{n(1-1/\eta)}} \sup_{|z-\zeta|<|z|^{1-1/\eta}} \left\{ \frac{e^{h(z)}(1+|\zeta|^2)^q}{(1+|z|^2)^q e^{h(z)}} \right\} \le$$

$$\le \sqrt{\frac{c}{\omega_{2n}}}\, e^{h(z)}(1+|z|^2)^N\cdot exp\left\{A\left(2|z|+|z|^{1-1/\eta}\right)^{1-\eta}|z|^{\eta(1-1/\eta)}+B\right\} =$$

$$= \text{const}\cdot e^{h(z)}(1+|z|^2)^N. \qquad (5.1.10)$$

Hence $F\in E_{H_N}$.

Note that (5.1.10) also implies the inequality

$$\mathcal{L}^*_f(z) \le h(z),\ \forall z\in\mathbb{C}^n.$$

At the same time, since $Rea = \tilde{h}(1) = h(z_0)$ and the function F coincides on Σ with the function $f = e^{\bar{a}w}$, we have $\mathcal{L}^*_F(z_0) \ge \mathcal{L}_F(z_0) = Rea = \tilde{h}(1) = h(z_0)$. Therefore $\mathcal{L}^*_f(z_0) = h(z_0)$ and hence $F\notin E_{u_k}$. Thus, the operator $I_{V_{k,N},H_N}$ is not surjective.

This finishes the proof of the lemma.

From the boundedness and non-surjectivity of the operator $I_{V_{k,N},H_N}$ it follows by the **open mapping Theorem** (see, for example, W.Rudin [1]) that the space $E_{V_{k,N}}$, regarded as a set in the space E_{H_N}, is a set of the first category. Hence the union $\bigcup_k E_{u_k}$ is a set of the first category too. Therefore $\bigcup_k E_{u_k} \ne E_{H_N}$. Let F be a function in E_{H_N}. Its indicator $\mathcal{L}^*_F$ satisfies the inequality $\mathcal{L}^*_F(z) \le h(z)$, $\forall z\in\mathbb{C}^n$. If the strict inequality $\mathcal{L}^*_F(z^0) \le h(z^0)$ takes place at at least one point z^0, then the function $\mathcal{L}^*_F(z)$ belongs to the set P_u with $u = h$. Recall that this set was defined in **Lemma 5.1.1**. Then, in accordance with the properties of the functions u_k, the inequalities

$$\left.\begin{aligned} &\mathcal{L}^*_F(z) \le u_{k'}(z),\ \forall z\in\mathbb{C}^n, \\ &\mathcal{L}^*_F(z) \le u_{k'}(z)-\delta|z|,\ \forall z\in K \overset{\text{def}}{=} \{\, z:\ u_{k'}(z) < h(z)\} \end{aligned}\right\} \qquad (5.1.11)$$

hold for some $k = k'$ and $\delta > 0$.

It is well-known, and was already mentioned above in the proof of **Lemma 2.3.1**, that for any $z^0\in S_1$ and $A > \mathcal{L}_F(z^0)$ there are numbers $r = r(z^0,A) > 0$ and $c = c(z^0,A)$ such that

$$|F(z)| \le ce^{A|z|}, \ \forall z: \frac{z}{|z|} \in B(z^0, r).$$

We set $A = u_{k'}(z^0) - \delta + \varepsilon$, $\varepsilon > 0$. Then let r be so small that

$$|u_{k'}(z) - u_{k'}(z^0)| \le \varepsilon, \ \forall z \in \bar{K} \cap B(z^0, r).$$

Now, taking into account the continuity of the functions u_k and the possibility of choosing a finite number of balls $B(z^0, r)$ covering $\bar{K}$, we conclude that for any $\varepsilon > 0$ there is a constant c_ε such that the inequality

$$|F(z)| \le c_\varepsilon \mathit{exp}\,\{ u_{k'}(z) - \delta|z| + 2\varepsilon|z|\}, \ \forall z \in \bar{K}, \qquad (5.1.12)$$

holds. In particular,

$$|F(z)| \le c_{\delta/2} e^{u_{k'}(z)}, \ \forall z: \frac{z}{|z|} \in K.$$

At the same time, since $F \in E_{H_N}$,

$$|F(z)| \le c(1+|z|^2)^N \cdot e^{u_{k'}(z)}, \ \forall z: u_{k'}(z) = u(z).$$

From this and (5.1.12) we find that the function $F(z)$, i.e. a function $F \in E_{H_N}$ having the property $\mathfrak{L}^*_F(z^0) < h(z^0)$, belongs to the space $E_{V_{k'}}$, $k' = k'(z^0)$. Hence, if $F \in E_h \setminus \bigcup_k E_{V_k}$, then $\mathfrak{L}_F(z) = h(z)$, $\forall z \in \mathbb{C}^n$. Thus, any function in $E_h \setminus \bigcup_k E_{V_k}$, which is a non-empty set of the second category, is as required, i.e. a function whose existence is stated in **Theorem 5.1.1.**

This finishes the proof of the theorem.

3. An Example of function which is bounded on $\mathbb{R}^2$ but which is not of c.r.g. In §3 of **Chapter 4** we showed that if a function $f \in H(\mathbb{C}, 1]$ satisfies the condition

$$\int_{-\infty}^{\infty} \frac{\mathit{ln}^+|f(x)|}{1+x^2}\,dx < \infty,$$

then f is of c.r.g. This condition is satisfied, in particular, if $\sup_{x\in\mathbb{R}}|f(x)| < \infty$, or $\sup_{x\in\mathbb{R}}|f(x)x^{-k}| < \infty$. It is natural to pose the question on the validity of similar statements in the multidimensional case. It was found that such statements are not true without additional assumptions on the function f. More exactly, the following theorem holds.

Theorem 5.1.4. There is a function $f \in H(\mathbb{C}^2, 1]$ satisfying the condition[37)]

$$\int_{\mathbb{R}^2} |f(x)|^2 d\omega_x < \infty \tag{5.1.13}$$

and not of c.r.g. in $\mathbb{C}^2$.

Proof. Note that the property of a function to be or not to be of c.r.g. does not change when it is multiplied by another function of c.r.g., in particular by the function $(z_1^{-1} z_2^{-1} \sin z_1 \sin z_2)^N$. Therefore, to prove the theorem it suffices to construct a function $g \in H(\mathbb{C}^2, 1]$ satisfying the condition

$$|g(x)| \leq c(1+|x|)^N, \ \forall x \in \mathbb{R}^2,$$

for some $c > 0$ and $N > 0$. We choose such a function in the form

$$g = \Psi \cdot f - u,$$

where the function $f \in H(\mathbb{C}^2, 1]$ is constructed, in accordance with **Theorem 5.1.3** from the function

$$h(z) = h(z_1, z_2) = \left\{ \left(Im\sqrt{z_1^2 + z_2^2} \right)^2 + (Im z_1)^2 + (Im z_2)^2 \right\}^{1/2}. \tag{5.1.14}$$

The possibility of using this function for the above purpose follows from the following lemma.

Lemma 5.1.4. Let the function h(z) be defined by (5.1.14). Then h(z) is plurisubharmonic in $\mathbb{C}^2$, strictly plurisubharmonic on $\Omega = \{ z = (z_1, z_2) \in \mathbb{C}^2 : z_1 \bar{z}_2 \neq z_2 \bar{z}_1, \ z_1^2 + z_2^2 \neq 0\}$, and satisfies condition 2) of **Theorem 5.1.3** with $\eta = 1/2$.

Proof. It is obvious that the function h(z) is infinitely

[37)] It is well-known (see, for example, L.Ronkin [1]) that if a function f satisfies the condition (5.1.13), then it also satisfies the condition $\sup_{\mathbb{R}^2} |f(x)| < \infty$. Besides, by **Theorem 5.1.1** such functions are Fourier transforms of functions of compact support.

differentiable in the domain $\Omega_1 = \{ z\in\mathbb{C}^2 : z_1^2+z_2^2 \neq 0\}$. We represent $h(z)$ in a small neighbourhood of an arbitrary point of the domain Ω_1 in the form

$$h(z) = \Phi(u_1,u_2,u_3) \overset{\text{def}}{=} (u_1+u_2+u_3)^{1/2},$$

where $u_1 = Im z_1$, $u_2 = Im z_2$, and u_3 is a single-valued branch of the function $Im\sqrt{z_1^2+z_2^2}$. In turn, the function $u_3(z_1,z_2)$ can be written in the form $u_3 = \Phi_1(w)$, $w = z_1^2 + z_2^2$, where $\Phi_1(w)$ is a single-valued branch of the function $Im\sqrt{w}$ in the corresponding neighbourhood on the plane $\mathbb{C}$ which does not contain the point $w = 0$. Since the function $\Phi_1(w)$ is harmonic,

$$\frac{\partial^2 u_3}{\partial z_i \partial \bar{z}_j} = 0, \; i,j = 1,2.$$

Taking this into account, we obtain the following representation for the Levi form of the function $h(z)$

$$\sum_{i,j=1}^{k} \frac{\partial^2 h}{\partial z_i \partial \bar{z}_j} \zeta_i \bar{\zeta}_j = \sum_{k,l=1}^{3} \frac{\partial^2 \Phi}{\partial u_k \partial u_l} \left(\sum_{i=1}^{2} \frac{\partial u_k}{\partial z_i} \zeta_i \right)\left(\sum_{j=1}^{2} \frac{\partial u_l}{\partial \bar{z}_j} \bar{\zeta}_j \right) +$$

$$+ \sum_{k=1}^{3} \frac{\partial \Phi}{\partial u_k} \sum_{i,j=1}^{k} \frac{\partial u_k}{\partial z_i \partial \bar{z}_j} \zeta_i \bar{\zeta}_j = \sum_{k,l=1}^{3} \frac{\partial^2 \Phi}{\partial u_k \partial u_l} W_k \bar{W}_l , \qquad (5.1.15)$$

where

$$W_k = \sum_{i=1}^{2} \frac{\partial u_k}{\partial z_i} \zeta_i .$$

Note that the function $\Phi(u_1,u_2u_3)$ is convex with respect to the variables u_1, u_2, u_3, and therefore

$$\sum_{k,l=1}^{3} \frac{\partial^2 \Phi}{\partial u_k \partial u_l} \xi_k \xi_l \geq 0, \; \forall(\xi_1,\xi_2,\xi_3)\in\mathbb{R}^3. \qquad (5.1.16)$$

This implies that

$$\sum_{k,l=1}^{3} \frac{\partial^2 \Phi}{\partial u_k \partial u_l} W_k \bar{W}_l \geq 0, \; \forall W = (W_1,W_2,W_3)\in\mathbb{C}^3. \qquad (5.1.17)$$

This inequality and (5.1.15) imply that the Levi form of the function $h(z_1,z_2)$ is non-negative on the domain Ω_1. Thus $h\in PSH(\Omega_1)$. Noting that $h(z) \geq 0$, $\forall z\in\mathbb{C}^2$, we obtain $h\in PSH(\mathbb{C}^2)$.

To prove that the function $h(z)$ is strictly plurisubharmonic on the domain Ω we first note that equality in (5.1.16) holds if and only if $\xi_1/u_1 = \xi_2/u_2 = \xi_3/u_3$. It is easy to see that this implies that equality in (5.1.17) holds only if $W_1/u_1 = W_2/u_2 = W_3/u_3$. We show that the equalities $W_1/u_1 = W_2/u_2 = W_3/u_3$ can be valid only if $z_1\bar{z}_2 = z_2\bar{z}_1$. Indeed, if we express the components of these equalities in the terms of ζ_j and z_j, then we obtain

$$\frac{W_1}{u_1} = \frac{\zeta_1}{2i\,Im z_1}, \quad \frac{W_2}{u_2} = \frac{\zeta_2}{2i\,Im z_2},$$

$$\frac{W_3}{u_3} = \frac{1}{2i} \frac{z_1\zeta_1 + z_2\zeta_2}{\sqrt{z_1^2+z_2^2}\cdot Im\sqrt{z_1^2+z_2^2}}.$$

This implies that for the solvability (with respect to ζ_1 and ζ_2) of the system $W_1/u_1 = W_2/u_2 = W_3/u_3$ it is necessary that

$$z_1 Im z_1 + z_2 Im z_2 = \sqrt{z_1^2+z_2^2}\cdot Im\sqrt{z_1^2+z_2^2}.$$

Squaring both sides of this equality and after obvious transformations we obtain the equality

$$2 = \frac{z_1}{\bar{z}_1}\cdot\frac{\bar{z}_2}{z_2} + \frac{z_2}{\bar{z}_2}\cdot\frac{\bar{z}_1}{z_1}.$$

It is clear that this equality is valid only if $z_1\bar{z}_2 - z_2\bar{z}_1 = 0$. Hence, for $(z_1,z_2)\in\Omega_1$, $z_1\bar{z}_2 - z_2\bar{z}_1 \neq 0$ we have

$$\sum_{i,j=1}^{2} \frac{\partial^2 h}{\partial z_i \partial \bar{z}_j}\zeta_i\bar{\zeta}_j = \sum_{k,l=1}^{3} \frac{\partial^2 \Phi}{\partial u_k \partial u_l} W_k \bar{W}_l > 0, \ \forall\zeta\in\mathbb{C}^2\setminus\{0\}.$$

Thus we have proved that the function $h(z)$ is strictly plurisubharmonic on the domain $\Omega = \Omega_1\setminus\{ z\in\mathbb{C}^2:\ z_1\bar{z}_2 - z_2\bar{z}_1 \neq 0\}$.

To complete the proof of the lemma we note that since $\Phi_1(w)$ clearly satisfies condition 2) of **Theorem 5.1.3**, the function $|u_3|$, which is uniquely defined in $\mathbb{C}^3$, also satisfies this condition. Further, since

$\Phi(u_1,u_2,u_3)$ is smooth on S_1, the function

$$h(z) = \Phi(u_1(z_1,z_2),u_2(z_1,z_2),u_3(z_1,z_2))$$

also satisfies condition 2) of **Theorem 5.1.3.**

This finishes the proof of the lemma.

Now we define the function Ψ figuring in the representation $F = \Psi\cdot f - u$.

We choose a point z^0 in the domain $\Omega = \mathbb{C}^2\setminus\{ z\in\mathbb{C}^2:\ z_1\bar{z}_2 - z_2\bar{z}_1 = 0,\ z_1^2+z_2^2 = 0\}$ such that $\mathcal{L}_f(z^0) = h(z^0)$. This is possible because $\mathcal{L}_f^*(z) \equiv h(z)$ and the equality $\mathcal{L}_f(z) = \mathcal{L}_f^*(z)$ is valid for almost all $z\in\mathbb{C}^2$ (see, for example, L.Ronkin [1], Lelong-Gruman [1]). Let $r > 0$ be such that $B(z^0,2r) \subset\subset \Omega$. Finally we choose a sequence $t_k\uparrow+\infty$ such that

$$\lim_{k\to\infty} \frac{\ln|f(t_k z^0)|}{t_k} = h(z^0) \tag{5.1.18}$$

and

$$B(t_k z^0,2t_k r) \cap B(t_{k+1}z^0,2t_{k+1}r) = \emptyset.$$

Then we define the function $\Psi(z)$ by

$$\Psi(z) = 1 - \sum_{k=1}^{\infty} \alpha\left(\frac{z-t_{2k}z^0}{2rt_{2k}}\right),$$

where $\alpha(z)\in C^\infty(\mathbb{C}^2)$ satisfies the conditions: $\alpha(z) = 1$ when $|z| < \frac{1}{2}$, $\alpha(z) = 0$ when $|z| > 1$, $0 \le \alpha(z) \le 1$ for any $z\in\mathbb{C}^2$. It is clear that

$$0 \le \Psi(z) \le 1,\ \forall z\in\mathbb{C}^2,$$

$$\Psi(z) = 1,\ \forall z\notin \bigcup_{k=1}^{\infty} B(t_{2k}z^0,t_{2k}r),$$

$$\Psi(z) = 0\ ,\ \forall z\in \bigcup_{k=1}^{\infty} B(t_{2k}z^0,t_{2k}r)$$

and

$$|\bar{\partial}\Psi(z)| = O\left(\frac{1}{|z|}\right) \text{ as } z \to \infty.$$

To define the function u figuring in the representation $F = \Psi\cdot f - u$, we note that the function F is holomorphic if and only if $\bar{\partial}F = 0$, and therefore u must be a solution of the equation $\bar{\partial}u = f\bar{\partial}\Psi$. To find such a solution, and moreover, a solution having some special properties, similarly as in the proof of the previous theorem we use **Hörmander's Theorem** on the solvability of the $\bar{\partial}$-problem in weighted spaces. As weight function we consider the function

$$P(z) = 2(1-\delta\kappa(z))h(z) + (N+3)ln(1+|z|^2),$$

where N is the as in **Theorem 5.1.3**,

$$\kappa(z) = \sum_{k=1}^{\infty}\alpha\left(\frac{z-t_k z^0}{rt_k}\right),$$

and $\delta > 0$ is such that $P\in PSH(\mathbb{C}^2)$. The existence of such a (sufficiently small) δ easily follows from the fact that h is positively homogeneous and strictly plurisubharmonic in Ω, and $B(z^0,2r) \subset\subset \Omega$.

We show that

$$\int_{\mathbb{C}^2}|\bar{\partial}\Psi|^2|f|^2e^{-P}d\omega_z < \infty.$$

Note that $\kappa(z) = 0$, $\forall z\in supp\bar{\partial}\Psi$, and by the choice of the function f the estimate

$$|f(z)| \le c(1+|z|^2)^N e^{h(z)}$$

holds. Hence

$$\int_{\mathbb{C}^2}|\bar{\partial}\Psi|^2|f|^2e^{-P}d\omega_z = \int_{supp\bar{\partial}\Psi}|\bar{\partial}\Psi|^2|f|^2e^{-2h(z)-(N+3)ln(1+|z|^2)}d\omega_z \le$$

$$\le \int_{\mathbb{C}^2}O\left(\frac{1}{|z|}\right)(1+|z|^2)^{-3}d\omega_z < \infty.$$

Thus, **Hörmander's Theorem** can be applied. Therefore, there is a function $u\in L^2_{loc}(\mathbb{C}^2)$ such that $\bar{\partial}u = f\bar{\partial}\Psi$ and

$$\int_{\mathbb{C}^2}|u|^2e^{-P(z)-2ln(1+|z|^2)}d\omega_z = c_1 < \infty. \tag{5.1.19}$$

If the functions u, Ψ and f are chosen as above, then the function $F(z) = \Psi(z)f(z)-u(z)$ is entire, and

$$\int_{\mathbb{C}^2}|F(z)|^2e^{-h(z)-(N+5)ln(1+|z|^2)}d\omega_z = c_2 < \infty. \tag{5.1.20}$$

From this it follows that $F\in H(\mathbb{C},1]$ and $\mathfrak{L}^*_F(z) \le h(z)$, $\forall z\in\mathbb{C}^2$. Besides, since obviously $h(x+iy) \le h(y)$, $\forall x \in \mathbb{R}^2$, $y \in \mathbb{R}^2$, (5.1.20) implies in a standard manner that

$$|F(x)| \le \text{const}\cdot(1+|x|)^{N+5}.$$

Now we show that the ray $l_{z^0} = \{z = tz^0 : t > 0\}$ is not a ray of c.r.g. of the function $F(z)$. Note that the function $u(z)$ is holomorphic outside $supp\bar{\partial}\Psi$, and hence for all sufficient large k inequality

$$|u(z)|^2 \le \frac{1}{\omega_4} \int_{B(z,1)} |u(z)|^2 d\omega_z \le$$

$$\le \frac{1}{\omega_4} sup\left\{e^{P(\zeta)+2ln(1+|\zeta|^2)} : \zeta \in B(z,1)\right\}\cdot \int_{\mathbb{C}^2} \frac{|u|^2 e^{-P(z)}}{(1+|z|^2)^2} d\omega_z =$$

$$= \text{const}\cdot exp\{(1-\delta/2)h(z)+(N+5)ln(1+|z|^2)\} \qquad (5.1.21)$$

holds in every ball $B(t_k z^0, \frac{1}{2}t_k r)$. This it clearly implies that

$$\lim_{r'\to 0} \overline{\lim_{k\to\infty}} \frac{1}{t_k} \mathfrak{M}_{ln|u|}(t_k z^0, t_k r') \le (1-\delta/2)h(z^0). \qquad (5.1.21)$$

Since $\Psi(z) = 0$ on $B(t_{2k}z^0, t_{2k}r)$, $k = 1, 2, \ldots,$ then $u(z) = F(z)$, $\forall z \in B(t_{2k}z^0, t_{2k}r)$. Therefore, (5.1.22) implies that

$$\lim_{r'\to 0} \underline{\lim_{t\to\infty}} \frac{1}{t} \mathfrak{M}_{ln|u|}(tz^0, tr') \le$$

$$\le \lim_{r'\to 0} \overline{\lim_{k\to\infty}} \frac{1}{t_k} \mathfrak{M}_{ln|u|}(t_k z^0, t_k r') \le (1-\delta/2)h(z^0). \qquad (5.1.23)$$

At the same time, $\Psi(z) = 1$ on $B(t_{2k+1}z^0, t_{2k+1}r)$, $k = 1, 2, \ldots,$ and therefore (5.1.19) and (5.1.20) imply that

$$\lim_{k\to\infty} \frac{ln|F(t_{2k+1}z^0)|}{t_{2k+1}} = h(z^0).$$

Hence $\mathfrak{L}_F(z^0) = h(z^0)$, and we can write the inequality (5.1.23) in the form

$$\lim_{r'\to 0}\lim_{t\to\infty} \frac{1}{t}\, \mathfrak{N}_{ln|F|}(tz^0,tr') \le (1-\delta/2)\mathfrak{L}_F(z^0) < \mathfrak{L}_f^*(z^0).$$

In accordance with the definition of ray of c.r.g., the last inequality means that l_{z^0} is not a ray of c.r.g.

This finishes the proof of the theorem.

4. Sufficient conditions for c.r.g. The existence of entire functions of exponential type in $\mathbb{C}^n$ having a polynomial growth on $\mathbb{R}^n$ and not being c.r.g. leads to the problem of the additional conditions that would guarantee c.r.g. Simple, geometrically easy-to-interpret conditions of this sort are contained in the following **theorem of Wiegerinck**.

Theorem 5.1.5. Let $\Phi\in\mathcal{D}'(\mathbb{R}^n)$, *supp*$\Phi \subset\subset \mathbb{R}^n$ and *convsupp*Φ[38] be a polyhedron.

Then the function

$$F(z) = \left(\frac{1}{\sqrt{2\pi}}\right)^n \int_{\mathbb{R}^n} \Phi(t)e^{-i\langle z,t\rangle}dt$$

which belongs to $H(\mathbb{C}^n,1]$, is of c.r.g.

Proof. In accordance with **Theorem 2.3.1**, the function F is of c.r.g. if the set $Fr\,ln|f|$ (the limit set of the function $ln|F|$) consists of one element. First we show that all the functions $v\in Fr\,ln|F|$ are equal to one another on the imaginary space, i.e. on the set $i\mathbb{R}^n = \{z\in\mathbb{C}^n\colon z = iy,\ y\in\mathbb{R}^n\}$. Indeed, by **Theorem 5.1.2** the function F(z) is of polynomial growth on $\mathbb{R}^n$, and hence for any $\lambda\in\mathbb{R}^n$ the function $F_\lambda(w) = F(\lambda w)$ is of polynomial growth on the real axis of the complex w-plane. Hence $F_\lambda(w)$ is a function of c.r.g.; therefore there is a set E_λ of relative measure zero such that

$$\lim_{\substack{t\to\infty\\ t\notin E_\lambda}} \frac{ln|F(\lambda e^{i\theta}t)|}{t} = \mathfrak{L}_F(\lambda e^{i\theta}) . \tag{5.1.24}$$

As noted in §1 of **Chapter 4**, the equality $\mathfrak{L}^*(i\lambda) = \mathfrak{L}(i\lambda)$ holds for almost all $\lambda\in\mathbb{R}^n$. Therefore, (5.1.24) implies that for almost all $\lambda\in\mathbb{R}^n$,

---

[38] *conv* means a convex hull

$$\lim_{\substack{t \to \infty \\ t \notin E_\lambda}} \frac{\ln|F(ti\lambda)|}{t} = \mathfrak{L}_F^*(i\lambda) .$$

From this, using the sufficient condition for c.r.g. on a ray given at the end of §3.1 of **Chapter 2**, we conclude that almost all rays $l_{i\lambda}$, $\lambda \in \mathbb{R}^n$, are rays of c.r.g. of the function $F(z)$. Further, note that in accordance with **Theorem 5.1.2** and the **Remark** following it (on the relationship between the indicators $\mathfrak{L}_F^*$ and P_F) we have that the function $\mathfrak{L}_f^*(z)$ coincides with the support function of the set *convsupp*Φ on $i\mathbb{R}^n$. Therefore $\mathfrak{L}_f^*(z)$ is continuous on $i\mathbb{R}^n$. Hence we can apply **Theorem 2.3.6**, from which it follows that any ray $l_{i\lambda}, \lambda \in \mathbb{R}^n$, is of c.r.g. Now, using **Theorem 2.3.2**, we obtain that

$$v(iy) = \mathfrak{L}^*(iy), \ \forall y \in \mathbb{R}^n, \ v \in Fr\ln|F|. \tag{5.1.25}$$

Then we show that this coincidence of all functions $v \in Fr\ln|F|$ on $i\mathbb{R}^n$ implies the coincidence of these functions on all of $\mathbb{C}^n$. Since the set *convsupp*Φ is a polyhedron, the support function $K(y)$ of this set, which coincides, by **Theorem 5.1.2**, with the function $\mathfrak{L}_F^*(i\lambda)$, can be represented in the form

$$K(\lambda) = \sup_{1 \le j \le N} \langle x^{(j)}, \lambda \rangle,$$

where $x^{(1)}, \ldots, x^{(N)}$ are the vertices of the polyhedron *convsupp*Φ. We set

$$K_j = int\{\lambda : K(\lambda) = \langle x^{(j)}, \lambda \rangle\}.$$

Every such set is an open cone in $\mathbb{R}^n$ and $\bigcup_j \overline{K}_j = \mathbb{R}^n$. Since $\mathfrak{L}_F^*(z) \le P_F(y)$, $\forall z = x+iy \in \mathbb{C}^n$, and $v(z) \le \mathfrak{L}_F^*(z)$, $\forall z \in \mathbb{C}^n$, $v \in Fr\ln|F|$, we have

$$v(x+iy) \le \mathfrak{L}_F^*(x+iy) \le P_F(y), \ \forall x \in \mathbb{R}^n, \ y \in \mathbb{R}^n.$$

From this taking into account that $P_F(y) = K(y) = \mathfrak{L}_F^*(iy)$ and that $K(y) = \langle x^{(j)}, y \rangle$ on K_j, we obtain

$$v(x+iy) \le \mathfrak{L}_F^*(x+iy) \le \langle x^{(j)}, y \rangle, \ \forall x \in \mathbb{R}^n, \ y \in K_j . \tag{5.1.26}$$

At the same time, in accordance with (5.1.25) we have

$$v(iy) = \mathcal{L}^*_F(iy) = \langle x^{(j)},y\rangle, \ \forall x\in\mathbb{R}^n, \ y\in K_j, \ j = 1, \ldots . \qquad (5.1.27)$$

Since the functions $v(z)$ and $\mathcal{L}^*_F(z)$ are subharmonic, and the function $\langle x^{(j)},y\rangle$ is harmonic on $\mathbb{R}^n\times K_j$, (5.1.26) and (5.1.27) and the maximum principle imply that

$$v(x+iy) = \mathcal{L}^*_F(x+iy) = \langle x^{(j)},y\rangle, \ \forall y\in K_j, \ x\in\mathbb{R}^n.$$

Since the complement of the union $\bigcup_j \{\mathbb{R}^n\times K_j\}$ has Lebesgue measure zero we obtain that

$$v(z) = \mathcal{L}^*_F(z), \ \forall z\in\mathbb{C}^n, \ v\in Frln|F|.$$

Thus the set $Frln|F|$ consists of a single element $\mathcal{L}^*_F$. In accordance with **Theorem 2.3.1** this means that the function F is of c.r.g.

This finishes the proof of the theorem.

§2 Discrete uniqueness sets

Let M be a family of functions which are holomorphic in a domain $G \subset \mathbb{C}^n$. A set $E \subset G$ is called a uniqueness set for the functions in M if the implication

$$f\in M, \ f(z) = 0, \forall z\in E \ \Rightarrow \ f(z) \equiv 0$$

is true; in other words, the set E is not contained in the support of the divisor Z_f for any function $f \not\equiv 0$ in M.

We can also interpret the concept of uniqueness set in more general manner. Namely, to each point $z\in E$ we associate a natural number $\gamma(z)$. We call E a uniqueness set with multiplicity $\gamma(z)$ (or, for brevity, simply a uniqueness set) for the class M if: the $f\in M$, $f(z) = 0$ and the multiplicity $\gamma_f(z)$ of zero of the function $f(z)$ at the point z is not less than $\gamma(z)$, $\forall z\in E$, imply $f(z) \equiv 0$.

The problem of the conditions under which E is a uniqueness set for $f\in M$ is a traditional problem of the theory of holomorphic functions. It is obvious that for entire functions of one variable, E must *a priori* be a discrete set. In this case the problem whether a set is a uniqueness set can,as a rule, be solved by constructing and studying the Weierstrass' canonical product with zeros on E. There are many studies along this way; they are reflected in the papers of N.Levinson [1], B.Levin [1], R.Boas [1], A.Leont'ev [1], and others. That is why we do not consider the one-dimensional case here.

The divisor of an entire function, more exactly, its support, is not discrete in the multidimensional situation. However, it is quite natural to consider discrete uniqueness sets in $\mathbb{C}^n$, $n > 1$, because problems on the completeness of function systems (in particular, of multidimensional systems of exponentials), problems on interpolation, etc., lead to such sets. It is clear that the methods connected with the application of the canonical Weierstrass' product and its analogues are not applicable for the study of discrete uniqueness sets of functions of several variables. The specific "multidimensional" method of study of these sets consists in obtaining and comparing upper and lower bounds on the volume of the zero set for a function from the class under consideration in certain specific neighbourhoods of these sets. The following **Lelong's Theorem** lays at the basis of this method.

Theorem 5.2.1. Let $f(z)$ be holomorphic in the ball $B_R \subset \mathbb{C}^n$ and vanish at the point $z = 0$ with multiplicity $\gamma_f(0)$. Then the function $t^{-2n+2}V_f(B_t)$, $0< t < R$, where $V_f(B_t)$ is the volume of the divisor[39] Z_f in the ball B_t, is a monotonically increasing function and

$$\lim_{t \to 0} \frac{1}{V_{2n-2}} t^{-2n+2}V_f(B_t) = \gamma_f(0),$$

where V_{2n-2} is the volume of the ball B_1 in the space $\mathbb{R}^{2n-2}$.

A proof of this theorem can be found in, for example, Lelong-Gruman [1], E.Čirka [1], L.Ronkin [1].

Before giving conditions of uniqueness for entire functions bounded on the real space that were obtained along the above-mentioned way, we introduce some necessary notation.

If $x \in \mathbb{R}^n$ and $z = x+iy \in \mathbb{C}^n$, then we set $\|x\|_\infty = \max\{|x_j|: 1 \le j \le n\}$ and $\|z\|_\infty = \max\{\|x\|_\infty, \|y\|_\infty\}$. We also denote by $T(x^0,\alpha)$ and $T_h(x^0,\alpha)$ the n-dimensional cube $\{x \in \mathbb{R}^n: \|x-x^0\|_\infty < \alpha\}$ and the 2n-dimensional parallelepiped $\{z=x+iy \in \mathbb{C}^n: \|x-x^0\|_\infty < \alpha, \|y\|_\infty < h\}$, respectively. Then

---

[39] See §1.1 of **Chapter 3**.

we consider a pair (E,γ), where E is a discrete set in $\mathbb{R}^n$ and $\gamma = \gamma(x)$ is a function on E taking on values in $\mathbb{N}$. We shall characterize the "largeness" of the set E, or, more exactly, of the pair (E,γ) in the direction of a ray l_{x^0} by the upper density $\delta_{E,\gamma}(x^0)$ in the direction l_{x^0}, which is defined as follows. Let $\mu_{E,\gamma}(r;x^0,\alpha)$ be the number of points of E in the "cube" $T(rx^0,r\alpha)$, counted with the multiplicities $\gamma(x)$, i.e.

$$\mu_{E,\gamma}(r;x^0,\alpha) = \sum_{x\in E\cap T(rx^0,r\alpha)} \gamma(x).$$

Then

$$\delta_{E,\gamma}(x^0) = \overline{\lim_{\alpha\to 0}}\ \overline{\lim_{r\to\infty}}\ \frac{1}{(2\alpha r)^n}\mu_{E,\gamma}(r;x^0,\alpha)\ .$$

If $x^0 = 0$, then $\delta_{E,\gamma}(x^0)$ is simply called the upper density, and is denoted by $\delta_{E,\gamma}$.

A set of points $x\in\mathbb{R}^n$ such that $f(x) = 0$, $f\in H(\mathbb{C}^n)$ fixed, can contain a continuum. Therefore, to characterize the conditions under which E is a uniqueness set for one or another class of functions in terms of $\mu_{E,\gamma}(r;x^0,\alpha)$ it is necessary to require that E satisfies some condition of "sparseness". In the present situation such a condition is the positivity of

$$h_E = \frac{1}{2}\ \inf\{\|x'-x''\|:\ x'\in E,\ x''\in E,\ x' \neq x''\}.$$

We set

$$W_n = \inf_{\alpha>0}\ \inf_{\substack{f\in H(T_\alpha(0,\alpha))\\ f(0)=0}}\ \frac{1}{\gamma_f(0)(2\alpha)^{2n-2}}\, V_f(T_\alpha(0,\alpha)).$$

From **Theorem 5.2.1** it follows that

$$W_n \geq 2^{-2n+2}V_{2n-2}=2^{-2n+2}\ \frac{\pi^{n-1}}{(n-1)!} > 0,\ \forall n = 2,\ 3,\ \ldots\ .$$

It is natural to conjecture that $W_n = 1$, $\forall n = 2, 3, \ldots$. However, this has been proved for $n = 2$ only (see V.Kac'nelson, L.Ronkin [1]). Let $\sigma = (\sigma_1,\ldots,\sigma_n)\in\mathbb{R}^n_+$. Then we denote by $B_\sigma(E,\gamma)$ the set of all functions $f\in H(\mathbb{C}^n)$ which vanish at the points of E with multiplicities $\gamma_f(x)$ not smaller than $\gamma(x)$ and which satisfy the condition

$$|f(z)| \le const\cdot exp\{\sigma_1|Imz_1| + \ldots + \sigma_n|Imz_n|\}.$$

The following theorem gives sufficient conditions for uniqueness for the class $B_\sigma(E,\gamma)$.

Theorem 5.2.2. Let a set E, $h_E = h > 0$, a function $\gamma(x)$ and positive numbers $\sigma_1, \ldots, \sigma_n$ be such that for some $x^0 \in R^n$ the inequality

$$\sigma_1 + \ldots + \sigma_n < W_n(h_E)^{n-1}\delta_{E,\gamma}(x^0)\cdot\pi \tag{5.2.1}$$

is valid. Then the class $B_\sigma(E,\gamma)$ consists of the single function $f \equiv 0$. In particular, if

$$\sigma_1 + \ldots + \sigma_n < W_n(2h_E)^{n-1}\delta_{E,\gamma}\frac{\pi}{2^{n-1}}$$

and $f \in B_\sigma(E,\gamma)$ then $f \equiv 0$.

Proof. We assume that $f \not\equiv 0$ and estimate from above and below the volume $V_f(T_h(rx^0,r\alpha))$ of the divisor Z_f in the domain $T_h(rx^0,r\alpha)$.

Lemma 5.2.1. Let $f \not\equiv 0$ be an entire function satisfying the condition

$$ln|f(x+iy)| \le \sigma_1|y_1| + \ldots + \sigma_n|y_n|, \ \forall x \in R^n, \ y \in R^n.$$

Then

$$\overline{lim}_{\alpha\to 0}\ \overline{lim}_{r\to\infty}\ \frac{1}{(2\alpha r)^n} V_f(T_h(rx^0,r\alpha)) \le \frac{\pi}{2^{n-1}}\ \delta_{E,\gamma}(x^0)h^{n-1}\sum_{j=1}^{n}\sigma_j.$$

Proof. Consider the function $ln|f(z)|$ as a function of one variable z_j for fixed $z^{(j)} = (z_1, \ldots, z_{j-1}, z_{j+1}, \ldots, z_n)$. We denote by $\mu^{(j)}(z_j;z^{(j)})$ the Riesz associated measure in $\mathbb{C}$ of the function $\varphi_{z^{(j)}}(z_j) = ln|f(z)|$. This measure is defined for all $z^{(j)}$ such that $\varphi_{z^{(j)}} \not\equiv 0$, is concentrated at the zeros of $\varphi_{z^{(j)}}$, coincides with its multiplicity and at each zero. Therefore the integral

$$\int_{G^{(j)}} d\omega_{z^{(j)}} \int_{G_{z^{(j)}}} d\mu^{(j)}(z_j;z^{(j)}),$$

where $G^{(j)}$ is the projection of the domain $G \subset \mathbb{C}^n$ to the hyperplane

$\{z\in\mathbb{C}^n:\ z_j = 0\}$ and $G_{z^{(j)}}$ is the intersection of G with the corresponding orthogonal plane, is naturally interpreted as the volume[40)] of the projection to $\{z\in\mathbb{C}^n:\ z_j = 0\}$ of the divisor in the domain G. It is known (see, for example, L.Ronkin [1]) that the volume of the divisor is equal to the sum of the volumes of its projections to the coordinate hyperplanes. Thus,

$$V_f(T_h(rx^0,r\alpha)) = \sum_{j=1}^{n} V_j\ , \tag{5.2.2}$$

where

$$V_j = \int\limits_{T_h^{(j)}(rx^0,r\alpha)} d\omega_{z^{(j)}} \int\limits_{\substack{|x_j-rx_j^0|<\alpha r\\ |y_j|<h}} d\mu^{(j)}(z_j;z^{(j)}),$$

$$T_h^{(j)}(rx^0,r\alpha) = \{z^{(j)}\in\mathbb{C}^{n-1}:\ |x_i-x_i^0r| < \alpha r,\ |y_i| < h,\ i \neq j\}.$$

To estimate V_j we define the auxiliary quantity

$$\tilde{V}_j = \int\limits_{T_h(rx^0,r\alpha)} d\omega_z \int\limits_0^{\varepsilon r} \frac{dt}{t} \int\limits_{|\zeta-z_j|<t} d\mu^{(j)}(\zeta;z^{(j)}) \tag{5.2.3}$$

We set, for compactness of notation, $j = 1$, $z^{(1)} = {'z} = {'x}+i{'y}$, ${'T}(rx^0,r\alpha) = \{{'x}\in\mathbb{R}^{n-1}:\ \|{'x}-r{'x}^0\|_\infty < \alpha r\}$, ${'T}_h(rx^0,r\alpha) = \{{'z}\in\mathbb{C}^{n-1}:\ \|{'x}-r{'x}^0\|_\infty < \alpha r,\ \|{'y}\|_\infty < h\}$, and use the ordinary **Jensen's formula** for the estimation of the inner integral in (5.2.3). Then we have

$$\tilde{V}_1 = \int\limits_{T_h(rx^0,r\alpha)} d\omega_z \left\{\frac{1}{2\pi} \int\limits_0^{2\pi} ln|f(z_1+\varepsilon re^{i\theta},{'z})|d\theta - ln|f(z)|\right\} \le$$

$$\le \int\limits_{T_h(rx^0,r\alpha)} \left\{\sigma_1|z_1| + \ldots + \sigma_n|z_n| + \sigma_1\cdot\frac{2}{\pi}\varepsilon r - ln|f(z)|\right\}d\omega_z \le$$

---

40) Here the volumes of the divisor and its projections are calculated counted with multiplicity.

$$\leq (2\alpha r)^n h^{n+1} 2^{n-1}\left(\sigma_1 + \dots + \sigma_n + \frac{4}{\pi h}\,\varepsilon r\right) +$$

$$- \int\limits_{T_h(rx^0,\, r\alpha)} \ln|f(z)|\}\,d\omega_z. \qquad (5.2.4)$$

At the same time, if we single out the integration over z_1 in the outer integral in (5.2.3), and replace the order of integration by integration over t and ζ, then for $o < \varepsilon < \alpha$ we find

$$\tilde{V}_1 = \int\limits_{'T_h(rx^0,\, r\alpha)} d\omega_{'z} \int\limits_{\substack{|x_1 - rx_1^0| < \alpha r \\ |y_1| < h}} d\omega_{z_1} \int_0^{\varepsilon r} \frac{dt}{t} \int\limits_{|\zeta - z_1| < t} d\mu^{(1)}(\zeta;\, 'z) \geq$$

$$\geq \int\limits_{'T_h(rx^0,\, r\alpha)} d\omega_{'z} \int_0^{\varepsilon r} \frac{dt}{t} \int\limits_{\substack{|x_1 - rx_1^0| < \alpha r \\ |y_1| < h}} d\omega_{z_1} \int\limits_{\substack{|\zeta - z_1| < t \\ |Im\zeta| < h}} d\mu^{(1)}(\zeta;\, 'z) \geq$$

$$\geq \int\limits_{'T_h(rx^0,\, r\alpha)} d\omega_{'z} \int_0^{\varepsilon r} \frac{dt}{t} \int\limits_{\substack{|Re\zeta - rx_1^0| < (\alpha - \varepsilon) r \\ |Im\zeta| < h}} d\mu^{(1)}(\zeta;\, 'z) \int\limits_{\substack{|\zeta - z_1| < t \\ |y_1| < h}} d\omega_{z_1} \geq$$

$$\geq \int\limits_{'T_h(rx^0,\, r\alpha)} d\omega_{'z} \int_{2h}^{\varepsilon r} \frac{dt}{t} \int\limits_{\substack{|Re\zeta - rx_1^0| < (\alpha - \varepsilon) r \\ |Im\zeta| < h}} d\mu^{(1)}(\zeta;\, 'z) \int\limits_{\substack{|x_1 - Re\zeta| < \sqrt{t^2 - 4h^2} \\ |y_1| < h}} d\omega_{z_1} \geq$$

$$\geq 4h \int_{2h}^{\varepsilon r} \sqrt{t^2 - 4h^2}\, \frac{dt}{t} \int\limits_{'T_h(rz^0,\, (\alpha - \varepsilon) r)} d\omega_{'z} \int\limits_{\substack{|Re\zeta - rx_1^0| < (\alpha - \varepsilon) r \\ |y_1| < h}} d\mu^{(1)}(\zeta;\, 'z) =$$

$$= 4h \int_{2h}^{\varepsilon r} \sqrt{t^2-4h^2}\,\frac{dt}{t} \cdot V_1. \tag{5.2.5}$$

Estimates similar to (5.2.4) and (5.2.5) are certainly also valid for $j = 2, 3, \ldots, n$. Comparing them and taking into account (5.2.2), we conclude that

$$V_f(T_h(rx^0, r(\alpha-\varepsilon))) \le \left(4h \int_{2h}^{\varepsilon r} \sqrt{t^2-4h^2}\,\frac{dt}{t} \right)^{-1} \times$$

$$\times \left\{ (2\alpha r)^n 2^{n-1} h^{n+1} \sum_{j=1}^{n} \sigma_j \left(n + \frac{4\varepsilon r}{\pi h}\right) - \int_{T_h(rz^0, r\alpha)} ln|f(z)|\,d\omega_z \right\}.$$

In view of

$$\lim_{r \to \infty} \frac{1}{\varepsilon r} \int_{2h}^{\varepsilon r} \sqrt{t^2-4h^2}\,\frac{dt}{t} = 1,$$

it follows that

$$\overline{\lim_{r \to \infty}} \frac{1}{(2(\alpha-\varepsilon)r)^n} V_f(T_h(rx^0, r(\alpha-\varepsilon))) \le$$

$$\le \left(\frac{\alpha}{\alpha-\varepsilon} \right)^n 2^{n-1} \frac{h^{n-1}}{\pi} (\sigma_1 + \ldots + \sigma_n) +$$

$$- \frac{1}{4h\varepsilon(2(\alpha-\varepsilon))^n} \lim_{r \to \infty} \frac{1}{r^{n+1}} \int_{T_h(rx^0, r\alpha)} ln|f(z)|\,d\omega_z ,$$

or, equivalently,

$$\overline{\lim_{r \to \infty}} \frac{1}{(2\alpha r)^n} V_f(T_h(rx^0, r\alpha)) \le$$

$$\le \left(\frac{\alpha+\varepsilon}{\alpha} \right)^n \frac{(2h)^{n-1}}{\pi} (\sigma_1 + \ldots + \sigma_n) +$$

$$-\frac{1}{4h\varepsilon(2\alpha)^n}\lim_{r\to\infty}\frac{1}{r^{n+1}}\int\limits_{T_h(rz^0,r\alpha)} ln|f(z)|d\omega_z\ , \qquad (5.2.6)$$

We show that the summand at the right-hand side of this inequality vanishes. Since $sup\{ln|f(z)|: x\in\mathbb{R}^n,\ \|y\|_\infty < h\} < \infty$, it suffices to show that

$$\lim_{r\to\infty}\frac{1}{r^{n+1}}\int\limits_{\substack{|x|<r\\ \|y\|_\infty<h}} ln|f(z)|d\omega_z = 0. \qquad (5.2.7)$$

We set

$$\Phi(w) = \int\limits_{\substack{|x|<1\\ \|y\|_\infty<h}} ln|f(xw+iy)|d\omega_z\ ,\quad w\in\mathbb{C}.$$

It is obvious that $\Phi\in SH(\mathbb{C},1]$ and $sup\{\Phi(t): -\infty < t < \infty\} < \infty$. Hence, in accordance with **Theorem 4.4.3**, more exactly, with its analogue for subharmonic functions, the function $\Phi(w)$ is of c.r.g. in $\mathbb{C}$ and $\mathfrak{L}_\Phi(1) = 0$. Therefore there is a set S of relative measure zero such that

$$\lim_{\substack{r\to\infty\\ r\notin S}}\frac{\Phi(r)}{r} = 0.$$

Noting that

$$\int\limits_{\substack{|x|<1\\ \|y\|_\infty<h}} ln|f(xr+iy)|d\omega_z = \frac{1}{r^n}\int\limits_{\substack{|x|<r\\ \|y\|_\infty<h}} ln|f(z)|d\omega_z\ ,$$

this implies

$$\lim_{\substack{r\to\infty\\ r\notin S}}\frac{1}{r^{n+1}}\int\limits_{\substack{|x|<r\\ \|y\|_\infty<h}} ln|f(z)|d\omega_z = 0. \qquad (5.2.8)$$

The condition $r\notin S$ figuring in this equality is unnecessary. Indeed, Since S is a set of relative measure zero, there is a function $\lambda(r) > 0$ such that $\lambda(r)+r = r'\in S$, $\forall r$, and $\lim\limits_{r\to\infty} r^{-1}\lambda(r) = 0$. Therefore, from the boundedness above of the function $ln|f(z)|$ (when $\|y\|_\infty < h$) and (5.2.8) it follows that

$$\underline{\lim}_{r \to \infty} \frac{1}{r^{n+1}} \int_{\substack{|x|<r \\ \|y\|_\infty<h}} ln|f(z)|d\omega_z =$$

$$= \underline{\lim}_{r \to \infty} \left(\frac{r'}{r}\right)^{n+1} \left\{ \frac{1}{r'^{\,n+1}} \int_{\substack{|x|<r' \\ \|y\|_\infty<h}} ln|f(z)|d\omega_z - \frac{1}{(r')^n} \int_{\substack{r<|x|<r' \\ \|y\|_\infty<h}} ln|f(z)|d\omega_z \right\} \geq$$

$$\geq \underline{\lim}_{r' \to \infty} \frac{1}{r'^{\,n+1}} \int_{\substack{|x|<r' \\ \|y\|_\infty<h}} ln|f(z)|d\omega_z = 0.$$

The opposite inequality is evident. Hence

$$\lim_{r \to \infty} \frac{1}{r^{n+1}} \int_{\substack{|x|<r \\ \|y\|_\infty<h}} ln|f(z)|d\omega_z = 0.$$

Further, from (5.2.6) it follows that for any $\alpha > 0$ and $\varepsilon > 0$,

$$\overline{\lim}_{r \to \infty} \frac{1}{(2\alpha r)^n} V_f(T_h(rx \overset{0}{,} r\alpha)) \leq \left(\frac{\alpha+\varepsilon}{\alpha} \right)^n \frac{(2h)^{n-1}}{\pi} (\sigma_1 + \ldots + \sigma_n) .$$

Taking into account that α and ε are arbitrary, we conclude that

$$\overline{\lim}_{\alpha \to 0} \overline{\lim}_{r \to \infty} \frac{1}{r^n} V_f(T_h(rx \overset{0}{,} r\alpha)) \leq \frac{2^{n-1}h^{n-1}}{\pi} (\sigma_1 + \ldots + \sigma_n) .$$

This finishes the proof of the lemma.

Lemma 5.2.2. Let $E \subset \mathbb{R}^n$ be a discrete set such that $h_E = h > 0$. Also let $\gamma(x)$ be a natural-valued function on E. Finally, let $f(z) \not\equiv 0$ be an entire function that vanishes at the points of E with multiplicity at least $\gamma(x)$. Then

$$\overline{\lim}_{\alpha \to 0} \overline{\lim}_{r \to \infty} \frac{1}{(2\alpha r)^n} V_f(T_h(rx \overset{0}{,} r\alpha)) \geq$$

$$\geq \delta_{E,\gamma}(x^0)(2h)^{2n-2}W_n\ ,\ x^0\in\mathbb{R}^n. \qquad (5.2.9)$$

Proof. We associate to each point $\xi^{(k)}\in E$ the cube $Q_k = \{z\in\mathbb{C}^n: \|z-\xi^{(k)}\|_\infty < \infty\}$. It is obvious that $Q_k\cap Q_l = \emptyset$ if $k \neq l$, and $Q_k \subset T_h(rx^0,r\alpha)$ if $\xi^{(k)}\in T(rx^0,r\alpha-h)$. Therefore

$$V_f(T_h(rx^0,r\alpha)) \geq \sum_{\xi^{(k)}\in T(rx^0,r\alpha-h)} V_f(Q_k).$$

Taking into account the definition of the constant W_n, we obtain

$$V_f(T_h(rx^0,r\alpha)) \geq W_n(2h)^{2n-2}\sum_{\xi^{(k)}\in T(rx^0,r\alpha-h)} \gamma(\xi^{(k)}).$$

This inequality and the definition of $\delta_{E,\gamma}(x^0)$ immediately imply (5.2.9).

The proof of the lemma is finished.

Now, to obtain the theorem it suffices to compare the estimates for

$$\overline{\lim_{\alpha\to 0}}\ \overline{\lim_{r\to\infty}}\ (2\alpha r)^{-1}V_f(T_h(rx^0,r\alpha))$$

that are contained in **Lemmas 5.2.1** and **5.2.2**.

The proof of the theorem is finished.

Remark 1. It is clear that in the conditions of **Theorem 5.2.1** we can replace (5.2.1) by: there exists an α such that the inequality

$$\sigma_1 + \ldots + \sigma_n < \pi(2h)^{n-1}W_n\ \overline{\lim_{r\to\infty}}\ \frac{1}{(2\alpha r)^n}\sum_{x\in E\cap T(rx^0,r\alpha)}\gamma(x)$$

is valid.

Remark 2. As already mentioned, $W_2 = 1$. Therefore, for $n = 2$ condition (5.2.1) may be concretized as follows:

$$\sigma_1 + \sigma_2 < 2\pi h_E\delta_{E,\gamma}(x^0).$$

This makes it possible to state that **Theorem 5.2.1** is exact in the class of all pairs (E,γ). Indeed, let $E = \mathbb{Z}^2$, $\gamma(x) \equiv 2$ and $f(z_1,z_2) = \sin\pi z_1 \sin\pi z_2$. It is obvious that in this case $\delta_{E,\delta} = 2$, $h_E = \frac{1}{2}$ and $f\in B_\sigma(E,\gamma)$, where $\sigma = (\sigma_1,\sigma_2) = (\pi,\pi)$. Thus, in this situation,

$$\sigma_1 + \sigma_2 = 2\pi,$$

but $\sin\pi z_1 \sin\pi z_2 \not\equiv 0$. Hence, if we replace the inequality (5.2.1) by the corresponding equality, then, generally speaking, the statement of **Theorem 5.2.1** would be false. Of course, this does not exclude the possibility of weakening condition (5.2.1) under certain additional assumptions on the class of sets E or for an individual set E. So, in the case $E = \mathbb{Z}^2$, $\gamma \equiv 1$, the sufficient uniqueness condition is as follows

$$\max(\sigma_1, \sigma_2) < \pi.$$

This condition is weaker than the one that follows from the theorem:

$$\sigma_1 + \sigma_2 < \pi.$$

Remark 3. If in **Lemma 5.2.2** we use the balls $B(\xi^{(k)}, h)$ instead of the cubes Q_k, the quantity

$$\tilde{h}_E = \frac{1}{2}\, \inf\left\{\frac{1}{2}|x'-x''|:\ x'\in E,\ x''\in E,\ x' \neq x''\right\}$$

instead of h_E, and the volume $\omega_{2n-2} = \dfrac{\pi^{n-1}}{(n-1)!}$ of the unit ball in $\mathbb{R}^2$ instead of W_n, then we obtain a statement differing from **Theorem 5.2.1** only by the replacement (5.2.1) by the condition

$$\sigma_1 + \dots + \sigma_n \le \frac{\pi^n}{2^{n-1}(n-1)}\, h_E^{n-1} \delta_{E,\gamma}(x^0).$$

§3. Norming sets

It follows from the definition of uniqueness set E for a class $M \subset H(\mathbb{C}^n)$ that every function $f\in M$ can be recovered from its values at the points of the uniqueness set E. Two problems naturally arise in this situation. The first one is the problem on the construction of a recovering algorithm. The solution of this problem is known for some trivial cases (for some special grids) only. The second problem is the one on the stability of recovery or, in other words, on the continuity of the dependence of a function on the totality of its values on E. Here we consider the latter problem in case the function f is itself completely characterized by the quantity

$$\|f\|_{\mathbb{R}^n} = \|f\|_{L^\infty(\mathbb{R}^n)} = \sup_{x\in\mathbb{R}^n} |f(x)|,$$

and the set of its values on E is characterized by

$$\|f\|_E = \sup_{x\in E}|f(x)|.$$

The set $E \subset \mathbb{R}^n$ is referred to as a norming (or Cartwright's) set for the class $M \subset H(\mathbb{C}^n)$ if there is a constant $c = c(M,E)$ such that

$$\|f\|_{\mathbb{R}^n} \le c\|f\|_E, \quad \forall f\in M. \tag{5.3.1}$$

Now we can formulate the problem on stability of recovery, described above in general terms, as the problem of conditions under which a set E is norming for a class M.

1. Discrete norming sets. We denote by $H(\mathbb{C}^n,1,\sigma]$, $\sigma \ge 0$, the set of all entire functions $f(z)$, $z\in\mathbb{C}^n$, that satisfy the condition

$$\overline{\lim_{|z|\to\infty}} \frac{\ln|f(z)|}{|z_1| + \dots + |z_n|} \le \sigma.$$

In the case of one variable, **Cartwright's Theorem** (see, for example, B.Levin [1]) was the first theorem about norming sets for such classes of the functions[41]. It states that if an entire function $f(z)$, $z\in\mathbb{C}$, is of exponential type $\sigma < \pi$ and its modulus is bounded by a constant A at the integer points of the real axis, then $|f(x)| \le cA$, $\forall x\in\mathbb{R}$, where the constant c is independent of f. In subsequent studies sets E of essentially more general nature were considered, the constant c was estimated and calculated, the formulation of the problem was changed, in particular, other norms as well as functions of arbitrary order and increasing on the axis were considered. The bibliography concerning this problem is quite voluminous and is partly represented in B.Levin [1]. Here we do not consider these investigations, but concentrate our attention on the first results in this direction for functions of several variables. Note that, as we already frequently have encountered, the traditional methods for proving of such theorems as Cartwright's, are useless in the case $n > 1$. The new method,

---

[41] **Cartwright's Theorem** was preceded by the solution of Polya's Problem by a number of mathematicians. **Polya's Problem** assumed that the conditions $f\in H(\mathbb{C},1,0]$ and $\sup_{k\in\mathbb{Z}}|f(k)| < \infty$ imply $f \equiv \text{const}$.

constructed by V.Logvinenko, is based on a special approximation procedure of functions belonging to $H(\mathbb{C}^n,1,\sigma]$. For $n = 1$ this method does not give the precise result as implied by the traditional methods. In particular, **Cartwright's Theorem** cannot be obtained in this way, but in the case of an arbitrary number of variables this method is useful.

Before stating and proving one of the results obtained in the above manner (namely, **Theorem 5.3.1**), we recall that a set $E \subset \mathbb{R}^n$ is called a δ-grid on $\mathbb{R}^n$ if

$$\bigcup_{x\in E} B(x,\delta) = \mathbb{R}^n,$$

or, equivalently, if

$$\forall x\in\mathbb{R}^n,\ \exists y\in E:\ |x-y| < \delta.$$

Theorem 5.3.1. Let a natural number p and positive number σ be such that

$$\frac{e^{\sigma}}{p\sigma} \sup_{0\le\tau\le\sigma} \left\{ e^{\tau/p}.\frac{|\sin\tau|}{\tau} \right\}^{n-1} < 1$$

and $p\sigma\delta < 1$. Then every δ-grid E on $\mathbb{R}^n$ is a norming set for the class $H(\mathbb{C}^n,1,\sigma]$. Besides, inequality (5.3.1) is valid with $c = (1-\delta\sigma)^{-1}$.

Proof. First we show that if we assume that a function f is bounded on $\mathbb{R}^n$, then inequality (5.3.1) is valid with a constant independent of f. More exactly, we prove the following lemma.

Lemma 5.3.1. Let a set E be a δ-grid on $\mathbb{R}^n$ and a function f belong to $H(\mathbb{C}^n,1,\sigma]$. Let also $\delta\sigma < 1$ and $\|f\|_{\mathbb{R}^n} < \infty$. Then $\|f\|_{\mathbb{R}^n} \le (1-\delta\sigma)^{-1}\|f\|_E$.

Proof. We use the classical inequality of S.Bernstein, connecting the norms on $\mathbb{R}$ of a function $f\in H(\mathbb{C},1,\sigma]$ and its derivative. In accordance with this inequality (see, for example, B.Levin [1]) we have

$$\|f\|_{\mathbb{R}} \le \sigma\|f'\|_{\mathbb{R}},\ \forall f\in H(\mathbb{C},1,\sigma]. \tag{5.3.2}$$

Applying (5.3.2) to the function $\Phi_{x,x'}(w) = f(x'+(x-x')w)$, $x\in\mathbb{R}^n$, $x'\in\mathbb{R}^n$, $w\in\mathbb{C}$, which clearly belongs to $H(\mathbb{C},1,\sigma']$, $\sigma' = |x-x'|\sigma$, we obtain

$$|f(x)-f(x')| = \left| \int_0^1 \frac{d}{dt} f(x'+t(x-x'))dt \right| \le$$

$$\leq \int_0^1 \sigma \|f\|_{\mathbb{R}^n} |x-x'|dt = \sigma|x-x'|\|f\|_{\mathbb{R}^n} . \tag{5.3.3}$$

Let $x \in \mathbb{R}^n$ be arbitrary; then we choose $x' \in E$ in such a way that $|x-x'| < \delta$. This is possible because E is a δ-grid. It then follows from (5.3.3) that

$$|f(x)| - \delta\sigma\|f\|_{\mathbb{R}^n} \leq |f(x')|,$$

whence we conclude that

$$\|f\|_{\mathbb{R}^n}(1-\delta\sigma) \leq \|f\|_E.$$

This finishes the proof of the lemma.

As already mentioned, the main point of the proof of **Theorem 5.3.1**, as well as of similar theorems not given here, is the use of some special approximation procedure for the functions $f \in H(\mathbb{C}^n,1,\sigma]$. This approximation procedure is based on the following lemma.

Lemma 5.3.2. Let

$$f(z) = \sum_{k \in \mathbb{Z}_+^n} \frac{a_k}{k!} z^k$$

belong to $H(\mathbb{C}^n,1,\sigma]$, and let p be a natural number such that $p > e\sigma n$. Then

$$\lim_{m \to \infty} \max_{\|z\|_\infty \leq m} | f(z) - \sum_{\|k\| \leq pm} \frac{a_k}{k!} z^k | = 0. \tag{5.3.4}$$

Proof. From the formulas defining the type of an entire function by its Taylor coefficients (see, for example L.Ronkin [1]), it follows that under the conditions of the lemma we have

$$\overline{\lim_{\|k\| \to \infty}} \sqrt[\|k\|]{|a_k|} \leq \sigma.$$

Therefore, for any $\varepsilon > 0$ the inequality

$$|a_k| \leq (\sigma+\varepsilon)^{\|k\|} \tag{5.3.5}$$

is valid for all k, except for an at most finite set. Hence

$$\max_{\|z\|_\infty < m} \left| f(z) - \sum_{\|k\|_\infty \le pm} \frac{a_k}{k!} z^k \right| = \max_{\|z\|_\infty < m} \left| \sum_{\|k\|_\infty > pm} \frac{a_k}{k!} z^k \right| \le$$

$$\le \sum_{l=pm+1} (\sigma+\varepsilon)^l m^l \sum_{\|k\|_\infty = l} \frac{1}{k!} = \sum_{l=pm+1}^{\infty} \frac{1}{l!} (\sigma+\varepsilon)^l m^l n^l \le$$

$$\le \sum_{l=pm+1}^{\infty} \left(\frac{(\sigma+\varepsilon)mne}{l}\right)^l . \qquad (5.3.6)$$

Since $p > en\sigma$, for $\varepsilon > 0$ sufficient small and $l > pm$ we have

$$\frac{(\sigma+\varepsilon)mne}{l} < \frac{(\sigma+\varepsilon)mne}{pm} = q < 1.$$

Therefore, letting $m \to \infty$ we obtain

$$\sum_{l=pm+1}^{\infty} \left(\frac{(\sigma+\varepsilon)mne}{l}\right)^l \to 0.$$

From this and (5.3.6) we conclude that (5.3.4) is valid.

The proof of the lemma is finished.

Now, using the number p and the function

$$f(z) = \sum_{k\in\mathbb{Z}_+^n} \frac{a_k}{k!} z^k,$$

we construct the following sequence of functions:

$$f_m(z) = \mathcal{P}_{pm}(z) \prod_{j=1}^{n} \left(\frac{\sin\frac{(\sigma+\varepsilon)}{m} z_j}{\frac{(\sigma+\varepsilon)}{m} z_j} \right)^{pm}, \qquad (5.3.7)$$

where

$$\mathcal{P}_l(z) = \sum_{\|k\| \le l} \frac{a_k}{k!} z^k,$$

and $\varepsilon > 0$ is such that $p(\sigma+\varepsilon)\delta < 1$ and

$$\frac{e^{\sigma+\varepsilon}}{p(\sigma+\varepsilon)} \sup_{0<\tau<\sigma+\varepsilon} \left\{ e^{\tau/p} \frac{\sin\tau}{\tau} \right\} < 1.$$

It is obvious that $f_m \in H(\mathbb{C}^n, 1, (\sigma+\varepsilon)p]$, $\|f_m\|_{R^n} < \infty$ and $f_m(x) \to f(x)$, as $m \to \infty$, uniformly on every compact set in $\mathbb{R}^n$. We show that

$$\|f_m\|_E < (1+o(1))\|f\|_E \ , \ m \to \infty.$$

We set $E_m = \{x\in E: \|x\|_\infty < m\}$. If $m \to \infty$, then by **Lemma 5.3.2** we have

$$\|\mathcal{P}_{pm}(x)\|_{E_m} \le (1+o(1))\|f\|_{E_m} .$$

From this it follows that

$$\|f_m\|_{E_m} \le (1+o(1))\|f\|_{E_m} \ , \ m \to \infty. \tag{5.3.8}$$

Now we estimate the function $f_m(x)$ for $\|x\|_\infty > m$. First of all we estimate $|\mathcal{P}_{pm}(z)|$ for arbitrary $(|z_1|, \ldots, |z_n|) = (r_1, \ldots, r_n) = r$. Note that since ε in (5.3.5) is fixed, without loss of generality we may assume that this estimate is valid for all $k\in\mathbb{Z}_+^n$. Hence

$$|\mathcal{P}_{mp}(z)| \le \sum_{\|k\|\le pm} \frac{(\sigma+\varepsilon)^{\|k\|} r^k}{k!} \le$$

$$\le \sum_{\|k\|_\infty \le pm} \frac{(\sigma+\varepsilon)^{\|k\|} r^k}{k!} = \prod_{j=1}^{n} P_{mp}(r_j), \tag{5.3.9}$$

where $P_\nu(t) = \sum_{l=0}^{\nu} \frac{(\sigma+\varepsilon)^l t^l}{l!}$. Note that for $t > 0$,

$$P_\nu(t) \le e^{(\sigma+\varepsilon)t},$$

and for $t > \nu$,

$$P_\nu(t) = \frac{t^\nu}{\nu!} \sum_{l=0}^{\nu} \frac{(\sigma+\varepsilon)^\nu \nu!}{l!\,t^{\nu-l}} \le \frac{t^\nu}{\nu!} \sum_{l=0}^{\nu} \frac{(\sigma+\varepsilon)^\nu}{l!} \frac{\nu!}{\nu^{\nu-l}} \le$$

$$\le \frac{t^\nu}{\nu!} (1+\gamma(\nu)) \sum_{l=0}^{\nu} \frac{(\sigma+\varepsilon)^\nu e^{-\nu} \nu^l \sqrt{2\pi\nu}}{l!} \le \frac{1+\gamma(\nu)}{\nu!} \sqrt{2\pi\nu} \left(\frac{t}{e} \right)^\nu e^{(\sigma+\varepsilon)\nu},$$

where $\gamma(\nu) = (\nu!\left(\frac{\nu}{e} \right)^\nu (2\pi\nu)^{-1/2} - 1) \to 0$ as $\nu \to \infty$. From these estimates it follows that for the function

$$Q_m(t) = P_{pm}(t)\left(\frac{m}{(\sigma+\varepsilon)t} \cdot \sin \frac{(\sigma+\varepsilon)t}{m} \right)^{mp}$$

the following inequalities are valid:

$$\max_{|t|>pm} |Q_m(t)| \le$$

$$\le \frac{1+\gamma(pm)}{(pm)!}\sqrt{2\pi pm}\left(\frac{pm}{e}\right)^{pm}\left(\frac{e^{\sigma+\varepsilon}}{p(\sigma+\varepsilon)}\right)^{pm} = \left(\frac{e^{\sigma+\varepsilon}}{p(\sigma+\varepsilon)}\right)^{pm} \qquad (5.3.10)$$

$$\max_{m\le|t|\le pm} |Q_m(t)| \le \max_{m<t<pm}\left\{ e^{(\sigma+\varepsilon)t}\left(\frac{m}{(\sigma+\varepsilon)t}\right)^{mp}\right\} =$$

$$= \max\left\{\left(\frac{e^{\sigma+\varepsilon}}{(\sigma+\varepsilon)^p}\right)^m, \left(\frac{e^{(\sigma+\varepsilon)p}}{(p(\sigma+\varepsilon)^p}\right)^m\right\} = \left(\frac{e^{\sigma+\varepsilon}}{p(\sigma+\varepsilon)}\right)^{pm}, \qquad (5.3.11)$$

$$\max_{|t|\le m} |Q_m(t)| \le \sup_{0<t\le m}\left\{ e^{(\sigma+\varepsilon)t}\left(\frac{m}{(\sigma+\varepsilon)t}\left|\sin\frac{(\sigma+\varepsilon)t}{m}\right|\right)^{mp}\right\} =$$

$$= \left\{\sup_{0<\tau\le\sigma+\varepsilon}\left\{e^{\tau}\left(\frac{\sin\tau}{\tau}\right)^p\right\}\right\}^m. \qquad (5.3.12)$$

Since $\|x\|_\infty \ge m$, for some i the inequality $|x_i| \ge m$ is valid Therefore, from (5.3.7), (5.3.9), (5.3.10)-(5.3.12) it follows that outside the cube $\|x\|_\infty < m$ the inequality

$$|f_m(x)| \le \left\{\frac{e^{\sigma+\varepsilon}}{p(\sigma+\varepsilon)} \sup_{0<\tau\le\sigma+\varepsilon}\left(e^{\tau/p}\left|\frac{\sin\tau}{\tau}\right|\right)^{n-1}\right\}^{pm} \overset{def}{=} q^{pm}$$

holds. In a view of the assumptions on p and ε we conclude that $q < 1$, and therefore

$$\lim_{m\to\infty} \sup_{\|x\|_\infty > m} |f_m(x)| = 0.$$

Comparing this equality with (5.3.8), we obtain

$$\|f_m\|_E \le (1+o(1))\|f\|_E, \quad m \to \infty. \qquad (5.3.13)$$

As already noted, $\|f_m\|_{\mathbb{R}^n} < \infty$ and $f_m \in H(\mathbb{C}^n,1,(\sigma+\varepsilon)\rho]$. Therefore we can apply **Lemma 5.3.1**, by which inequality (5.3.13) implies

$$\|f_m\|_{\mathbb{R}^n} \le (1-\delta(\sigma+\varepsilon)p)^{-1}\|f\|_E(1+o(1)), \quad m \to \infty.$$

Passing to the limit in this inequality as $m \to \infty$, we obtain

$$\|f\|_{\mathbb{R}^n} \le (1-\delta(\sigma+\varepsilon)p)^{-1}\|f\|_E.$$

Again using **Lemma 5.3.1**, we conclude that

$$\|f\|_{\mathbb{R}^n} \le (1-\delta\sigma)^{-1}\|f\|_E.$$

Thus the proof of the theorem is finished.

Every statement on norming sets for a class $H(\mathbb{C}^n,1,\sigma]$ with σ fixed evidently implies the corresponding statement for $H(\mathbb{C}^n,1,\sigma)$ with arbitrary σ. Using this remark we can weaken, at least formally, the requirement on σ and δ in **Theorem 5.3.1**. We set

$$p(\sigma) = \min\left\{ p\in\mathbb{N}: \frac{e^\sigma}{p\sigma}\cdot \sup_{0<\tau\le\sigma}\left(e^{\tau/p}\,\frac{|\sin\pi\tau|}{\tau}\right)^{n-1} < 1\right\},$$

$$p_0 = \inf_{0<\sigma<\infty} p(\sigma).$$

Note that $p_0 \le 3+2\sqrt{n}$. This can be easily proved if we estimate $p(\sigma)$ when $\sigma = \dfrac{2}{1+\sqrt{n}}$.

Theorem 5.3.2. Let a set $E \subset \mathbb{R}^n$ be a δ-grid in $\mathbb{R}^n$. Then E is a norming set for any class $H(\mathbb{C}^n,1,\sigma]$ with $\sigma < \dfrac{1}{p_0\delta}$.

Proof. Let $\sigma_0\in(0,\infty)$ be such that $p(\sigma_0) = p_0$. It is easy to see that such σ_0 exists. Then we construct the function $\tilde{f}(z) = f\left(\dfrac{\sigma_0}{\sigma}\, z\right)$ and the set $\tilde{E} = \{x = \dfrac{\sigma}{\sigma_0}\, y\colon\ y\in E\}$ from a function $f\in H(\mathbb{C}^n,1,\sigma]$ and set E, respectively. Note that $\tilde{f}\in H(\mathbb{C}^n,1,\sigma_0]$, and that $\tilde{E}$ is a $\tilde{\delta}$-grid in $\mathbb{R}^n$ with $\tilde{\delta} = \dfrac{\sigma}{\sigma_0}\delta$. Therefore, in accordance with **Theorem 5.3.1** the inequality

$$\|\tilde{f}\|_{\mathbb{R}^n} \le \|\tilde{f}\|_{\tilde{E}}(1-\tilde{\delta}\sigma_0)^{-1}$$

holds when $p_0\sigma_0\tilde{\delta} < 1$. From this, going back to the function f and the set E and taking into account that $\|\tilde{f}\|_{\tilde{E}} = \|f\|_E$, $\|f\|_{\mathbb{R}^n} = \|\tilde{f}\|_{\mathbb{R}^n}$, $p_0\sigma_0\tilde{\delta} = p_0\sigma\delta$, we conclude that if $p_0\sigma\delta < 1$, then

$$\|f\|_{\mathbb{R}^n} \le (1-\sigma\delta)^{-1}\|f\|_E,\quad \forall f\in H(\mathbb{C}^n,1,\sigma].$$

The proof of the theorem is finished.

2. Norming sets of positive measure. We can state the problem of the conditions under which a set E is a norming set for a class $H(\mathbb{C}^n,1,\sigma]$ in the case when E is an essentially more massive set than a δ-grid; in particular, if E is of positive measure in $\mathbb{R}^n$. For example, the following problem seems natural: what are the classes $H(\mathbb{C}^n,1,\sigma]$, for which the set of balls with fixed radius and with the centres as a δ-grid is a norming set. This problem is natural in a more general situation; namely, for the sets that are called relatively dense with respect to Lebesgue measure. These are defined as follows.

A set $E \subset \mathbb{R}^n$ is called relatively dense in $\mathbb{R}^n$ with respect to n-dimensional Lebesgue measure if there are numbers $L > 0$, $\delta > 0$ (the parameters of relative density) such that

$$mes_n\{E \cap T(y,L)\} \geq \delta, \ \forall y \in \mathbb{R}^n,$$

where, as above,

$$T(y,L) = \{x \in \mathbb{R}^n: \ \|x-y\|_\infty \leq \infty\}.$$

The solution of this problem is contained in the following theorem of B.Levin.

Theorem 5.3.3. Let $E \subset \mathbb{R}^n$ be a relatively dense set in $\mathbb{R}^n$ with respect to Lebesgue measure with parameters L and δ. Then E is a norming set for every class $H(\mathbb{C}^n,1,\sigma]$, $0 \leq \sigma \leq \infty$, and the constant c in inequality (5.3.1) can be chosen equal to $\frac{\sigma}{\delta}(4\pi)^n L^{n+1} c_1$, where c_1 is an absolute constant.

Proof. Let $f \in H(\mathbb{C}^n,1,\sigma]$ and $\|f\|_E < \infty$. For simplicity of notation we assume that $\|f\|_E < 1$. We set:

$$u(z) = ln|f(z)|,$$

$$T_R = T(0,R) = \{x \in \mathbb{R}^n: \ \|x\|_\infty < R\},$$

$$m_0(R) = sup\{u(x): \ x \in T_R\},$$

$$m_h(R) = sup\{u(x+ihI): \ x \in T_R\}, \ h > 0, \ I = (1, \ldots, 1) .$$

The statement of the theorem on the boundedness of the function $|f(x)|$ and the corresponding estimate will be obtained as a rather simple consequence of the following lemma, which contains estimates relating the quantities $m_0(R)$ and $m_h(R)$.

Lemma 5.3.3. Let an entire function $f(z)$ satisfy the conditions

$$|f(z)| \le ce^{\sigma|z|}, \ \forall z\in\mathbb{C}^n,$$

and

$$|f(x)| \le 1, \ \forall x\in E,$$

where the set E is relatively dense in $\mathbb{R}^n$ with parameters L and δ.

Then there is a function $D(\nu)$, $0 < \nu < 1$, independent of f, such that for any ν the inequalities

$$m_L(\nu R) \le \left(1 - \frac{\delta}{(2L)^n}\right)m_0(R) + \sigma D(\nu)2^{n-1}L + O(1), \tag{5.3.14}$$

and

$$m_0(R) \le m_L(R) + \sigma D(\nu)2^{n-1}L + O(1) \tag{5.3.15}$$

are valid as $R \to \infty$.

Proof. To obtain estimates leading to inequalities (5.3.14) and (5.3.15) we consider some auxiliary functions. We denote by $\omega(\zeta)$, $\zeta = \xi+i\eta\in\mathbb{C}$, a function which is bounded and continuous on the set $\tilde{U}^+ = \{\zeta: |\zeta| < 1, \eta \ge 0, \zeta \ne \pm1\}$, harmonic in the semidisc $U^+ = \{\zeta: |\zeta| < 1, \eta > 0\}$, and such that $\omega(\xi) = 0$ when $-1 < \xi < 1$, and $\omega(e^{i\theta}) = 1$ when $0 < \theta < \pi$. In other words $\omega(\zeta)$ is the harmonic measure of the arc $S^+ = \{\zeta: |\zeta| = 1, \xi > 0\}$ with respect to U^+. We can give $\omega(\xi)$ in explicit form, but this is unnecessary. It is important only that $\omega\in C^1(\tilde{U}^+)$ and hence for any $\nu\in(0,1)$ there is a constant $D(\nu)$ such that

$$\sup_{-\nu < \xi < \nu} \omega(\xi+i\eta) \le D(\nu)\eta, \ \forall\eta\in(0,1-\nu). \tag{5.3.16}$$

Then we consider the function

$$\Omega^{(1)}(z) = 1 - (1-\omega(z_1))(1-\omega(z_2))\cdot\ldots\cdot(1-\omega(z_n)).$$

This function is defined on the set $\tilde{U}^{+n} = \tilde{U}^+\times\ldots\times\tilde{U}^+$. It is n-harmonic on the domain $U^{+n} = int\tilde{U}^{+n} = U^+\times\ldots\times U^+$, i.e. it is harmonic in every variable z_j when the other variables are fixed. The function $\Omega^{(1)}(z)$ vanishes when $z = x\in T_1$, and equals 1 for $z\in\tilde{S}_1\setminus\tilde{T}_1$, where $\tilde{S}_1 = \partial U^+\times\ldots\times\partial U^+$ is the skeleton of the domain U^{+n}. Besides, $0 < \Omega^{(1)}(z) <$

1, $\forall z\in U^{+n}$, and it follows from (5.3.16) that

$$\sup_{\xi\in T_\nu} \Omega^{(1)}(\zeta+ihI) \le 2^{n-1}D(\nu)h, \quad \forall h\in(0,1-\nu). \tag{5.3.17}$$

Finally we set

$$\Omega_R^{(2)}(z) = \int_{T_R\setminus E} P(x-t,y)dt,$$

where $P(x,y)$ is the product of the Poisson kernels for the half-plane, i.e.

$$P(x,y) = \frac{1}{\pi^n}\prod_{j=1}^{n} \frac{y}{x^2+y^2} .$$

It follows from the well-known properties of the Poisson kernel that the function $\Omega_R^{(2)}(z)$ is n-harmonic in the product of half-planes $\mathbb{C}^{+n} = \mathbb{C}^+\times\ldots\times\mathbb{C}^+$ and satisfies the conditions

$$\lim_{z'\to x,\, z'\in\mathbb{C}^{+n}} \Omega_R^{(2)}(z') = 0 \quad \text{for almost all } x\in E\cap T_R$$

and

$$\lim_{z'\to x,\, z'\in\mathbb{C}^{+n}} \Omega_R^{(2)}(z') = 1 \quad \text{for almost all } x\in T_R\setminus E.$$

Besides, $0 < \Omega_R^{(2)}(z) < 1$, $\forall z\in\mathbb{C}^{+n}$.

Since $u(z)\in PSH(\mathbb{C}^n)$, then function $u(z)$ is n-subharmonic, i.e. it is subharmonic in every variable z_j. It is clear that for such functions the n-harmonic majorant principle holds. This principle states that if an n-harmonic function v majorizes an n-subharmonic function u on $\partial G_1\times\ldots\times\partial G_n$, where G_1, ..., G_n are bounded domains in $\mathbb{C}$, then v majorizes u on $G_1\times\ldots G_n$ too. To estimate the function $u(z)$ in this manner we take as corresponding n-harmonic majorant the function

$$(\ln c+\sigma R)\Omega^{(1)}(z/R) + m_0(R)\Omega_R^{(2)}(z),$$

where c and σ are the same as in the lemma. From the above-mentioned properties of the functions $\Omega^{(1)}(z)$ and $\Omega_R^{(2)}(z)$ and from the estimates of the function f (or, equivalently, of the function $u(z)$) given in the lemma, it follows that the inequality

$$u(z) \le (\ln c+\sigma R)\Omega^{(1)}(z/R) + m_0(R)\Omega_R^{(2)}(z)$$

is valid on the set $\tilde{S}_R = \{z\colon\ z/R\in\tilde{S}_1\}$. By the n-harmonic majorant principle this inequality implies

$$u(z) \le (\ln c+\sigma R)\Omega^{(1)}(z/R) + m_0(R)\Omega_R^{(2)}(z),\ \forall z\in U_R^{+n}, \tag{5.3.18}$$

where $U_R^{+n} = U_R^{+}\times\ldots\times U_R^{+}$, $U_R^{+} = \{\zeta\in\mathbb{C}\colon\ |\zeta| < R,\ Im\zeta > 0\}$.

We estimate the quantity $m_L(\nu R)$, $0 < \nu < 1$, using the last inequality. From (5.3.17) it follows that

$$(\ln c+\sigma R)\sup_{x\in T_{\nu R}} \Omega^{(1)}\left(\frac{x+iLI}{R}\right) \le 2^{n-1}D(\nu)\,\frac{L}{R}\,(\ln c+\sigma R) =$$

$$= \sigma D(\nu)2^{n-1}L + o(1),\ R\to\infty. \tag{5.3.19}$$

Then we estimate $\Omega_R^{(2)}(x+iLI)$. We have

$$\Omega_R^{(2)}(x+iLI) = \int_{T_R\setminus E} P(x-t,LI)dt \le$$

$$\le \int_{\mathbb{R}^n\setminus(T_R\cap E)} P(x-t,LI)dt = 1 - \int_{T_R\cap E} P(x-t,LI)dt. \tag{5.3.20}$$

Note that if $|x_i-t_i| \le L$, then

$$\frac{L}{(x_i-t_i)^2 + L^2} \ge \frac{1}{2L},$$

and hence

$$P(x-t,LI) \ge \frac{1}{(2\pi L)^n} \quad \text{for } \|x-t\|_\infty \le L.$$

Therefore, from (5.3.20) it follows that for $x\in T_{R-L}$,

$$\Omega_R^{(2)}(x+iLI) \le 1 - \int_{T_R\cap E} P(x-t,LI)dt \le 1 - \int_{T(x,L)\cap E} P(x-t,LI)dt \le$$

$$\le 1 - \left(\frac{1}{2\pi L}\right)^n \int_{T(x,L)\cap E} dt \le 1 - \frac{\delta}{(2\pi L)^n}. \tag{5.3.21}$$

Comparing the estimates (5.3.18), (5.3.19) and (5.3.21), we find that as $R \to \infty$ the inequality

$$m_L(\nu R) \le \sigma D(\nu) 2^{n-1} L + m_0(R)\left(1 - \frac{\delta}{(2\pi L)^n}\right) + o(1)$$

is valid. Thus we have proved inequality (5.3.14).

To prove inequality (5.3.15) it is clearly sufficient to set $E = \emptyset$ and to apply the estimates obtained to the function $v(z) = u(-z+iLh)$.

The proof of the lemma is finished.

It follows from this lemma that for any $\nu \in (0,1)$ the inequality

$$m_0(\nu^2 R) \le m_0(R)\left(1 - \frac{\delta}{(2\pi L)^n}\right) + 2^n D(\nu)\sigma L + o(1) \qquad (5.3.22)$$

is valid as $R \to \infty$. Using this inequality we find that

$$\overline{lim}_{R\to\infty} \frac{m_0(R)}{R} = \frac{1}{\nu^2}\, \overline{lim}_{R\to\infty} \frac{m_0(\nu^2 R)}{R} \le \frac{1}{\nu^2}\left(1 - \frac{\delta}{(2\pi L)^n}\right) \overline{lim}_{R\to\infty} \frac{m_0(R)}{R} .$$

Setting $\nu^2 > 1 - \dfrac{\delta}{(2\pi L)^n}$, we conclude that

$$\lim_{R\to\infty} \frac{m_0(R)}{R} = 0. \qquad (5.3.23)$$

Now we suppose that $\sup_R m_0(R) = \infty$ or, equivalently, $m_0(R)\uparrow\infty$ as $R \to \infty$. Then (5.3.22) implies that for any $\kappa > 1$, any $\nu \in (0,1)$ and all sufficiently large R the inequality

$$m_0(R) \le m_0\left(\frac{1}{\nu^2} R\right)\left(1 - \frac{\delta}{\kappa(2\pi L)^n}\right)$$

is true. Iterating this inequality we obtain

$$m_0(R) \le m_0\left(\frac{1}{\nu^{2l}} R\right)\left(1 - \frac{\delta}{\kappa(2\pi L)^n}\right)^l.$$

Taking into account (5.3.23), we pass to the limit as $l \to \infty$. Assuming $\nu^2 > 1 - \dfrac{\delta}{\kappa(2\pi L)^n}$, we obtain that for all sufficiently large R

$$m_0(R) \le \lim_{l\to\infty} \frac{m_0(\nu^{-2l}R)}{\nu^{-2l}R}\; \lim_{R\to\infty} R\nu^{-2l}\left(1 - \frac{\delta}{\kappa(2\pi L)^n}\right) = 0.$$

This contradicts the assumption that $m_0(R)\uparrow\infty$. Hence

$$\sup_{0<R<\infty} m_0(R) = c_1 < \infty.$$

To estimate c_1 we use inequality (5.3.22) again, which clearly implies

$$c_1 \le c_1\left(1 - \frac{\delta}{(2\pi L)^n}\right) + 2^n D(\nu)\sigma L,$$

and therefore

$$c_1 \le \frac{(2\pi L)^n 2^n D(\nu)\sigma L}{\delta}.$$

The proof of the theorem is finished.

In conclusion of this section we note that in the proof of **Theorem 5.3.3** the special form of the function $u(z)$, i.e. that $u(z) = \ln|f(z)|$ where $f \in H(\mathbb{C}^n, 1, \sigma]$, was not used. We used only that the function $u(z)$ is n-harmonic. Therefore **Theorem 5.3.3** can be formulated in a more general form.

Theorem 5.3.4. Let E be a relatively dense set in $\mathbb{R}^n$ with respect to Lebesgue measure with parameters L and δ. Let the upper semicontinuous function $u(z)$ be n-harmonic in $\mathbb{C}^n$ and satisfy the conditions

$$\overline{\lim_{|z|\to\infty}} \frac{u(z)}{|z_1| + \ldots + |z_n|} < \infty$$

and

$$\sup_{x\in E} u(x) < \infty.$$

Then

$$\sup_{x\in\mathbb{R}^n} u(x) \le \frac{\sigma(4\pi)^n L^{n+1}}{\delta} c\cdot\sup_{x\in E} u(x).$$

Notes

Theorem 5.1.1 is the multidimensional analogue of the **Paley-Wiener Theorem** (see, for example, B.Levin [1]). It was obtained by G.Polya and M.Plancherel [1]. As already mentioned, **Theorem 5.2.2** is contained in the monographs of V.Napalkov [1] and L.Hörmander [1]. **Theorem 5.1.3** is due to J.Wiegerinck [1]. The example that is the subject of **Theorem 5.1.4** was first constructed by J.Vauthier [1], [2]. Here we have given the simplified construction due to J.Wiegerinck [1], who also obtained the sufficient conditions of c.r.g. containing in **Theorem 5.1.5**.

The study of discrete uniqueness sets was started by J.Korevaar and S.Hellerstein [1]. They considered functions $f\in H(\mathbb{C}^{+n})$ and sets $E\subset i\mathbb{Z}^n_+$. The method of studying discrete uniqueness sets given here was developed in the papers of L.Ronkin [6] - [11]. **Theorem 5.2.2**, with a constant at the right hand side of inequality (5.2.1) that differs from that given here by the factor $\frac{1}{e}$, was obtained by L.Ronkin [9] - [11]. Later B.Berndtsson [1] refined this theorem, and get rid of the above-mentioned factor. Earlier, without factor 1/e but with another definition of $\delta_{E,\gamma}(x^0)$, **Theorem 5.2.2.** was proved by L.Ronkin [9]-[11].

All the statements of §3 concerning discrete norming sets are due to V.Logvinenko [2] - [3]. These papers, as well as [4] contain also other statements on these sets. **Theorem 5.3.3** was given by B.Levin in his report at the Conference on Function Theory (Kharkov 1971). The first published proof of this theorem (in the more general situation of n-subharmonic functions) is due to V.Kac'nelson [1]. B.Levin and V.Logvinenko [1] obtained a generalization of **Theorem 5.3.3** on subharmonic functions for sets of variables. Our proof of **Theorem 5.3.3** is an adaptation of their proof.

Note that the problem of equivalent norms in the space $W^p_{\sigma,n}$ of entire functions $f(z)\in H(\mathbb{C}^n,1]$ that belong to L^p on $\mathbb{R}^n$ is close to that of the norming sets of positive measure. More exactly, this problem is as follows: what are the conditions on a set $E\subset\mathbb{R}^n$ such that if they are satisfied, then the inequality

$$\int_{\mathbb{R}^n}|f(x)|^p dx \le c\int_E |f(x)|^p dx,\ \forall f\in W^p_{\sigma,n},$$

is valid with a constant c independent of f. Here we have not considered this problem, but one can find it in B.Panejah [1], [2], Logvinenko-Sereda [1], V.Kac'nelson [1].

CHAPTER 6

QUASIPOLYNOMIALS

Among the several classes of special f.c.r.g.,next to the class considered in the previous chapter, the class of functions called quasipolynomials is of great interest. A function is called a quasipolynomial if it is a finite linear combination of exponentials with certain coefficients (for the exact definition, see § 6.1). Quasipolynomials are encountered in various part of mathematics, in applied problems, and in various fields of physics. An account of a number of general facts concerning quasipolynomials will be given in this chapter. Main attention will focus on quasipolynomials of several variables.

§1 M-quasipolynomials. Growth and zero distribution

1. Basic definition. We set, for brevity, $M_0 = H(\mathbb{C}^n,1,0]$. In other words, M_0 is the set of all entire functions of at most minimal type with respect to the order 1. Now let $M \subset M_0$.

An entire function f(z) is called to as a M-quasipolynomial if it can be represented in the form

$$f(z) = \sum_{i=1}^{m} a_i(z)e^{\langle\lambda_i,z\rangle}, \qquad (6.1.1)$$

where $m < \infty$, $a_i \in M$, $a_i \not\equiv 0$, $\lambda_i \in \mathbb{C}^n$, $\forall i = 1,\dots, m$, and $\lambda_i \neq \lambda_j$ for $i \neq j$. We denote by Q_M the set of all M-quasipolynomials.

The case $M = P$, where P is the set of all polynomials in $z = (z_1,\dots,z_n)$, is of utmost interest, because P-quasipolynomials (or exponential polynomials) are the solutions of linear partial differential equations with constant coefficients. The case $M = \mathbb{C}$, i.e. when all a_i are constant, is of great interest too. Usually, $\mathbb{C}$-quasipolynomials are simply called quasipolynomials or exponential sums. Note also that when discussing M-quasipolynomials with arbitrary, non-concretized. M for brevity we sometimes call them quasipolynomials too.

It is easy to see that if

$$\sum_{i=1}^{m} a_i e^{\langle\lambda_i, z\rangle} \equiv 0,$$

where $a_i \in M_0$, $\forall i$, and $\lambda_i \neq \lambda_j$ for $i \neq j$, then $a_i \equiv 0$. Therefore, the coefficients a_i and exponents λ_i in the representation (6.1.1) are uniquely determined[42] by f.

The set of all points (exponents) λ_j in the representation (6.1.1) will be referred to as the spectrum of the quasipolynomial f, and will be denoted by $\Lambda = \Lambda(f)$.

In the sequel we will see that both the growth of an M-quasipolynomial f and its zero distribution are, in essence, determined by its spectrum only, namely, by the convex hull of the spectrum, denoted $\tilde{\Lambda} = \tilde{\Lambda}(f)$. It is obvious that $\tilde{\Lambda}$ is a convex polyhedron in $\mathbb{C}^n = \mathbb{R}^{2n}$. We denote by $\tilde{\Lambda}_0$ the set of its vertices and by $\tilde{\lambda}_j$ the vertices themselves. Similarly, we denote by $\tilde{\Lambda}_p$, $p = 1, \ldots, 2n-1$, the set of p-dimensional sides and by $\Gamma_{j,p}$, $j = 1, \ldots, m_p < \infty$, the sides themselves. For simplicity of notation we set $\Gamma_{j,0} = \tilde{\lambda}_j$. Note that each side $\Gamma_{j,p}$ is a closed convex p-dimensional polyhedron.

We denote by $H_\Lambda(\zeta)$, $\zeta \in \mathbb{C}^n$, the support function of the polyhedron $\tilde{\Lambda}$, i.e. we set

---

[42] Sometimes more general M-quasipolynomials are considered: namely, without the assumption that $M \subset H(\mathbb{C}^n)$, in particular, when M consists of meromorphic functions or functions holomorphic in a cone. However, M must be such that the coefficients a_i and the exponents λ_i are uniquely determined by the function f as in (6.1.1). In particular, this requirement will be fulfilled if $M = \mathfrak{R}$, where $\mathfrak{R}$ is the set of all rational functions in z.

$$H_\Lambda(\zeta) = \sup_{z\in\tilde\Lambda} Re\langle\bar\zeta, z\rangle.$$

Since $\tilde\Lambda$ is a polyhedron, $H_\Lambda(\zeta)$ is a piecewise-linear function. We single out sets on which this function is linear. For this we associate to each side $\Gamma_{j,p}$, $j = 1, \dots \omega_p$, $p = 0, 1, \dots, 2n-1$, the $(2n-p)$-dimensional cone

$$\Gamma^*_{j,p} = \{\zeta\in\mathbb{C}^n\colon H_\Lambda(\zeta) = Re\langle z,\bar\zeta\rangle,\ \forall z\in\Gamma_{j,p}\}.$$

The function $H_\Lambda(\zeta)$ is linear in each such cone ($H_\Lambda(\zeta) = Re\langle z,\bar\zeta\rangle$, $z\in\Gamma_{j,p}$) and if $\zeta\in\Gamma^*_{j,p}$, then the hyperplane

$$\{z\in\mathbb{C}\colon Re\langle z,\bar\zeta\rangle = H_\Lambda(\zeta)\}$$

contains the side $\Gamma_{j,p}$.

The set

$$\mathfrak{Z}(\tilde\Lambda) = \bigcup_{j=1}^{m} \Gamma^*_{j,1}$$

is referred to as the star of the polyhedron $\tilde\Lambda$. We also set

$$\mathfrak{Z}_0(\tilde\Lambda) = \bigcup_{j=1}^{m_0} int\Gamma^*_{j,0}\ .$$

Note that $int\Gamma^*_{j,0} \cap int\Gamma^*_{i,0} = \emptyset$, $\forall i,j$, $i\neq j$ and $\bar{\mathfrak{Z}}_0(\tilde\Lambda) = \mathbb{C}^n$, $\mathfrak{Z}_0(\tilde\Lambda) \cup \mathfrak{Z}(\tilde\Lambda) = \mathbb{C}^n$. We write $\mathfrak{Z}(f)$ and $\mathfrak{Z}_0(f)$ instead of $\mathfrak{Z}(\tilde\Lambda)$ and $\mathfrak{Z}_0(\tilde\Lambda)$ if the polyhedron $\tilde\Lambda$ is constructed from the spectrum of an M-quasipolynomial f. These sets will be used for the description of the growth and zero distribution of M-quasipolynomials.

2. Growth of M_0-quasipolynomials. It follows directly from the definition that an M-quasipolynomial is an entire function of exponential type. We can get more refined information on the asymptotic behaviour of an M_0-quasipolynomial if we take into account that there is a main term for almost all directions in $\mathbb{C}^n$. The main term is a component at the right-hand side of (6.1.1) that grows faster than the sum of the other terms. The following theorem contains a description of the growth of M-quasipolynomials in terms of the indicator $\mathfrak{L}^*_f$.

Theorem 6.1.1. Let f(z) be an M_0-quasipolynomial with spectrum Λ. Then

$$\mathfrak{L}^*_f(z) = H_\Lambda(\bar z), \tag{6.1.2}$$

and f(z) is a function of c.r.g.

Proof. First of all we show that

$$\mathfrak{L}_f(z) \le H_\Lambda(\bar{z}), \ \forall z \in \mathbb{C}^n. \tag{6.1.3}$$

Indeed, since f(z) is an M_0-quasipolynomial, the representation (6.1.1) holds, and hence

$$\mathfrak{L}_f(z) = \overline{\lim_{t\to\infty}} \frac{\ln|f(tz)|}{t} = \overline{\lim_{t\to\infty}} \ln\left|\sum_{j=1}^m a_j(tz)e^{t\langle z,\lambda_j\rangle}\right|\cdot\frac{1}{t} \le$$

$$\le \overline{\lim_{t\to\infty}} \frac{1}{t} \max_j(\ln|a_j(tz)| + tRe\langle z,\lambda_j\rangle) \le$$

$$\le \max_j Re\langle z,\lambda_j\rangle = \max_{\zeta\in\tilde{\Lambda}} Re\langle\zeta,z\rangle = H_\Lambda(\bar{z})$$

(recall that the a_j are functions of minimal type).

Then we let z be a point such that $\bar{z}\in\mathfrak{Z}_0(f)$. In accordance with the definition of $\mathfrak{Z}_0(f)$, the point $\bar{z}$ belongs to one and only one set $int\Gamma^*_{j,0}$. This means that the hyperplane $\{\zeta\in\mathbb{C}^n: Re\langle\zeta,z\rangle = H_\Lambda(\bar{z})\}$ contains one and only one point of the spectrum Λ. For the sake of being specific, we assume that this point is λ_1. Thus

$$Re\langle\lambda_1,z\rangle = H_\Lambda(\bar{z})$$

and

$$Re\langle\lambda_j-\lambda_1,z\rangle < 0, \ \forall j = 2, \dots .$$

Therefore, at z we have

$$f(tz) = a_1(tz)e^{t\langle\lambda_1,z\rangle}\left(1 + \sum_{j=2}^m a_j(tz)e^{t\langle\lambda_j-\lambda_1,z\rangle}\right) =$$

$$= a_1(tz)e^{t\langle\lambda_1,z\rangle}(1 + o(1)) \tag{6.1.4}$$

as $t\to\infty$.

Since $a_1(z)$ is a function of minimal type, it is a function of c.r.g. in $\mathbb{C}^n$ and, moreover, its non-zero restriction to any complex line is a f.c.r.g. of one variable. Therefore, if $a_1(z) \ne 0$, then $t^{-1}\ln|a_1(tz)| \to 0$ as $t\to\infty$ outside a set E of relative measure zero. From this and (6.1.4) it follows that

$$\lim_{\substack{t \to \infty \\ t \notin E}} \frac{\ln|f(tz)|}{t} = Re\langle\lambda_1, z\rangle = H_\Lambda(\bar{z}), \ \forall z: \ \bar{z}\in\mathfrak{Z}_0(f), \ \prod_{j=1}^{m} a_j(z) \neq 0.$$

In view of (6.1.3) This means that for $\bar{z}\in\mathfrak{Z}_0(f)$, $\prod_{j=1}^{m} a_j(z) \neq 0$,

$$\mathfrak{L}_f(z) = H_\Lambda(\bar{z}),$$

and each ray $l_z = \{\zeta\in\mathbb{C}^n: \ \zeta = tz, \ t>0\}$ is a ray of c.r.g. of the function f. Taking into account the continuity of $H_\Lambda(z)$ and the coincidence of the closure of the set $\left\{z\in\mathbb{C}^n: \ \bar{z}\in\mathfrak{Z}_0(f), \ \prod_{j=1}^{m} a_j(z) \neq 0\right\}$ with $\mathbb{C}^n$, we conclude that

$$\mathfrak{L}^*_f(z) = H_\Lambda(\bar{z})$$

and that f(z) is a function of c.r.g.

This finishes the proof.

In the case of $\mathbb{C}$-quasipolynomials **Theorem 6.1.1** can be refined as follows.

Theorem 6.1.2. Let

$$f(z) = \sum_{j=1}^{m} a_j e^{\langle\lambda_j, z\rangle}, \ a_j\in\mathbb{C}.$$

Then there are positive numbers c_1, c_2, d such that

$$c_1 \le |f(z)|e^{-H_\Lambda(\bar{z})} < c_2, \forall z: \ dist(z,\mathfrak{Z}(f)) > d. \qquad (6.1.5)$$

Proof. Let $z^{(p)}$ be any point belonging to $int\Gamma^*_{p,0}$, where p is arbitrary but fixed, $1 \le p \le \omega_0$. Then

$$H_\Lambda(\bar{z}^{(p)}) = Re\langle z^{(p)}, \tilde{\lambda}_p\rangle.$$

Further, the definitions of $\Gamma^*_{p,0}$ and $\tilde{\lambda}_p$ imply

$$Re\langle\lambda_j-\tilde{\lambda}_p, z\rangle < 0, \ \forall\lambda_j \neq \tilde{\lambda}_p, \ z\in int\Gamma^*_{p,0}.$$

It is obvious that the theorem will be proved if we show that for some $c_1 > 0$, $c_2 > 0$, $d > 0$ the inequality

$$c_1 \le |f(z)|e^{-H_\Lambda(\bar{z})} \le c_2, \forall z = d\cdot z^{(p)}+w, \ w\in\Gamma^*_{p,0}$$

is satisfied.

As in the proof of the previous theorem, for simplicity of notation we assume $\tilde{\lambda}_p = \lambda_1$ and, correspondingly,

$$H_\Lambda(\bar{z}^{(p)}) = Re\langle z^{(p)}, \lambda_1\rangle,$$

$$Re\langle \lambda_j - \lambda_1, z\rangle < 0, \forall z \in int\Gamma^*_{p,0}, \ j = 2, \ldots, m. \tag{6.1.6}$$

We choose d large enough so that the inequality

$$\sum_{j=2}^{m} |a_j| e^{Re\langle \lambda_j - \lambda_1, z^{(p)}\rangle} < \frac{1}{2}|a_1|$$

holds. This is possible in view of (6.1.6). Then for $z = d \cdot z^{(p)} + w$, where $w \in \Gamma^*_{p,0}$, we obtain

$$|f(z)| e^{-H_\Lambda(\bar{z})} = |f(z)| e^{-Re\langle z, \lambda_1\rangle} = \left| \sum_{j=1}^{m} a_j e^{\langle z, \lambda_j\rangle - Re\langle z, \lambda_1\rangle} \right| \geq$$

$$\geq |a_1| - \sum_{j=2}^{m} |a_j| e^{Re\langle \lambda_j - \lambda_1, z\rangle} = |a_1| - \sum_{j=2}^{m} |a_j| e^{Re\langle \lambda_j - \lambda_1, z^{(p)}\rangle d} e^{Re\langle \lambda_j - \lambda_1, w\rangle} \geq$$

$$\geq |a_1| - \sum_{j=2}^{m} |a_j| e^{Re\langle \lambda_j - \lambda_1, z^{(p)}\rangle} \geq |a_1| - \frac{1}{2}|a_1| = c_1.$$

Similarly, if d is chosen as above, then

$$|f(z)| e^{-H_\Lambda(\bar{z})} \leq \frac{3}{2}|a_1| = c_2, \ \forall z = z^{(p)} d + w, w \in \Gamma^*_{p,0}.$$

The proof is finished.

3. The zero distribution of M_0-quasipolynomials. Since M_0-quasipolynomials are f.c.r.g. (see **Theorem 6.1.1**), in accordance with the general properties of f.c.r.g. the divisor Z_f of each M_0-quasipolynomial f is regularly distributed, and hence its density exists and can be represented by means of the Riesz associated measure of the indicator $\mathcal{L}^*_f(z)$. In the situation under consideration this measure, and therefore the density of the divisor, can be calculated clearly sufficiently simply directly in terms of the spectrum of the quasipolynomial. We introduce the notation necessary for formulating the corresponding result.

Let f, $\tilde{\Lambda}$, $\tilde{\zeta}_j$, $\Gamma_{j,p}$, $\Gamma^*_{j,p}$, $\mathfrak{Z}_0(f)$, and $\mathfrak{Z}(f)$ be as above. We associate to each edge $\Gamma_{1,1}$ the vertices $\tilde{\zeta}_{j'(1)}$ and $\tilde{\zeta}_{j''(1)}$ defining it. We set

$$\mathfrak{F}_\Lambda(z) = \frac{1}{2\pi} |\tilde{\zeta}_{j'(1)} - \tilde{\zeta}_{j''(1)}|$$

if the point z belongs to the cone $\Gamma^*_{1,1}$ and not to any other cone $\Gamma^*_{j,1}$, $j \neq 1$. Such points form the relative interior $int\Gamma^*_{1,1}$ of the cone $\Gamma^*_{1,1}$. The set $\mathfrak{Z}(f)\setminus\bigcup_j \Gamma^*_{j,1}$ has real dimension $2n-2$, and we prescribe the function $\mathfrak{F}_\Lambda(z)$ on it arbitrarily. We denote by $d\sigma_{2n-1}$ the $(2n-1)$-dimensional volume element on $\mathfrak{Z}(f)$ induced by the metric of $\mathbb{C}^n$. As above (see **Chapter 4**), we denote by $V_f(G)$ the $(2n-2)$-dimensional volume of the divisor Z_f in a domain G, $G \subset\subset \mathbb{C}^n$.

Theorem 6.1.3. Let f be a M_0-quasipolynomial with spectrum Λ, and let K be an open cone in $\mathbb{C}^n$ with vertex at the origin. Then if $mes_{2n-1}(\partial K \cap \mathfrak{Z}(f)) = 0$ then

$$\lim_{R\to\infty} \frac{1}{R^{2n-1}} V_f(K_R) = \int_{\tilde{K}} \mathfrak{F}_\Lambda(z) d\sigma_{2n-1}, \tag{6.1.7}$$

where $K_R = K \cap B_R$, $\tilde{K} = K_1 \cap \mathfrak{Z}(f)$.

Proof. Theorems 3.1.1 and 6.1.1 imply

$$\lim_{R\to\infty} \frac{1}{R^{2n-1}} V_f(K_R) = \frac{\theta_{2n}}{2\pi} \int_{K_1} d\mu,$$

where $\mu = \frac{1}{\theta_{2n}} \Delta H_\Lambda(\bar{z})$ is the Riesz measure associated with the function $H_\Lambda(\bar{z}) = \mathfrak{L}_f(z)$, and K is an open cone with vertex at the origin such that $\mu(\partial K) = 0$. Therefore, to prove equality (6.1.7) it suffices to show that the measure μ is concentrated on the star $\mathfrak{Z}(f)$ with density $\mathfrak{F}_\Lambda$. We use the properties of the function $H_\Lambda(z)$, mentioned at the beginning of this paragraph; in particular, $H_\Lambda(z) = Re\langle\bar{z},\tilde{\zeta}_j\rangle$, $\forall z\in\Gamma^*_{j,0}$. Letting $\varphi\in C^\infty_0(\mathbb{C}^n)$ and taking into account the rules of operation with distributions, we obtain

$$\frac{\theta_{2n}}{2\pi}\int_{\mathbb{C}^n}\varphi d\mu = \sum_{j=1}^{\omega_0}\int_{\Gamma^*_{j,0}\cap\, supp\varphi} H_\Lambda(\bar{z})\Delta\varphi d\omega_z =$$

$$= -\frac{1}{2\pi}\sum_{j=1}^{\omega_0}\left\{\int_{\partial\Gamma^*_{j,0}\cap\, supp\varphi} H_\Lambda(\bar{z})\frac{\partial\varphi}{\partial n}d\omega_z + \int_{\partial\Gamma^*_{j,0}\cap\, supp\varphi}\varphi\frac{\partial}{\partial n}Re\langle z,\tilde{\zeta}_j\rangle d\omega_z\right\}.$$

The first sum on the right vanishes because of the following arguments. The sets $\partial\Gamma^*_{j,0}$, which form the star $\mathfrak{Z}(f)$, consist of (2n-1)-dimensional cones $\Gamma^*_{1,1}$. Each $\Gamma^*_{1,1}$ is included in exactly two sets $\partial\Gamma^*_{j',0}$ and $\partial\Gamma^*_{j'',0}$, $j' = j'(1)$, $j'' = j''(1)$, with opposite orientations. By analogy, taking into account that μ is a positive measure, we conclude that the second sum equals

$$\frac{1}{2\pi}\sum_{1=1}^{\omega_1}\int_{\Gamma^*_{1,1}\cap\, supp\varphi}\varphi\left|\frac{\partial}{\partial n}Re\langle\tilde{\zeta}_{j'},z\rangle - \frac{\partial}{\partial n}Re\langle\tilde{\zeta}_{j''},z\rangle\right|d\sigma_{2n-1},$$

where $\frac{\partial}{\partial n}$ is the derivative with respect to the normal to the plane $Re\langle z,\tilde{\zeta}_{j'}-\tilde{\zeta}_{j''}\rangle = 0$ containing $\Gamma^*_{1,1}$.

Thus,

$$\frac{\theta_{2n}}{2\pi}\int_{\mathbb{C}^n}\varphi d\mu =$$

$$= \frac{1}{2\pi}\sum_{1=1}^{\omega_1}\int_{\Gamma^*_{1,1}\cap\, supp\varphi}\varphi\left|\frac{\partial}{\partial n}Re\langle z,\tilde{\zeta}_{j'}\rangle - \frac{\partial}{\partial n}Re\langle z,\tilde{\zeta}_{j''}\rangle\right|d\sigma_{2n-1}. \quad (6.1.8)$$

Now we calculate $\frac{\partial}{\partial n}Re\langle z,\tilde{\zeta}_{j'}\rangle - \frac{\partial}{\partial n}Re\langle z,\tilde{\zeta}_{j''}\rangle$. It is obvious that $grad\{Re\langle z,\tilde{\zeta}_j\rangle\} = \bar{\tilde{\zeta}}_j$, and the normal to the plane $Re\langle z,\tilde{\zeta}_{j'}-\tilde{\zeta}_{j''}\rangle = 0$ is defined by the equality $\bar{z} = (\tilde{\zeta}_{j'}-\tilde{\zeta}_{j''})t$, $t\in\mathbb{R}$. Therefore,

$$\frac{\partial}{\partial n}Re\langle z,\tilde{\zeta}_{j'}\rangle = Re\langle\bar{\tilde{\zeta}}_{j'},\frac{\tilde{\zeta}_{j'}-\tilde{\zeta}_{j''}}{|\tilde{\zeta}_{j'}-\tilde{\zeta}_{j''}|}\rangle,$$

$$\frac{\partial}{\partial n} Re\langle z, \tilde{\zeta}_{j''}\rangle = Re\langle \bar{\tilde{\zeta}}_{j''}, \frac{\tilde{\zeta}_{j'} - \tilde{\zeta}_{j''}}{|\tilde{\zeta}_{j'} - \tilde{\zeta}_{j''}|}\rangle ,$$

and hence

$$\left| \frac{\partial}{\partial n} Re\langle z, \tilde{\zeta}_{j'}\rangle - \frac{\partial}{\partial n} Re\langle z, \tilde{\zeta}_{j''}\rangle \right| = |\tilde{\zeta}_{j'} - \tilde{\zeta}_{j''}|.$$

From this and (6.1.8) it follows that

$$\frac{\theta_{2n}}{2\pi} \int_{\mathbb{C}^n} \varphi d\mu = \frac{1}{2\pi} \sum_{l=1}^{\omega_1} \int_{\Gamma^*_{1,1} \cap \, supp\varphi} \varphi(z) | \tilde{\zeta}_{j'} - \tilde{\zeta}_{j''} | d\sigma_{2n-1} =$$

$$= \int_{\mathfrak{Z}(f)} \varphi(z) \mathfrak{F}_\Lambda(z) d\sigma_{2n-1}.$$

This means that the measure μ is concentrated on $\mathfrak{Z}(f)$ with density $\mathfrak{F}_\Lambda(z)$. Thus,

$$\frac{\theta_{2n}}{2\pi} \int_{K_1} d\mu = \int_{K_1 \cap \mathfrak{Z}(f)} \mathfrak{F}_\Lambda(z) d\sigma_{2n-1} = \int_{\tilde{K}} \mathfrak{F}_\Lambda(z) d\sigma_{2n-1}.$$

Therefore if $mes_{2n-1}(\partial K \cap \mathfrak{Z}(f)) = 0$ or, equivalently,

$$\int_{\partial K \cap \mathfrak{Z}(f)} \mathfrak{F}_\Lambda(z) d\sigma_{2n-1} = 0,$$

then $\mu(\partial K) = 0$.

This finishes the proof.

Theorem 6.1.3 can be refined for $\mathbb{C}$-quasipolynomials by giving an estimate of the rate of convergence of the value $R^{-2n+1} V_f(K_R)$ to its limit. Here we give two results of the above type without proofs (which are rather unwieldy). A cone $K \subset \mathbb{C}^n$ with piecewise smooth boundary is said to be transversal to the star $\mathfrak{Z}(f)$ if its boundary ∂K is transversal to the (2n-1)-dimensional planes containing the sets $\Gamma^*_{j,1}$. Similarly we define transversality of a cone $K \subset \mathbb{R}^n$ to the "real" star $\mathfrak{Z}(f) \cap \mathbb{R}^n$.

Theorem 6.1.4. Let a function $f(z)$ be a $\mathbb{C}$-quasipolynomial with spectrum Λ. Let $K \subset \mathbb{C}^n$ be an open cone with vertex at the origin and piecewise smooth boundary which is transversal to the star $\mathfrak{Z}(f)$.

Then

$$V_f(K_R) = R^{2n-1}\int_{\tilde{K}} \mathfrak{F}_\Lambda(z)d\sigma_{2n-1} + Q_1R^{2n-2} + Q_2, \ \forall R > 0,$$

where the quantities Q_1 and Q_2 are bounded on $\mathbb{R}_+$.

Theorem 6.1.5. Let $f(z)$ be a $\mathbb{C}$-quasipolynomial with spectrum $\Lambda \subset \mathbb{R}^n$. Let $K \subset \mathbb{R}^n$ be an open cone with vertex at the origin and piecewise smooth boundary which is transversal to $\mathfrak{Z}(f) \cap \mathbb{R}^n$. Then

$$V_f(\{z = x+iy\in\mathbb{C}^n: |x| < R, |y| < h, x\in K\}) =$$

$$= R^{n-1}h^nV_n \int_{\mathfrak{Z}(f) \cap \mathbb{R}^n} \mathfrak{F}_\Lambda(z)d\sigma_{n-1} +$$

$$+ O_1R^{n-2}h^n + O_2R^{n-1}h^{n-1} + O_3, \ R > 0, \ h > 0,$$

where σ_{n-1} is the volume element on $\mathfrak{Z}(f)\cap\mathbb{R}^n$ and O_1, O_2, O_3 are bounded on $\mathbb{R}_+$.

In conclusion of the section we state (as above without proof) one useful result on the zero distribution of quasipolynomials of one variable.

Theorem 6.1.6. Let

$$f(w) = \sum_{l=1}^{m} P_l(w)e^{\lambda_l w},$$

where $w\in\mathbb{C}$, $\lambda_l\in\mathbb{R}$, $\lambda_1 < \lambda_2 < \ldots < \lambda_m$, and the $P_l(w)$ are polynomials of degrees $\deg P_l$. Let also $n_f(R,h)$ be the number of zeros of the function f in the rectangle $\{w\in\mathbb{C}: |Re w| < R, \ 0 < Im w < h\}$ counted with multiplicities. If the inequalities

$$|P_1(-R+iu)e^{\lambda_1(-R+iu)}| > \left|\sum_{l=2}^{m} P_l(-R+iu)e^{\lambda_l(-R+iu)}\right|, \ \forall u: 0 < u < h,$$

and

$$|P_m(R+iu)e^{\lambda_m(R+iu)}| > \left|\sum_{l=1}^{m-1} P_l(R+iu)e^{\lambda_l(R+iu)}\right|, \ \forall u: 0 < u < h,$$

are valid, then

$$n_f(R,h) = \frac{\lambda_m-\lambda_1}{2\pi} R + c(R),$$

where

$$c(R) < \frac{3}{2} \sum_{l=1}^{m} (1+degP_l).$$

In the case $degP_l = 0$, $\forall l = 1,\dots, m$ (i.e. when f is a $\mathbb{C}$-quasipolynomial), the estimate $c(R) \le m-1$ is true.

§2. Entire functions that are quasipolynomials in every variable

As already mentioned in **Chapter 4**, concerning there is the traditional problem in the theory of functions of several complex variables concerning the properties of a function $f \in H(\mathbb{C}^n)$ if it is known that its restrictions to some family of complex lines satisfy (as functions of one variable) certain conditions. In the situation considered in this chapter, i.e. for M-quasipolynomials, it is natural to study the problem of the general form of functions which are quasipolynomials in each variable if the other variables are fixed. The solution of this problem is also important, because it gives the possibility to obtain in a relatively simple manner properties of quasipolynomials of several variables from corresponding properties of quasipolynomials of one variable. The problem mentioned above will be considered here in the case of *P*-quasipolynomials and $\mathbb{C}$-quasipolynomials. To state the main result (**Theorem 6.2.1**) we need some properties of *P*-quasipolynomials of one variable, as well as properties of holomorphic functions of several variables that are *P*-quasipolynomials in one variable.

1. Auxiliary knowledge. Let

$$f(w) = \sum_{l=1}^{m} P_l(w)e^{\lambda_l w}, \quad w \in \mathbb{C}, \tag{6.2.1}$$

where the $P_l(w)$ are polynomials in w of degree $degP_l$, $P_l(w) \not\equiv 0$, $\forall l$, and the λ_l are complex numbers such that $\lambda_l \ne \lambda_j$ if $j \ne l$. Thus, the function f(w) is a *P*-quasipolynomial. We call the number

$$degf \overset{\text{def}}{=} \sum_{l=1}^{m} (degP_l+1)$$

the degree of this *P*-quasipolynomial. It is clear that *degf* coincides with the minimal order of a linear differential equation with constant coefficients such that the *P*-quasipolynomial belongs to the set of its solutions. This fact gives the possibility to formulate a simple effective criterion for entire function to belong to the class of *P*-quasipolynomials. The determinants

$$D_N(w;f) = \begin{vmatrix} f & f' \dots f^{(N)} \\ f' & \dots\dots\dots f^{(N+1)} \\ \dots\dots\dots\dots\dots\dots\dots \\ f^{(N)} & \dots\dots\dots f^{(2N)} \end{vmatrix}$$

will figure in the formulation of this criterion (**Lemma 6.2.1**).

Lemma 6.2.1. In order that an entire function $f(w)$, $w \in \mathbb{C}$, be a *P*-quasipolynomial of degree N, it is necessary and sufficient that $D_N(w;f) \equiv 0$ and $D_{N-1}(w;f) \not\equiv 0$.

Proof. Let $D_N(w;f) \equiv 0$ and $D_{N-1}(w;f) \not\equiv 0$. Consider the linear differential equation (with respect to y)

$$\begin{vmatrix} f & f' \dots f^{(N)} \\ f^{(N-1)} & \dots\dots\dots f^{(2N-1)} \\ y & y' \dots y^{(N)} \end{vmatrix} = 0. \tag{6.2.2}$$

Since $D_{N-1}(w;f) \not\equiv 0$, the order of this equation is equal to N and the functions $y_1 = f, \dots, y_N = f^{(N-1)}$, which are obviously its solutions, form linearly independent system. On the other hand, since $D_N(w;f) \equiv 0$, the function $y_{N+1} = f^{(N)}$ is also a solution of equation (6.2.2). Hence there are constants $c_1, \dots, c_N$ such that $y_{N+1} = c_1y_1 + \dots + c_Ny_N$ or, equivalently,

$$f^{(N)} - c_Nf^{(n-1)} - \dots - c_1f = 0.$$

Thus, the function f is a solution of a linear differential equation of order N with constant coefficients. This is equivalent to the fact that f is a *P*-quasipolynomial of the degree $\textit{degf} = \tilde{N} \le N$.

Now we assume that $f(w)$ is of the form (6.2.1) and $\textit{degf} = \tilde{N}$. Then

$$f^{(\tilde{N})} + c_{\tilde{N}}f^{(\tilde{N}-1)} + \dots + c_1f = 0, \tag{6.2.3}$$

where $c_1, \ldots, c_{\tilde N}$ are the corresponding coefficients of the characteristic polynomial

$$Y^{\tilde N} + c_{\tilde N} Y^{\tilde N - 1} + \ldots + c_1 = (Y-\lambda_1)^{degP_1} \cdot \ldots \cdot (Y-\lambda_m)^{degP_m}.$$

We differentiate equality (6.2.3) step by step to obtain the systems

$$\begin{cases} f^{(\tilde N)} + c_{\tilde N} f^{(\tilde N - 1)} + \ldots + c_1 f = 0 \\ \ldots\ldots\ldots\ldots\ldots\ldots\ldots\ldots\ldots\ldots \\ f^{(2\tilde N - 1)} + c_{\tilde N} f^{(2\tilde N - 2)} + \ldots + c_1 f^{(\tilde N - 1)} = 0 \end{cases}$$

and

$$\begin{cases} f^{(\tilde N)} + c_{\tilde N} f^{(\tilde N - 1)} + \ldots + c_1 f = 0 \\ \ldots\ldots\ldots\ldots\ldots\ldots\ldots\ldots\ldots\ldots \\ f^{(2\tilde N)} + c_{\tilde N} f^{(2\tilde N - 1)} + \ldots + c_1 f^{(\tilde N)} = 0 \end{cases}$$

The solvability conditions for this systems are as follows: $D_{\tilde N - 1}(w;f) \not\equiv 0$ and $D_{\tilde N}(w;f) \equiv 0$. From this, taking into account the evident implication $D_N(w;f) \equiv 0 \Rightarrow D_{N+1}(w;f) \equiv 0$, we conclude that $\tilde N = N$.

This finishes the proof.

The following lemma is an obvious consequence of the **Liouville-Ostrogradskiǐ Theorem** on the Wronskian of a system of fundamental solutions of a linear differential equation.

Lemma 6.2.2. If f is a *P*-quasipolynomial of degree N, then

$$D_{N-1}(w;f) = D_{N-1}(0;f)exp\left\{w \sum_{l=1}^{m} (degP_l + 1)\lambda_l\right\},$$

where the P_l are the polynomials figuring in the representation of f in the form (6.2.1).

Now we consider some "semi-local" statements concerning functions of several variables which are quasipolynomials in one variable.

Lemma 6.2.3. Let, in some domain $\{(z,w): z \in G \subset \mathbb{C}^n,\ w \in G' \subset \mathbb{C}\}$, the equality

$$\sum_{l=1}^{m}\sum_{j=0}^{k_l} a_{j,l}(z)w^j e^{\lambda_l(z)w} \equiv 0$$

be valid for certain holomorphic functions $a_{j,l}(z)$, $\lambda_l(z)$ in G. Then for each l, either $a_{j,l}(z) \equiv 0$, $\forall j$, or there is an index $\tilde{l} \neq l$ such that $\lambda_l(z) \equiv \lambda_{\tilde{l}}(z)$.

Proof. Since the coefficients $a_{j,l}(z)$ and the exponents $\lambda_l(z)$ are holomorphic, the lemma clearly follows from the uniqueness theorem for holomorphic functions and the corresponding well-known statement on functions of one variable.

Lemma 6.2.4. Let a function $f(z,w)$, $z\in\mathbf{C}^n$, $w\in\mathbf{C}$, be holomorphic in the domain $\Omega = B_\delta \times U_r$. Let also $D_N(w;f(z,w)) \equiv 0$ for some N. Then for some set $\omega = B_\varepsilon(z^0)\times U_\varepsilon(w^0) \subset \Omega$ the following representation holds:

$$f(z,w) = \sum_{l=1}^{m}\sum_{j=0}^{k_l} c_{j,l}(z)w^j e^{\lambda_l(z)w},$$

where $c_{j,l}(z)\in H(B_\varepsilon(z^0))$, $\lambda_l(z)\in H(U_\varepsilon(w^0))$, $\lambda_j \not\equiv \lambda_i$ for $i \neq j$, $c_{k_l,l} \not\equiv 0$, $\forall l$, and $\sum_{l=1}^{m}(k_l+1) \le N$.

Proof. Without loss of generality we may assume that $D_{N-1}(w;f) \not\equiv 0$. Then (see **Lemma 6.2.1**) the function $f(w,z)$ satisfies for almost all $z\in B_r$ the equation

$$\frac{\partial^N f}{\partial w^N} + c_{N-1}\frac{\partial^{N-1} f}{\partial w^{N-1}} + \dots + c_0 f = 0 \tag{6.2.4}$$

with coefficients c_1, ..., c_{N-1}, $c_N = 1$, that depend (generally speaking) on z, but not on w. These coefficients are of the form

$$c_j = \frac{A_j}{D_{N-1}} = \frac{A_j(z,w)}{D_{N-1}(w;f(z,w))},\ j = 0, \dots, N,$$

where A_j, j = 0, ..., N, are the algebraic cofactors of the elements of the last row of the determinant $D_N = D_N(w;f(z,w))$. Therefore the functions $c_j(z)$ are holomorphic at the points z^0 for which, for some w^0, $D_N(w;f(z^0,w^0)) \neq 0$. The set E of such points is obviously open and everywhere dense in B_r. Since the function $f(z,w)$ is a solution of equation (6.2.4), it is a P-quasipolynomial in w and can be represented in the form

$$f(z,w) = \sum_{l=1}^{m} \sum_{j=0}^{k_l} c_{j,l} w^j e^{\lambda_l w}, \qquad (6.2.5)$$

where the $\lambda_l = \lambda_l(z)$ are the distinct zeros of the characteristic polynomial (in λ)

$$\mathfrak{K}(\lambda,z) = \sum_{j=0}^{N} c_j(z)\lambda^j.$$

The multiplicities of the zeros λ_l are equal to $k_l + 1$, where the k_l depend on z. Since the coefficients c_j are holomorphic on E, $\mathfrak{K}(\Lambda,z)$ is a pseudopolynomial[43]. We factor it into irreducible components. We have

$$\mathfrak{K}(\lambda,z) = \mathfrak{K}_1^{\nu_1}(\lambda,z)\cdot\ldots\cdot\mathfrak{K}_p^{\nu_p}(\lambda,z), \qquad (6.2.6)$$

where p, ν_1, ..., ν_p are positive integers, and $\mathfrak{K}_1$, ..., $\mathfrak{K}_p$ are irreducible polynomials in λ with coefficients which are holomorphic functions in z on E (i.e. the pseudopolynomial $\mathfrak{K}_j$ is irreducible in the ring of pseudopolynomials). Consider the polynomial

$$\tilde{\mathfrak{K}}(\lambda,z) = \mathfrak{K}_1(\lambda,z)\cdot\ldots\cdot\mathfrak{K}_p(\lambda,z).$$

It is obvious that the discriminant of this polynomial is not identically equal to zero. Hence its zero set $\tilde{E}$ is (n-1)-dimensional analytic set in B_r. In accordance with the well-known properties of pseudopolynomials and their discriminants (see, for example, L.Ronkin [12]) in every sufficient small neighbourhood $B_\varepsilon(z^0)$ of a point $z^0 \notin E\setminus\tilde{E}$ the equation $\tilde{\mathfrak{K}}(\lambda,z) = 0$ has m single-valued analytic solutions $\tilde{\lambda}_1(z)$, ..., $\tilde{\lambda}_m(z)$, where $m = deg\tilde{\mathfrak{K}}$ and $\tilde{\lambda}_i(z) \neq \tilde{\lambda}_j(z)$ if $i \neq j$. Therefore. if we consider the representation (6.2.5) on $B_\varepsilon(z^0)$ then the exponents $\lambda_l(z)$ can be assumed holomorphic and k_l can be assumed constant (each k_l equals one of the numbers ν_j). In this situation the coefficients $c_{j,l}$ are holomorphic too. Indeed, for $l = 1$ the fact that the

---

[43] A function of the form $\sum_{j=0}^{N} a_j(z)w^j$, where $a_j(z)$ are analytic functions, is called here as pseudopolynomial in to w.

highest coefficient $c_{k_1,1}$ is holomorphic obviously follows from the equality

$$c_{k_1,1} = \frac{1}{k!}\prod_{j=2}^{m}(\lambda_1(z)-\lambda_1(z))^{-k_j-1} \times \frac{\partial^{k_1}}{\partial w^{k_1}}\Bigg(e^{(\lambda_2(z)-\lambda_1(z))w} \times$$

$$\times \frac{\partial^{k_2+1}}{\partial w^{k_2+1}}\Bigg(e^{(\lambda_3(z)-\lambda_2(z))w} \cdots \frac{\partial^{k_m+1}}{\partial w^{k_m+1}}\Big(e^{-\lambda_m(z)w} f(z,w)\Big)\cdots\Bigg)\Bigg).$$

Similarly the coefficient $c_{k_1-1,1}$, that is the leading one of the function

$$f(z,w) - c_{k_1,1} w^{k_1} exp\{\lambda_1(z)w\},$$

is holomorphic. By analogy we establish that all the other coefficients are also holomorphic.

This finishes the proof of the lemma.

Side by side with the above properties of P-quasipolynomials of one variable (or in one of the variables) we need an elementary property of special systems of linear (algebraic) equations. This property is stated in the following lemma.

Lemma 6.2.5. If the non-negative numbers $c_{i,j}$, $i, j = 1, 2, \ldots, m$, satisfy the conditions

$$\sum_{j=1}^{m} c_{i,j} = 1$$

and $c_{j,j} = 0$, $\forall j = 2, \ldots, m$, then every solution $(\lambda_1, \ldots, \lambda_m)$ of the system

$$\begin{cases} \lambda_1 = \sum_{j=1}^{m} c_{1,j}\lambda_j \\ \cdots\cdots\cdots\cdots\cdots\cdots \\ \lambda_m = \sum_{j=1}^{m} c_{m,j}\lambda_j \end{cases} \tag{6.2.7}$$

has at least two equal components.

Proof. We prove this by induction with respect to the number m. If $m = 2$, then the system is necessarily of the form

$$\begin{cases} \lambda_1 = \lambda_2 \\ \lambda_2 = \lambda_1 \end{cases}.$$

Therefore the lemma is true for m = 2. Now we assume that the lemma is true for (n-1)-dimensional systems (consisting of m-1 equations). Then we substitute the right hand side of the first equation for λ_1 in every other equation of the system (6.2.7). Thus, we obtain the system

$$\begin{cases} \lambda_1 = \sum_{j=1}^{m} c_{1,j}\lambda_j \\ \lambda_2 = \sum_{j=2}^{m} \tilde{c}_{2,j}\lambda_j \\ \dots\dots\dots\dots \\ \lambda_m = \sum_{j=2}^{m} \tilde{c}_{m,j}\lambda_j \end{cases}, \tag{6.2.8}$$

which is equivalent to the first one. It is obvious that the numbers $\tilde{c}_{i,j}$, i, j = 2, ..., m, are non-negative and satisfy the condition $\sum_{j=1}^{m} \tilde{c}_{i,j} = 1$. Besides, $\tilde{c}_{j,j} = c_{j,1}c_{1,j}$, $\forall j = 2, \dots, m$.

If $\tilde{c}_{j,j} = 1$ for some j then for the same j we obviously have $\lambda_j = \lambda_1$. This means that the lemma is true. In the case $\tilde{c}_{j,j} \neq 1$ for all j, we replace the system (6.2.8) by the equivalent system

$$\begin{cases} \lambda_1 = \sum_{j=1}^{m} c_{1,j}\lambda_j \\ \lambda_2 = \sum_{j=2}^{m} c^{*}_{2,j}\lambda_j \\ \dots\dots\dots\dots \\ \lambda_m = \sum_{j=2}^{m} c^{*}_{m,j}\lambda_j \end{cases}, \tag{6.2.9}$$

where $c^{*}_{j,j} = 0$, $\forall j = 2, \dots,$ and $c^{*}_{i,j} = \dfrac{\tilde{c}_{i,j}}{1 - \tilde{c}_{j,j}}$ for $i \neq j$. It is obvious that $\sum_{j=2}^{m} c^{*}_{i,j} = 1$, $\forall i$. Thus, if we consider all the equations of (6.2.9) without the first one, we obtain a system of m-1 equation satisfying the condition of the lemma. In accordance with the induction hypothesis, there are p and q, $2 \le p < q \le m$, such that $\lambda_p = \lambda_q$.

The proof is finished.

2. The general form of functions that are *P*-quasipolynomial in every variable. Let $f(\zeta,w)$ be a $\mathbb{C}$-quasipolynomial in every variable with constant spectrum, i.e.

$$f(\zeta,w) = \sum_{j=1}^{p} a_j(\zeta)e^{\lambda_j w}, \ \forall(\zeta,w)\in\mathbb{C}^2, \tag{6.2.10}$$

and

$$f(\zeta,w) = \sum_{i=1}^{q} b_i(w)e^{\mu_i \zeta}, \ \forall(\zeta,w)\in\mathbb{C}^2, \tag{6.2.11}$$

where p, q, λ, μ are constants. Choose numbers w_1, ..., w_p in such a way that

$$\begin{vmatrix} e^{\lambda_1 w_1} & \dots & e^{\lambda_1 w_p} \\ \dots & \dots & \dots \\ e^{\lambda_p w_1} & \dots & e^{\lambda_p w_p} \end{vmatrix} \neq 0 .$$

Then consider the system of equations (with respect to $a_1(\zeta)$, ..., $a_p(\zeta)$)

$$\begin{cases} f(\zeta,w_1) = a_1(\zeta)e^{\lambda_1 w_1} + \dots + a_p(\zeta)e^{\lambda_p w_1} \\ \dots\dots\dots\dots\dots\dots \\ f(\zeta,w_p) = a_1(\zeta)e^{\lambda_1 w_p} + \dots + a_p(\zeta)e^{\lambda_p w_p} \end{cases}$$

following from (6.2.10). By solving this system we find that each function $a_j(\zeta)$ is a linear combination of the functions $f(\zeta,w_1)$, ..., $f(\zeta,w_p)$. In view of (6.2.11) this means that the $a_j(\zeta)$ are $\mathbb{C}$-quasipolynomials with spectrum $\{\mu_1, \ \dots, \ \mu_q\}$. Substituting the corresponding representations of $a_j(\zeta)$, $j = 1$, ..., p, in (6.2.10), we find that $f(\zeta,w)$ is a $\mathbb{C}$-quasipolynomial in ζ, w with spectrum $\{(\lambda_j,\mu_i)\}$. A similar statement is clearly also true for an functions of arbitrary number of variables and in the case of *P*-quasipolynomials instead of $\mathbb{C}$-quasipolynomials.

The situation becomes essentially more complicated if we assume that the exponents λ_1, ..., λ_p and the number p itself depend on ζ, and μ_1, ..., μ_q and q depend on w. In this case the function $f(\zeta,w)$ is not necessarily a $\mathbb{C}$-quasipolynomial. The function $f(\zeta,w) = e^{\zeta w}$ is the simplest example. This situation turned out to be standard. All functions having the above property are of similar form (in a certain respect). More exactly, the following theorem holds.

Theorem 6.1.1. Let $f(z)$, $z\in\mathbb{C}^n$, be a *P*-quasipolynomial in each variable z_j when the other ones are arbitrarily fixed. Then

$$f(z) = \sum_{l=1}^{m} P_l(z)e^{\lambda_l(z)}, \tag{6.2.12}$$

where the $P_l(z)$ are polynomials in z and the $\lambda_l(z)$ are polynomials in z that are linear in each variable z_j (i.e. they are multilinear functions of z).

Proof. We proceed stepwise from simple particular cases to more complicated ones. In the final step we obtain the general case, i.e. the theorem itself. First we consider the case of one-point spectra.

Lemma 6.2.6. Let a function $f(z_1,z_2)$ be holomorphic in a domain $U^2 = U\times U$ and in this domain the following representations[44]:

$$f(z_1,z_2) = a(z_2)e^{\mu(z_2)z_1}, \tag{6.2.13}$$

$$f(z_1,z_2) = e^{\lambda(z_1)z_2}\sum_{l=0}^{m} b_l(z_1)z_2^l, \tag{6.2.13'}$$

with functions a, λ, μ and b_l, $l = 0, \ldots, m$ which are holomorphic in U. Then $\lambda(z_1)$ is a linear function.

Proof. We set

$$\sum_{l=0}^{m} b_l(z_1)z_2^l = P(z_1,z_2) = P,$$

$$z_2\lambda(z_1) = A(z_1,z_2) = A.$$

Then it follows from (6.2.13') that

---

[44] Certainly, if these representations hold then by **Hartogs' Theorem** the function f is holomorphic in U^2. However, it seems to be more preferable (for convenience of exposition) to include the condition that f be holomorphic directly in the corresponding statements.

$$\frac{\partial f}{\partial z_1} = \left(\frac{\partial P}{\partial z_1} + P \frac{\partial A}{\partial z_1} \right) e^A,$$

$$\frac{\partial^2 f}{\partial z_1^2} = \left[\frac{\partial^2 P}{\partial z_1^2} + 2 \frac{\partial P}{\partial z_1} \cdot \frac{\partial A}{\partial z_1} + P \frac{\partial^2 A}{\partial z_1^2} + P\left(\frac{\partial A}{\partial z_1}\right)^2 \right] e^A.$$

Substitute these expressions in the equality

$$\begin{vmatrix} f & \frac{\partial f}{\partial z_1} \\ \frac{\partial f}{\partial z_1} & \frac{\partial^2 f}{\partial z_1^2} \end{vmatrix} = 0,$$

which follows from (6.2.13) by **Lemma 6.2.1**. We have:

$$\begin{vmatrix} Pe^A & \left(\frac{\partial P}{\partial z_1} + P \frac{\partial A}{\partial z_1} \right) e^A \\ \left(\frac{\partial P}{\partial z_1} + P \frac{\partial A}{\partial z_1} \right) e^A & \left[\frac{\partial^2 P}{\partial z_1^2} + 2 \frac{\partial P}{\partial z_1} \cdot \frac{\partial A}{\partial z_1} + P \frac{\partial^2 A}{\partial z_1^2} + P\left(\frac{\partial A}{\partial z_1}\right)^2 \right] e^A \end{vmatrix} = 0 .$$

By elementary transformations this equality can be transformed to the form

$$\begin{vmatrix} P & \frac{\partial P}{\partial z_1} \\ \frac{\partial P}{\partial z_1} & \frac{\partial^2 P}{\partial z_1^2} + P \frac{\partial^2 A}{\partial z_1^2} \end{vmatrix} = 0 .$$

Then, substituting the representations of P and A in the last equality, we obtain

$$\left(\sum_{l=0}^{m} b_l(z_1) z_2^l \right) \cdot \left(\sum_{l=0}^{m} b_l''(z_1) z_2^l \right) + \\ + \sum_{l=0}^{m} \left(b_l(z_1) z_2^l \right)^2 \lambda''(z_1) z_2 - \left(\sum_{l=0}^{m} b_l'(z_1) z_2^l \right)^2 = 0.$$

Since this equality is valid for any $z_1 \in U$, $z_2 \in U$, the coefficients of the polynomial in z_2 at stays on the right-hand side must be equal to zero. In particular, the coefficient of the term z_2^{2m+1} is equal to zero. Hence

$$b_m(z_1)\lambda''(z_1) = 0, \ \forall z_2 \in U. \tag{6.2.14}$$

Without loss of generality we may assume that $b_m \not\equiv 0$. Therefore, from (6.2.14) it follows that $\lambda''(z_1) \equiv 0$. This means that $\lambda(z_1)$ is a linear function.

This finishes the proof of the lemma.

Let us complicate the situation by assuming the existence of a variable such that the spectrum with respect to it has more than one point.

Lemma 6.2.7. Let $f(z_1, z_2)$ be a holomorphic function in a domain U^2 and have in this domain the representations:

$$f(z_1, z_2) = a(z_2)e^{\mu(z_2)z_1}, \tag{6.2.15}$$

$$f(z_1, z_2) = \sum_{l=1}^{m} \sum_{j=0}^{k_l} b_{l,j}(z_1)z_2^j e^{\lambda_l(z_1)z_2}.$$

Here the functions a, μ, $b_{l,j}$, λ_l are taken holomorphic in U, and $\lambda_{l'} \not\equiv \lambda_l$ if $l' \neq l$. Then $\mu(z_2)$ is a linear function.

Proof. Without loss of generality we may assume that $b_{l,k_l} \not\equiv 0$. Thus, if z_1 is fixed in such a way that $b_{l,k_l}(z_1) \neq 0$, then the function $f(z_1, z_2)$ is a *P*-quasipolynomial in z_2 of the degree $N = \sum_{l=1}^{m}(k_l+1)$. Hence, by **Lemma 6.2.2** we obtain

$$D_{N-1}(z_2; f(z_1, z_2)) =$$

$$= D_{N-1}(0; f(z_1, z_2))exp\left\{\sum_{l=1}^{m}(k_l+1)\lambda_l(z_1)z_2\right\}, \ \forall z_2 \in U. \tag{6.2.16}$$

Since all functions figuring in (6.2.16) are analytic, this equality is also valid when $b_{l,k_l}(z_1) = 0$. Hence (6.2.15) is true for all z_1.

Further, note that (6.2.15) implies the equalities

$$\frac{\partial^\nu f}{\partial z_2^\nu} = e^{\mu(z_2)z_1} \sum_{j=0}^{\nu} c_{j,\nu}(z_2) z_1^j$$

for any $\nu = 0, 1, 2, \ldots$ (where the $c_{j,\nu}$ are functions holomorphic in U).

Substituting these expressions for $\frac{\partial^\nu f}{\partial z_2^\nu}$ in the determinant $D_{N-1}(z_2;f)$, we obtain

$$D_{N-1}(z_2;f) = \left(\sum_{j=0}^{(N-1)^2} c_j(z_2) z_1^j \right) e^{N\mu(z_2)z_1},$$

where the c_j are functions holomorphic in U. Comparing this equality with (6.2.16), we conclude that the function $D_{N-1}(z_2;f(_1,z_2))$ (as a function of z_1, z_2) satisfies the conditions of **Lemma 6.2.5**. Applying this lemma we obtain that $\mu(z_2)$ is a linear function.

This finishes the proof of the lemma.

In the next step we assume that the function $a(z_2)$ figuring in **Lemma 6.2.7** is a pseudopolynomial.

Lemma 6.2.8. Let f be a holomorphic function on U^2 that can be represented in the forms

$$f(z_1,z_2) = \sum_{j=0}^{\nu} a_j(z_2) z_1^j e^{\mu(z_2)z_1}, \tag{6.2.17}$$

and

$$f(z_1,z_2) = \sum_{l=1}^{m} \sum_{j=0}^{k_l} c_{l,j}(z_1) z_2^j e^{\lambda_l(z_1)z_2}, \tag{6.2.18}$$

where the functions a_j, μ, λ_l, $c_{l,j}$ are holomorphic, and $\lambda_j \neq \lambda_i$ if $i \neq j$. Then $\mu(z_2)$ is a linear function.

Proof. Without loss of generality we may assume $a_\nu \not\equiv 0$. Then, applying **Lemma 6.2.2**, we find that (6.2.17) implies the following equality:

$$D_\nu(z_1;f) = \tilde{a}(z_2) e^{(\nu+1)\mu(z_2)z_1},$$

where, as is easily seen, $\tilde{a}(z_2) \in H(U)$.

On the other hand, the representation (6.2.18) implies that

$$D_\nu(z_1;f) = \sum_{l=1}^{\tilde{m}} \sum_{j=0}^{\tilde{k}_l} \tilde{b}_{l,j}(z_1)z_2^j e^{\tilde{\lambda}_l(z_1)z_2},$$

where $\tilde{b}_{l,j}$ and $\tilde{\lambda}_l$ are analytic functions.

Now, regarding $D_\nu(z_1,f(z_1,z_2))$ as a function of z_1 and z_2, we conclude that according to **Lemma 6.2.7** the function $(1+\nu)\mu(z_2)$ is linear.

This finishes the proof of the lemma..

Now we pass to the case when both spectra are multi-pointed.

Lemma 6.2.9. Let f be a holomorphic function on U^2 that can be represented in such forms

$$f(z_1,z_2) = \sum_{l=1}^{p} \sum_{j=0}^{k_l} a_{l,j}(z_2)z_1^j e^{\mu_l(z_2)z_1}, \tag{6.2.19}$$

and

$$f(z_1,z_2) = \sum_{l=1}^{q} \sum_{j=0}^{m_l} b_{l,j}(z_1)z_2^j e^{\lambda_l(z_1)z_2},$$

where the functions $a_{l,j}$, μ_l, $b_{l,j}$, λ_l are holomorphic, and $\lambda_j \not\equiv \lambda_i$, $\mu_j \not\equiv \mu_i$ if $i \neq j$. Then $\mu_l(z_2)$ and $\lambda_l(z_1)$ are linear functions.

Proof. We assume that $a_{l,k_l} \not\equiv 0$, $b_{l,m_l} \not\equiv 0$ and set $N = \sum_{l=1}^{p}(k_l+1)$. According to **Lemma 6.2.1** we have

$$D_N(z_1;f(z_1,z_2)) \equiv 0. \tag{6.2.20}$$

We set

$$f_l = \sum_{j=0}^{m_l} b_{l,j}(z_1)z_2^j e^{\lambda_l(z_1)z_2}. \tag{6.2.21}$$

Then

$$\frac{\partial^\nu f}{\partial z_1^\nu} = \sum_{l=1}^{q} \frac{\partial^\nu f_l}{\partial z_1^\nu} = \sum_{l=1}^{q} \sum_{j=0}^{\nu+m_l} c_{l,j}(z_1)z_2^j e^{\lambda_l(z_1)z_2}, \quad \nu = 0,\ 1,\ \dots,$$

where the $c_{l,j}$ are holomorphic functions. Substitute these expressions for $\frac{\partial^\nu f}{\partial z_1^\nu}$ in the determinant $D_N(z_1;f)$. After a simple computation, equality (6.2.20) is transformed to the form

$$\sum_{j,i} g_{j,i}(z_1)z_2^j e^{\gamma_i(z_1)z_2} \equiv 0, \tag{6.2.22}$$

where the summation is over a finite set of indices, and the $g_{i,j}, \gamma_i(z_1)$ are holomorphic on U. Moreover, the functions γ_i are defined by the equalities

$$\gamma_i(z_1) = \sum_{l=1}^{q} \alpha_{i,l}\lambda_l(z_1),$$

with $\alpha_{i,l} \in \mathbb{Z}_+$ satisfying the condition

$$\sum_{l=1}^{q} \alpha_{i,l} = N+1.$$

Note that for any l, $1 \le l \le q$, there is a $i(l)$ such that $\gamma_{i(l)}(z_1) = (N+1)\lambda_l(z_1)$. Applying **Lemma 6.2.3** we obtain that either there is another index $j \ne i(l)$ such that $\gamma_j(z_1) = (N+1)\lambda_l(z_1)$ or

$$\sum_j g_{j,i(l)}(z_1)z_2^j e^{(N+1)\lambda_l(z_1)z_2} \equiv 0. \tag{6.2.23}$$

Note that the left-hand side of this equality is equal to $D_N(z_1;f_j)$. Indeed, substitute

$$\frac{\partial^\nu f}{\partial z_1^\nu} = \sum_{l=1}^{q} \frac{\partial^\nu f_l}{\partial z_1^\nu}, \quad \nu = 0, 1, \dots,$$

into $D_N(z_1;f)$ and represent $D_N(z_1;f)$ as a sum of determinants. Each of these determinants is the product of some exponential $e^{\gamma_i(z_1)z_2}$ and a pseudopolynomial in z_2. It is easy to see that the exponential $e^{(N+1)\lambda_l(z_1)z_2} = e^{\gamma_{i(l)}(z_1)z_2}$ belongs to a single determinant $D_N(z_1;f_l)$ (in the case when $\gamma_k \ne (N+1)\lambda_k$ for $k \ne i(l)$). Taking into account how we have obtained the functions $g_{i,j}$ and γ_i (see (6.2.1)), we conclude that

$$\sum_j g_{j,i(l)}(z_1)z_2^j e^{(N+1)\lambda_l(z_1)z_2} = D_N(z_1;f_l).$$

Hence we can rewrite equality (6.2.23) in the form $D_N(z_1;f_l) \equiv 0$. Therefore, according to **Lemma 6.2.4** we have the representation

$$f_1(z_1,z_2) = \sum_{k=1}^{\kappa} \sum_{j=1}^{\gamma_k} h_{k,j}(z_2) z_1^j e^{\beta_k(z_2) z_1}$$

in some polydisc $G \subset U^2$. Here the $h_{k,j}(z_2)$, $\beta_j(z_2)$ are holomorphic. From this representation and (6.2.21), by **Lemma 6.2.8** it follows that $\lambda_1(z_1)$ is a linear function. In the case when there is an index $j \neq i(1)$ such that $(N+1)\lambda_1(z_1) = \gamma_j(z_1)$ (taking into account the form of the functions $\gamma_j(z_1)$) we have the equality

$$\lambda_1(z_1) = \sum_{j=1}^{q} \tilde{\alpha}_{j,1} \lambda_j(z_1),$$

where the $\tilde{\alpha}_{j,1}$ are positive numbers satisfying the conditions $\sum_{j=1}^{q} \tilde{\alpha}_{j,1} = 1$, $\tilde{\alpha}_{j,j} = 0$.

Thus, after renumbering of the functions λ_1 we find that they satisfy some system

$$\begin{cases} \lambda_1(z_1) = \sum_{j=1}^{q} \hat{\alpha}_{j,1} \lambda_j(z_1) \\ \cdots\cdots\cdots\cdots\cdots\cdots\cdots \\ \lambda_{\tilde{q}}(z_1) = \sum_{j=1}^{q} \hat{\alpha}_{j,\tilde{q}} \lambda_j(z_1) \\ \lambda_{\tilde{q}+1}(z_1) = a_{\tilde{q}+1} z_1 + b_{\tilde{q}+1} \\ \cdots\cdots\cdots\cdots\cdots\cdots\cdots \\ \lambda_q(z_1) = a_q z_1 + b_q \end{cases}, \qquad (6.2.24)$$

where $\tilde{q} \leq q$, and the coefficients $\hat{\alpha}_{j,i}$, a_j, b_j are constants such that $\hat{\alpha}_{j,i} \geq 0$, $\hat{\alpha}_{i,i} = 0$, $\forall i,j$ and $\sum_{j=1}^{q} \hat{\alpha}_{j,1} = 1$, $\forall l$. We show that always $\tilde{q} < q$. Indeed, otherwise, in accordance with **Lemma 6.2.5**, for each $z_1 \in U$ there are indices $j(z_1)$ and $i(z_1)$ such that $\lambda_{j(z_1)}(z_1) = \lambda_{i(z_1)}(z_1)$. Using the uniqueness theorem for holomorphic functions, we conclude that $\lambda_j(z_1) \equiv \lambda_i(z_1)$ for some j and i, $i \neq j$. This contradicts the conditions of the lemma. Hence $\tilde{q} < q$. We, now reduce system (6.2.24) to the form

$$\begin{cases} \lambda_1(z_1) = \sum\limits_{j=2}^{q} \hat{\beta}_{j,1}\, \lambda_j(z_1) \\ \lambda_1(z_1) = \sum\limits_{j=3}^{q} \hat{\beta}_{j,1}\, \lambda_j(z_1) \\ \dots\dots\dots\dots\dots\dots\dots \\ \lambda_{\tilde{q}}(z_1) = \sum\limits_{j=\tilde{q}+1}^{q} \hat{\beta}_{j,\tilde{q}}\, \lambda_j(z_1) \\ \lambda_{\tilde{q}+1}(z_1) = a_{\tilde{q}+1} z_1 + b_{\tilde{q}+1} \\ \dots\dots\dots\dots\dots\dots\dots \\ \lambda_q(z_1) = a_q z_1 + b_q \end{cases}$$

It now immediately follows that all the functions $\lambda_j(z_1)$ are linear. Similarly, the functions $\mu_j(z_2)$ are linear.

This finishes the proof of the lemma.

In the following lemma we state the form of the coefficients $a_{j,1}$ and $b_{j,1}$ figuring in the representations (6.2.19).

Lemma 6.2.10. Let $f(z_1,z_2)$ bee a holomorphic function in U^2 that can be represented in the forms

$$f(z_1,z_2) = \sum_{l=1}^{p} \sum_{j=0}^{k_l} a_{l,j}(z_2) z_1^j e^{\mu_l(z_2) z_1},$$

and

$$f(z_1,z_2) = \sum_{l=1}^{q} \sum_{j=0}^{m_l} b_{l,j}(z_1) z_2^j e^{\lambda_l(z_1) z_2},$$

with functions $a_{l,j}$, $b_{l,j}$ holomorphic in U and linear functions λ_l, μ_l such that $\lambda_l \neq \lambda_{l'}$, $\mu_l \neq \mu_{l'}$ if $l \neq l'$. Then all the functions $a_{l,j}$ and $b_{l,j}$ are *P*-quasipolynomials.

Proof. We use the equality

$$k_1!\, b_{1,m_1} \prod_{j=2}^{p} (\lambda_1(z_1)-\lambda_1(z_1))^{k_j+1} =$$

$$= \frac{\partial^{k_1}}{\partial z_2^{k_1}}\left(e^{(\lambda_2(z_1)-\lambda_1(z_1))z_2} \frac{\partial^{k_2+1}}{\partial z_2^{k_2+1}} \left(e^{(\lambda_3(z)-\lambda_2(z))z_2} \times \ldots \right.\right.$$

$$\left.\left. \ldots \frac{\partial^{k_p+1}}{\partial z_2^{k_p+1}}\left(e^{-\lambda_p(z_1)z_2} f(z_1,z_2)\right)\ldots\right)\right),$$

which was already mentioned in the proof of **Lemma 6.2.4**. As the λ_l are linear functions, this equality implies that

$$b_{1,m_1}(z_1) = \frac{Q_{1,m_1}(z_1)}{P_{1,m_1}(z_1)},$$

where $Q_{1,m_1}(z_1)$ is a *P*-quasipolynomial and $P_{1,m_1}(z_1)$ is a polynomial.

Further, considering in a similar way the function

$$\tilde{f}(z_1,z_2) = P_{1,m_1}(z_1)\cdot(f(z_1,z_2) - b_{1,m_1}(z_1)e^{\lambda_1(z_1)z_2})$$

instead of f, we find that

$$b_{1,m_1-1}(z_1) = \frac{Q_{1,m_1-1}(z_1)}{P_{1,m_1-1}(z_1)},$$

where $Q_{1,m_1-1}(z_1)$ is a *P*-quasipolynomial and $P_{1,m_1-1}(z_1)$ is a polynomial. Continuing this process, we find that for any l and j,

$$b_{l,j}(z_1) = \frac{Q_{l,j}(z_1)}{P_{l,j}(z_1)},$$

where $Q_{l,j}(z_1)$ is a *P*-quasipolynomial and $P_{l,j}(z_1)$ is a polynomial.

Since after multiplication by polynomial a *P*-quasipolynomial is still a *P*-quasipolynomial, we may assume that all polynomials $P_{l,j}$ are equal, say to $P(z_1)$. Then

$$b_{l,j}(z_1) = \sum_j \frac{g_{l,j,i}(z_1)}{P(z_1)} e^{\beta_{l,j,i}z_1},$$

where the $g_{l,j,i}$ are polynomials and the $\beta_{l,j,i}$ are distinct numbers for distinct i. In turn, this implies that in U^2 we have

$$f(z_1,z_2) = \sum_{l,j,i} \frac{g_{l,j,i}(z_1)z_2^j}{P(z_1)} e^{\lambda_l(z_1)z_2+\beta_{l,j,i}z_1} . \qquad (6.2.25)$$

Similarly,

$$f(z_1,z_2) = \sum_{l,j,i} \frac{h_{l,j,i}(z_2)z_1^j}{Q(z_2)} e^{\mu_l(z_2)z_1+\gamma_{l,j,i}z_2} , \qquad (6.2.26)$$

where $h_{l,j,i}(z_2)$, $Q(z_2)$ are polynomials and $\gamma_{l,j,i}$ are distinct numbers for distinct i.

Note that the functions $e^{az_1z_2+bz_1+cz_2}$, $a\in\mathbb{C}$, $b\in\mathbb{C}$, $c\in\mathbb{C}$, are linearly independent over the field rational functions. Therefore, comparing (6.2.25) with (6.2.26) we obtain the following: for any indices l, j, i there are $\tilde{l}$, $\tilde{j}$, $\tilde{i}$ such that

$$\frac{g_{l,j,i}(z_1)z_2^j}{P(z_1)} = \frac{h_{\tilde{l},\tilde{j},\tilde{i}}(z_2)z_1^{\tilde{i}}}{Q(z_2)} .$$

This immediately implies that the function

$$\frac{g_{l,j,i}(z_1)}{P(z_1)}$$

is a polynomial, and hence $b_{l,j}$ is a *P*-quasipolynomial. Similarly, every coefficient $a_{l,j}(z_2)$ is a *P*-quasipolynomial.

This finishes the proof of the lemma.

Now we pass on directly to the proof of the theorem. Let $f(z)$, $z\in\mathbb{C}^n$, be a function satisfying in its condition. Since it is a *P*-quasipolynomial, and therefore holomorphic in every variable z_j, by **Hartogs' Theorem** $f(z)$ is a holomorphic function with respect to all variables $z_1, \ldots, z_n$ jointly. Thus $f(z)$ is an entire function. Besides, by **Lemma 6.2.1** for every $\hat{z}_j = (z_1,\ldots,z_{j-1},z_{j+1},\ldots,z_n)$ there is a number $\tilde{N} = \tilde{N}(\hat{z}_j)$ such that $D_{\tilde{N}}(z_j;f) = 0$, $\forall z_j\in\mathbb{C}$. We set $E_{N,j} = \{\hat{z}_j : D_N(z_j;f) = 0, \forall z_j\in\mathbb{C}\}$. Every set $E_{N,j}$ is closed, and their union $\bigcup_N E_{N,j}$ coincides with $\mathbb{C}^{n-1}$. Therefore there is at least one $N = N_j$ such that the Lebesgue measure of the set $E_{N_j,j}$ is positive. By the uniqueness theorem it now follows that $D_{N_j}(z_j;f) \equiv 0$. From this and by **Lemma 6.2.4** we conclude that in some polydisc $G = U_{r_1}(z_1^0)\times\ldots\times U_{r_n}(z_n^0)$ the representations

$$f(z) = \sum_{l=1}^{p_j} \sum_{i=0}^{k_{l,j}} a_{l,i,j}(\hat{z}_j)z_j^i e^{\mu_{l,j}(\hat{z}_j)z_j}, \quad j = 1, \ldots, n, \tag{6.2.27}$$

hold, where $a_{l,i,j}$ and $\mu_{l,j}$ are holomorphic functions.

Now we proceed by induction with respect to the number of variables. Let $n = 2$. Then we have

$$f = f(z_1,z_2) = \sum_{l=1}^{p_1} \sum_{i=0}^{k_{l,1}} a_{l,i,1}(z_2)z_1^i e^{\mu_{l,1}(z_2)z_1} =$$

$$= \sum_{l=1}^{p_2} \sum_{i=0}^{k_{l,2}} a_{l,i,2}(z_1)z_2^i e^{\mu_{l,2}(z_1)z_2}.$$

From this and **Lemmas 6.2.9** and **6.2.10** it follows that the $\mu_{l,1}$, $\mu_{l,2}$ are linear functions and the $a_{l,i,1}$, $a_{l,i,2}$ are quasipolynomials. Hence

$$f(z_1,z_2) = \sum c_{l,i,j}z_1^l z_2^j e^{\kappa_i z_1 z_2 + \kappa_i' z_1 + \kappa_i'' z_2},$$

where the $c_{l,i,j}$, κ_i, κ_i', κ_i'' are constants. Thus we have proved the theorem in case $n = 2$.

Now we assume that theorem is true for functions of n-1 variables. We consider the representations (6.2.27) for $j = 1$ and $j = \nu$. From **Lemmas 6.2.9** and **6.2.10** it follows that, firstly, the functions $a_{l,i,1}(\hat{z}_1)$ are *P*-quasipolynomials in every variable z_ν, $\nu = 2, \ldots, n$, and, secondly, the functions $\mu_{l,1}(\hat{z}_1)$ are multilinear. By the induction assumption, the first fact implies that for all l, i,

$$a_{l,i,1}(\hat{z}_1) = \sum_k P_{l,i,k}(\hat{z}_1)e^{\lambda_{l,i,k}(\hat{z}_1)},$$

where the $P_{l,i,k}(\hat{z}_1)$ are polynomials and the $\lambda_{l,i,k}(\hat{z}_1)$ are multilinear functions. Substituting these representations for the coefficients $a_{l,i,1}(\hat{z}_1)$ in the equality (6.2.7) (in the case $j = 1$), we find that

$$f(z_1,\hat{z}_1) = \sum_{l,i,k} P_{l,i,k}(\hat{z}_1)z_1^j exp\{\lambda_{l,i,k}(\hat{z}_1) + \mu_{l,1}(\hat{z}_1)z_1\} . \qquad (6.2.28)$$

Since $\lambda_{l,i,k}(\hat{z}_1)$ and $\mu_{l,1}(\hat{z}_1)$ are multilinear functions in $\hat{z}_1$, the function $\lambda_{l,i,k}(\hat{z}_1) + \mu_{l,1}(\hat{z}_1)z_1$ is multilinear in z for any l, i, k. Thus the representation (6.2.8) is the required one, i.e. it coincides with (6.2.12).

This finishes the proof of the theorem.

3. Corollaries, generalizations, applications. The following three statements are evident consequences of **Theorem 6.2.1**.

Corollary 1. If a function f(z), $z\in\mathbf{C}^n$, is a **C**-quasipolynomial in each variable z_j when the others are fixed, then

$$f(z) = \sum_{j=1}^{m} a_j e^{\lambda_j(z)} , \qquad (6.2.29)$$

where the a_j are constants and the $\lambda_j(z)$ are multilinear functions of $z_1, \ldots, z_n$.

Corollary 2. If a function f(z), $z\in\mathbf{C}^n$, is of at most minimal type with respect to the order $\rho = 2$ and is a *P*-quasipolynomial (a **C**-quasipolynomial) in each variable z_j when the others are fixed, then f(z) is a *P*-quasipolynomial (respectively, a **C**-quasipolynomial) in z.

Corollary 3. If the restriction of an entire function f(z) to any complex line $\{z: z = \lambda w, w\in\mathbf{C}\}$, $\lambda\in\mathbf{C}^n$, is a *P*-quasipolynomial (a **C**-quasipolynomial), then f(z) is a *P*-quasipolynomial (respectively, a **C**-quasipolynomial) in z.

Under the additional assumption that f(z) be holomorphic in $\mathbf{C}^n$ **Theorem 6.2.1** is also true if the requirement that f(z) be a quasipolynomial in z_j is not a priori satisfied for all $\hat{z}_j$. Namely, the following theorem holds.

Theorem 6.2.2. Let sets $E_1, \ldots, E_n \subset \mathbf{C}^n$ be such that each of them is not a subset of the countable union of analytic sets $A_j \subset \mathbf{C}^{n-1}$, $A_j \neq \mathbf{C}^{n-1}$. Let also the entire function f(z) be a *P*-quasipolynomial (a **C**-quasipolynomial) in z_j for each fixed $\hat{z}_j\in E_j$, j = 1, 2, ..., n. Then f(z) can be represented in the form (6.2.12) (respectively, (6.2.29)).

By **Lemma 6.2.1** this theorem can be easily reduced to **Theorem 6.2.1**.

We note also the "real" analogue of **Theorem 6.2.1**.

Theorem 6.2.3. Let a function $f(x)$, $x\in R^n$, be a *P*-quasipolynomial in x_j for each fixed $\hat{x}_j = (x_1,\ldots,x_{j-1},x_{j+1},\ldots,x_n)\in R^n$, i.e.

$$f(x) = \sum_{l=1}^{m_j(\hat{x}_j)} \left(\sum_{i=0}^{k_{l,j}(\hat{x}_j)} a_{l,i,j}x^i \right)\cdot e^{\lambda_{l,j}(\hat{x}_j)x_j}, \quad j = 1, \ldots, n.$$

Then there are polynomials $P_1(x)$, ..., $P_m(x)$ and multilinear functions $\lambda_1(x)$, ..., $\lambda_m(x)$ such that

$$f(x) = \sum_{l=1}^{m} P_l(x)e^{\lambda_l(x)}, \quad \forall x\in R^n.$$

Besides, if $k_{l,j}(\hat{x}_j) \equiv 0$, $\forall l,j$ then $degP_l = 0$, $\forall l$.

This theorem can easily be obtained from **Theorem 6.2.1** with the help of **Lemma 6.2.1** and Bernstein-Siciak theorem on separately holomorphic functions (see, for example, Ahiezer-Ronkin [1], [2]).

In conclusion of this section we give a typical example of application of **Theorem 6.2.1**.

Theorem 6.2.4. Let $f(z)$, $z\in \mathbb{C}^n$, be an entire function such that $f^m(z)$, $m\in \mathbb{N}$, is a $\mathbb{C}$-quasipolynomial. Then the function $f(z)$ itself is also a $\mathbb{C}$-quasipolynomial.

Proof. First we consider the case $n = 1$. We have

$$f^m(w) = \sum_{j=1}^{m} a_j e^{\lambda_j w}, \quad w\in \mathbb{C},$$

where $a_j\in \mathbb{C}$, $a_j\neq 0$, $\forall j$, $\lambda_j\in \mathbb{C}$ and $\lambda_j \neq \lambda_i$ for $j \neq i$. We denote by $\tilde{\Lambda}$ the polyhedron based on the points λ_1, ..., λ_m (i.e. $\tilde{\Lambda}$ is the convex hull of the spectrum of the function $f^m(w)$). As we have already seen the indicator of the function $f^m(w)$ coincides with the function $H(-\theta)$, where

$$H(\theta) = \max_{w\in\tilde{\Lambda}}\{Re\, we^{-i\theta}\}$$

is the support function of the polyhedron $\tilde{\Lambda}$. Therefore the indicator $h_f(\theta)$ of the function $f(w)$ is equal to $\frac{1}{m} H(-\theta)$. For sake of being specific we assume that $\lambda_1 = 0$ is one of the vertices of $\tilde{\Lambda}$. Then, for sufficiently small $\varepsilon > 0$ the points λ_2, ..., λ_n belong to the sector $|argw-\pi| < \frac{\pi}{2} - 2\varepsilon$, and, correspondingly, $h_f(\theta) = 0$ for $|\theta| < 2\varepsilon$. We choose $R > 0$ large enough so that $\left| e^{\lambda_j} \frac{a_j}{a_1} \right| < \frac{1}{2m}$ for $w\in G = \{ w\in \mathbb{C}:$

$|w| > R,\ |argw| < \varepsilon\}$. Then in G the following equalities hold

$$f(w) = \left(\sum_{j=1}^{m} a_j e^{\lambda_j w} \right)^{1/m} = \sqrt[m]{a_1} \left(1 + \sum_{j=2}^{m} \frac{a_j}{a_1} e^{\lambda_j w} \right)^{1/m} =$$

$$= \sqrt[m]{a_1} \sum_{l=0}^{\infty} c_l \left(\sum_{j=2}^{m} \frac{a_j}{a_1} e^{\lambda_j w} \right)^l = \sqrt[m]{a_1} + \sum_{\nu=1}^{\infty} b_\nu e^{\gamma_\nu w}, \qquad (6.2.30)$$

where the c_l are the corresponding binomial coefficients and the numbers γ_ν are of the form

$$\gamma_\nu = \sum_{j=2}^{m} k_{j,\nu} \lambda_j$$

and are not equal to one another. Here $k_{j,\nu} \in Z_+$, $\forall j,\ \nu$, and $\sum_{j=2}^{m} k_{j,\nu} \to \infty$ as $\nu \to \infty$. Note that $|arg\gamma_\nu - \pi| < \frac{\pi}{2} - 2\varepsilon$, $\forall \nu$; and from the condition $|arg\lambda_j - \pi| < \frac{\pi}{2} - \varepsilon$ it follows that

$$\lim_{\nu \to \infty} Re\gamma_\nu = \infty.$$

Without loss of generality we may assume that the γ_j are numbered in such a way that $Re\gamma_{\nu+1} \geq Re\gamma_\nu$, $\forall \nu$.

We show that the set of non-zero coefficients is at most finite. Let the converse be true. Then we can choose ν_0 in such a way that $b_{\nu_0+1} \neq 0$ and $Re\gamma_{\nu_0+1} < Re\gamma_{\nu_0}$. Since

$$h_f(\pi) = H(-\pi) = \frac{1}{m} \min_j Re\lambda_j,$$

the indicator $h_F(\theta)$ of the entire function

$$F(w) = f(w) - \sqrt[m]{a_1} - \sum_{\nu=1}^{\nu_0} e^{\gamma_\nu w}$$

satisfies the condition

$$h_F(\pi) = -Re\gamma_{\nu_0}.$$

From this and the well-known inequality $h_F(\pi+0) + h_F(\theta) \geq 0$, $\forall F \in H(\mathbb{C},1]$, we conclude that

$$h_F(0) \geq Re\gamma_{\nu_0}. \tag{6.2.31}$$

At the same time, in accordance with (6.2.30) we have the representation

$$F(w) = \sum_{\nu=\nu_0+1}^{\infty} b_\nu e^{\gamma_\nu w}$$

in the domain G. Therefore,

$$h_f(0) = Re\gamma_{\nu_0+1}.$$

Since $Re\gamma_{\nu_0+1} < Re\gamma_{\nu_0}$, this contradicts (6.2.30). Hence $b_\nu = 0$ for all sufficient large ν, and therefore (see (6.2.30)) f(w) is a C-quasipolynomial.

Thus we have proved the theorem for functions of one variable.

Now to obtain the theorem for an arbitrary number of variables it is sufficient to refer to **Corollary 3.**

This finishes the proof of the theorem.

Similarly, but somewhat more complicated, we can prove the corresponding theorem for *P*-quasipolynomials.

Theorem 6.2.5. Let f(z), $z \in \mathbb{C}^n$, be an entire function such that $f^m(z)$, $m \in \mathbb{N}$, is a *P*-quasipolynomial. Then the function f(z) itself is also a *P*-quasipolynomial.

§3 Factors of quasipolynomials

Here we consider some problems concerned with the form of factors of quasipolynomials and their existence.

1. Division of quasipolynomials. The simplest (in formulation) problem concerning factors of quasipolynomials is as follows. What can we say about the form of a factor f_1 if $f = f_1 \cdot f_2$ and f, f_2 are quasipolynomials? In other words, it is the problem of the form of an entire function that is a quotient of two quasipolynomials. We consider the case of *P*-quasipolynomials.

Theorem 6.3.1. Let *P*-quasipolynomials f(z), g(z), $z \in \mathbb{C}^n$, and an entire function h(z) be such that $f = g \cdot h$. Then h(z) can be represented in the form

$$h(z) = \frac{\varphi(z)}{P(z)}, \tag{6.3.1}$$

where $\varphi(z)$ is a P-quasipolynomial and $P(z)$ is a polynomial.

If, in addition, $g(z)$ is a $\mathbb{C}$-quasipolynomial, then $h(z)$ is a P-quasipolynomial. Finally, if both f and g are $\mathbb{C}$-quasipolynomials, then $g(z)$ is also a $\mathbb{C}$-quasipolynomial.

Proof. For the sake of brevity we call functions of the form (6.3.1) or, equivalently, quasipolynomials with coefficients rational in z (see footnote in the beginning of this chapter) R-quasipolynomials. All notions (such as the spectrum, the coefficients, the convex hull of the spectrum, etc.) which we have defined for M-quasipolynomials, $M \subset M_0$, can be easily extended to R-quasipolynomials. The main point of the proof of **Theorem 6.3.1** is the use of a property of R-quasipolynomials contained in the following lemma.

Lemma 6.3.1. Let R-quasipolynomials $f(z)$ and $g(z)$ be such that their quotient $f(z)/g(z)$ is an entire function. Then the support functions $H_{\tilde{\Lambda}(f)}$ and $H_{\tilde{\Lambda}(g)}$ of the convex hulls $\tilde{\Lambda}(f)$ and $\tilde{\Lambda}(g)$ of their spectra $\Lambda(f)$ and $\Lambda(g)$ satisfy the condition

$$H_{\tilde{\Lambda}(f)}(\zeta) + H_{\tilde{\Lambda}(f)}(-\zeta) \geq H_{\tilde{\Lambda}(g)}(\zeta) + H_{\tilde{\Lambda}(g)}(-\zeta), \ \forall \ \zeta \in \mathbb{C}^n. \tag{6.3.2}$$

Proof. By the definition of R-quasipolynomial, the functions $f(z)$ and $g(z)$ can be represented in the form $f(z) = f_1(z)/P_1(z)$, $g(z) = g_1(z)/Q_1(z)$, where f_1, g_1 are P-quasipolynomials and P_1, Q_1 are polynomials. It is obvious that the spectra of the P-quasipolynomials f_1 and g_1 coincide with the spectra of the R-quasipolynomials f and g, respectively. Therefore, the regularized indicators of the functions f_1 and g_1 are defined by the equalities

$$\left.\begin{aligned} \mathfrak{L}^*_{f_1}(z) &= H_{\tilde{\Lambda}(f)}(-z), \\ \mathfrak{L}^*_{g_1}(z) &= H_{\Lambda(g)}(-z). \end{aligned}\right\} \tag{6.3.3}$$

Set $h(z) = f(z)/g(z)$. Then

$$h(z)P_1(z)g_1(z) = f_1(z)Q_1(z). \tag{6.3.4}$$

By the condition of the lemma $h(z)$ is an entire function. Besides, since the functions f_1 and g_1 are *P*-quasipolynomials, they are entire functions of c.r.g. Noting also that the indicator of a polynomial is identically equal to zero, we obtain from (6.3.4) that $h \in H(\mathbb{C}^n, 1]$ and

$$\mathfrak{L}_h^*(z) = \mathfrak{L}_{f_1}^*(z) - \mathfrak{L}_{g_1}^*(z). \tag{6.3.5}$$

It is well-known (see, for example, L.Ronkin [1]) that

$$\mathfrak{L}_h^*(z) + \mathfrak{L}_h^*(-z) \geq 0, \ \forall z \in \mathbb{C}^n,$$

for any function $h \in H(\mathbb{C}^n, 1]$. Hence from (6.3.5) it follows that

$$\mathfrak{L}_{f_1}^*(z) - \mathfrak{L}_{g_1}^*(z) + \mathfrak{L}_{f_1}^*(-z) - \mathfrak{L}_{g_1}^*(-z) \geq 0.$$

From this, taking into account (6.3.3), we find that

$$H_{\tilde{\Lambda}(f)}(z) + H_{\tilde{\Lambda}(f)}(-z) \geq H_{\tilde{\Lambda}(g)}(z) + H_{\tilde{\Lambda}(g)}(-z), \ \forall z \in \mathbb{C}^n.$$

Thus the proof is finished.

Now let f and g be the functions figuring in the condition of the theorem, i.e.

$$f(z) = \sum_{l=1}^{m} a_l(z) e^{\langle \lambda_l, z \rangle},$$

$$g(z) = \sum_{l=1}^{k} b_l(z) e^{\langle \mu_l, z \rangle},$$

where $a_l \not\equiv 0$, $b_l \not\equiv 0$ are polynomials, $\lambda_l = (\lambda_{l,1}, \dots, \lambda_{l,n}) \in \mathbb{C}^n$, $\mu_l = (\mu_{l,1}, \dots, \mu_{l,n}) \in \mathbb{C}^n$, $\lambda_{l'} \neq \lambda_{l''}$, $\mu_{l''} \neq \mu_{l'}$ if $l'' \neq l'$, and $f/g = h \in H(\mathbb{C}^n)$.

Let $\zeta \in \mathbb{C}^n$ be such that the supporting hyperplane $L = \{z: Re\langle z, \bar{\zeta} \rangle - H_{\tilde{\Lambda}(g)}(\zeta) = 0\}$ contains only one of the points μ_l. Without loss of generality we may assume that the ζ and μ_1 are such that $\zeta = e_1 = (1, 0, \dots, 0)$, $\mu_l = \mu_1$. Thus, $H_{\tilde{\Lambda}(g)}(e_1) = Re\mu_{1,1} > Re\mu_{l,1}$, $\forall l = 2, \dots, k$. We renumber the exponents μ_l, $l = 2, \dots, k$ and λ_l, $l = 1, \dots, m$ in such a manner that $Re\mu_{2,1} \geq Re\mu_{3,1} \geq \dots \geq Re\mu_{k,1}$ and $Re\lambda_{1,1} \geq Re\lambda_{2,1} \geq \dots \geq Re\lambda_{m,1}$. We set $d = Re\mu_{1,1} - Re\mu_{2,1}$, $\kappa(f) = H_{\tilde{\Lambda}(f)}(e_1) + H_{\tilde{\Lambda}(f)}(-e_1)$. Note that $d > 0$, $\kappa(f) = Re\lambda_{1,1} - Re\lambda_{m,1}$ and by **Lemma 6.3.1** the inequality $\kappa(f) \geq \kappa(g)$ holds.

Now we choose the index l_1 in such a way that $Re\mu_{l_1,1} > Re\mu_{1,1} - d$ and $Re\mu_{l_1+1,1} \leq Re\mu_{1,1} - d$. Then we consider the *R*-quasipolynomials

$$g^{(1)}(z) = \frac{1}{b_1(z)} g(z)$$

and

$$f^{(1)}(z) = f(z) - a_1(z)g^{(1)}(z)e^{\langle\lambda_1-\mu_1,z\rangle}.$$

Note that the intersection of the spectrum $\Lambda(f^{(1)}(z))$ of the quasipolynomial $f^{(1)}$ with the half-space $\{z\in\mathbb{C}^n\colon\ Rez_1 > Re\lambda_{1,1} - d\}$ consists of the points $\lambda_2, \ldots, \lambda_{l_1}$ and, in view of the inequality $\kappa(f)\geq \kappa(g)$, the whole spectrum is contained in the half-space $\{z\in\mathbb{C}^n\colon\ Rez_1 > Re\lambda_{m,1}\}$. Besides, it is obvious that $f^{(1)}/g^{(1)}$ is an entire function and, therefore, by **Lemma 6.3.1** the inequality $\kappa(f^{(1)}) \geq \kappa(g^{(1)})$ is valid. Thus, the pair of functions $f^{(1)}$, $g^{(1)}$ is similar to the pair f, g. Hence it is possible to repeat the constructing of *R*-quasipolynomial with the above-mentioned properties (i.e. of $f^{(1)}$). Repeating this procedure step-by-step we obtain *R*-quasipolynomials

$$f^{(2)}(z) = f^{(1)}(z) - a_2(z)g^{(1)}(z)e^{\langle\lambda_2-\mu_1,z\rangle},$$

....

$$f^{(j)}(z) = f^{(j-1)}(z) - a_j(z)g^{(1)}(z)e^{\langle\lambda_j-\mu_1,z\rangle},$$

....

....

$$f_1(z)=f^{(l_1)}(z)=f^{(l_1-1)}-a_{l_1}(z)g^{(1)}(z)e^{\langle\lambda_{l_1}-\mu_1,z\rangle}=f(z)-h_1(z)g^{(1)}(z),$$

where $h_1(z)$ is a *P*-quasipolynomial.

It follows from the construction of these *R*-quasipolynomials that if $f_1 \not\equiv 0$, then:

i) $\Lambda(f_1) \subset \{z\in\mathbb{C}^n\colon\ Rez_1 \geq Re\lambda_{m,1}\}$;

ii) $\Lambda(f_1) \cap \{z\in\mathbb{C}^n\colon Rez_1 > \lambda_{1,1}-d\} = \emptyset$;

iii) $\kappa(g) = \kappa(g^{(1)}) \le \Lambda(f_1) \le \kappa(f)$;

iiii) the quotient f_1/g_1 is an entire function.

From i) and ii) it follows that $\kappa(f_1) \le \kappa(f) - d$. Repeating the reasoning above with the functions f_1 and $g^{(1)}$ instead of the functions f and g, we obtain on *R*-quasipolynomial $f_2(z) = f_1 - h_2 g^{(1)}$, where h_2 is a *P*-quasipolynomial such that $\kappa(g^{(1)}) \le \kappa(f_2) \le \kappa(f) - 2d$. Similarly we obtain an *R*-quasipolynomial $f_3 = f_2 - h_3 g^{(1)}$ such that h_3 is a *P*-quasipolynomial and $\kappa(g^{(1)}) \le \kappa(f_3) \le \kappa(f) - 3d$, etc. In view of the inequality $\kappa(g^{(1)}) \le \kappa(f_\nu) \le \kappa(f) - \nu\cdot d$, this process, i.e. the process of constructing the functions f_ν, $\nu = 1, 2, \ldots$, terminates after at most $\tilde{\nu}$ step, where $\tilde{\nu} = [d^{-1}(\kappa(f)-\kappa(g))] + 1$. Note that the process terminates only if the corresponding *R*-quasipolynomial $f_\nu \equiv 0$. Therefore there is an index $\hat{\nu} \ge \tilde{\nu}$ such that $f_{\hat{\nu}} \equiv 0$. Hence $f_{\hat{\nu}-1} = h_{\hat{\nu}} g^{(1)}$, $f_{\hat{\nu}-2} = f_{\hat{\nu}-1} + h_{\hat{\nu}-1} g^{(1)} = g^{(1)}(h_{\hat{\nu}} + h_{\hat{\nu}-1})$, ..., $f = g^{(1)}(h_{\hat{\nu}} + \ldots + h_1)$. Thus

$$\frac{f}{g} = \frac{h_{\hat{\nu}} + \ldots + h_1}{b_1}. \tag{6.3.6}$$

Here, $b_1(z)$ is a polynomial and the function $h_{\hat{\nu}} + \ldots + h_1$ is a *P*-quasipolynomial with spectrum consisting of the points $k_1\lambda_1 + \ldots + k_m\lambda_m + l_1\mu_1 + \ldots + l_k\mu_k$, $k_j\in\mathbb{Z}$, $l_j\in\mathbb{Z}$, $\forall j$. If we assume that, in addition, g is a $\mathbb{C}$-quasipolynomial, then b_1 is a constant and it follows from (6.3.6) that f is a *P*-quasipolynomial. It is also plain that if both f and g are $\mathbb{C}$-quasipolynomials then $h_{\hat{\nu}} + \ldots + h_1$ is also a $\mathbb{C}$-quasipolynomial, as is the function $f/g=(h_{\hat{\nu}} + \ldots + h_1)/b_1$.

This finishes the proof of the theorem.

One possible of generalization of **Theorem 6.3.1** is extension of the class of functions under consideration. We give a result obtained in this manner. To formulate it we need some definitions and notation.

A meromorphic function $f(z)$, $z\in\mathbb{C}^n$, is said to be a function of degree zero if it is the quotient of two entire functions of at most minimal type with respect to the order $\rho = 1$.

We denote by MS_n^p, $1 \le p \le n$, the set of all meromorphic functions of the form

$$f(z) = \sum_{l=1}^{m} a_l(z)e^{\lambda_l(z)},$$

where the coefficients $a_l(z) \not\equiv 0$ are meromorphic functions of degree zero and the exponents $\lambda_l(z)$ are polynomials of degree at most p which are linear in every variable $z_1, \dots, z_n$.

Note that all the quasipolynomials considered above belong to the class MS_n^1, while the functions figuring in **Theorem 6.2.3** belong to the class MS_n^n.

We denote by $n_f^-(r)$ the (2n-2)-dimensional volume of the set of poles of the function f in the ball B_r, counted with multiplicities.

Theorem 6.3.2. Let the functions

$$f(z) = \sum_{l=1}^{m} a_l(z)e^{\lambda_l(z)}$$

and

$$g(z) = \sum_{l=1}^{k} b_l(z)e^{\mu_l(z)}$$

belong to the class MS_n^p, $1 \le p \le n$. Let also the meromorphic function

$$h(z) = \frac{f(z)}{g(z)}$$

be such that

$$\lim_{r \to \infty} \frac{1}{r^{2n-1}} n_h^-(r) = 0.$$

Then $h \in MS_n^p$ and

$$h(z) = \sum_{l=1}^{d} c_l(z)e^{\alpha_l(z)},$$

where the $c_1, \dots, c_d$ are rational functions of $a_1, \dots, a_m, b_1, \dots, b_k$ with integer coefficients and $\alpha_1, \dots, \alpha_m$ are linear combinations of the functions $\lambda_1, \dots, \lambda_m, \mu_1, \dots, \mu_k$ with integer coefficients.

The proof of this theorem, which we omit, is in certain part conceptually similar to the proof of **Theorem 6.3.1**, but it is essentially more complicated.

Another generalization of **Theorem 6.3.1** consist of the consideration of operations that are more general than division. To understand which operations must be studied, it is sufficient to compare **Theorems 6.2.4**

and 6.3.1 and to note that they can be regarded as theorems on the form of an entire solution of a binomial algebraic equation with quasipolynomial coefficients. It is therefore natural to consider general algebraic equations.

Theorem 6.3.3. Let a meromorphic function $f(z)$, $z\in\mathbb{C}^n$, be such that

$$\lim_{r\to\infty} \frac{1}{r^{2n-1}} n_{\bar{f}}(r) = 0.$$

Let also

$$a_0 f^m + a_1 f^{m-1} + \ldots + a_m = 0,$$

where $a_0, \ldots, a_m$ are *P*-quasipolynomials. Then

$$f(z) = \frac{g(z)}{Q(z)},$$

where $Q(z)$ is polynomial and $g(z)$ is a *P*-quasipolynomial with exponents which are linear combinations of points belonging to the spectra $\Lambda(a_j)$ with coefficients of the form $\frac{\nu}{q}$, $\nu\in\mathbb{Z}$, $q\in\mathbb{Z}$. Here q depends on the spectra $\Lambda(a_j)$ only.

We omit the proof of this theorem too. It is not only unwieldy, but also needs a special apparatus, namely, the theory of asymptotic series of functions of one variable which are holomorphic in sector. Note only that in view of **Theorem 6.2.3.** it is sufficient to prove **Theorem 6.3.3** for functions of one variable.

2. Algebraic factors of *P*-quasipolynomials. In connection with the theorem on the division of quasipolynomials, the problem arises on the general form of polynomials which are (more exactly, can be) factors of a quasipolynomial of some class. The example of the polynomial $P(z_1,z_2) = z_1^2-z_2$ indicates that not every polynomial need be such a factor. Indeed, if some $\mathbb{C}$-quasipolynomial

$$f(z_1,z_2) = \sum_{l=1}^{m} c_k e^{\lambda_1 z_1+\mu_1 z_2}$$

can be divided by $z_1^2-z_2$, then

$$\sum_{l=1}^{m} c_k e^{\lambda_1 z_1+\mu_1 z_1^2} \equiv 0,$$

which is possible only if $c_l = 0$, $\forall l$. The following theorem shows that this example represents the general situation.

Theorem 6.3.4. Let a *P*-quasipolynomial

$$f(z) = \sum_{l=1}^{m} a_l(z)e^{\langle\lambda_l,z\rangle}$$

and an irreducible polynomial $P(z)$ be such that $\frac{f(z)}{P(z)} \in H(\mathbb{C}^n)$. Then either $\frac{a_l(z)}{P(z)} \in H(\mathbb{C}^n)$, $\forall l$, or there are indices l', l'' and numbers γ, $c_l \in \mathbb{C}$ such that

$$P(z) = (\langle\lambda_{l'}-\lambda_{l''},z\rangle - \gamma)c_l.$$

Proof. We prove this by induction with respect to the number of points of the spectrum. To obtain the induction base, i.e. the theorem in the case of *P*-quasipolynomials with two-point spectrum, we need the following lemmas.

Lemma 6.3.2. Let χ be an algebraic set[45] in $\mathbb{C}^2$ of pure dimension $dim\chi = 1$ which does not contain components of the form $z_1 = \gamma$, $\gamma\in\mathbb{C}$. Let also

$$ln|z_2| \le c_1 ln(1+|z_1|),\ \forall(z_1,z_2)\in\chi,\ |z_1| > R_0. \tag{6.3.7}$$

Then

$$\overline{\lim_{\substack{|z_1|+|z_2| \to\infty \\ (z_1,z_2)\in\chi}}} \frac{|Imz_1|}{ln(|z_1|+|z_2|)} = \infty. \tag{6.3.8}$$

Proof. Assume the opposite, i.e. that these are constants $c_2 > 0$ and $R_1 > 0$ such that

$$|Imz_1| \le c_2 ln(|z_1|+|z_2|),\ \forall(z_1,z_2)\in\chi,\ |z_1|+|z_2| > R_1.$$

Then consider the entire function $\frac{sin\varepsilon z_1}{z_1}$ in $\mathbb{C}^2$, where $\varepsilon > 0$ is such that $\varepsilon c_2 c_3 < 1$, $c_3 = max\{1,c_1\}$. Note that $|sin\varepsilon z_1| \le e^{\varepsilon|Imz_1|}$. Besides, from (6.3.7) it follows that

---

[45] Recall that a set $\chi \subset \mathbb{C}^n$ is called algebraic if $\chi = \{z\in\mathbb{C}^n\colon P_1(z) = 0, \dots, P_l(z) = 0\}$, where $P_1(z), \dots, P_l(z)$ are polynomial in z.

$$|z_1|+|z_2| \le 2(1+|z_1|)^{c_3}$$

on χ for $|z_1| > R_0$. Therefore,

$$\overline{\lim_{\substack{|z_1|+|z_2| \to \infty \\ (z_1,z_2)\in\chi,\, |z_1|>R_0}}} \left| \frac{\sin\varepsilon z_1}{z_1} \right| \le \overline{\lim_{\substack{|z_1|+|z_2| \to \infty \\ (z_1,z_2)\in\chi,\, |z_1|>R_0}}} \frac{e^{\varepsilon|Im z_1|}}{|z_1|} \le$$

$$\le \overline{\lim_{\substack{|z_1|+|z_2| \to \infty \\ (z_1,z_2)\in\chi,\, |z_1|>R_0}}} \frac{(|z_1|+|z_2|)^{\varepsilon c_2}}{|z_1|} \le$$

$$\le \overline{\lim_{|z_1| \to \infty}} \frac{2^{\varepsilon c_2}(1+|z_1|)^{\varepsilon c_2 c_3}}{|z_1|} = 0. \tag{6.3.9}$$

This and the inequality

$$\sup_{(z_1,z_2)\in\chi,\, |z_1|<R_1} \left| \frac{\sin\varepsilon z_1}{z_1} \right| <\infty$$

implies that

$$\sup_{(z_1,z_2)\in\chi} \left| \frac{\sin\varepsilon z_1}{z_1} \right| <\infty. \tag{6.3.10}$$

Since the analogue of **Liouville's Theorem** is true for functions holomorphic on an algebraic set, (6.3.10) implies that

$$\frac{\sin\varepsilon z_1}{z_1} \equiv const$$

on χ. Taking into account (6.3.9) we obtain

$$\frac{\sin\varepsilon z_1}{z_1} = 0,\ \forall (z_1,z_2)\in\chi.$$

Hence the irreducible components of χ must be of the form $\left\{(z_1,z_2):\ z_1 = \frac{\pi l}{\varepsilon},\ z_2\in\mathbb{C}\right\}$, $l\in\mathbb{Z}$. This is impossible.

The proof of the lemma is finished.

Lemma 6.3.3. Let

$$P(z_1,z_2) = a_0(z_1)z_2^m + a_1(z_1)z_2^{m-1} + \dots + a_m(z_1),\ a_0(z_1) \not\equiv 0,\ m > 1,$$

be an irreducible polynomial (in z_1 and z_2) which is not a factor[46] of a polynomial (in z_1 and z_2)

$$Q(z_1,z_2) = b_0(z_2)z_1^k + b_1(z_2)z_1^{k-1} + \dots + b_k(z_2).$$

Then there are constants $c_1 > 0$, $R_0 > 0$ and $\alpha > 0$ such that

$$\frac{1}{c_1|z_1|^\alpha} \le |Q(z_1,z_2)| \le c_1|z_1|^\alpha, \forall(z_1,z_2)\in|Z_P|,\ |z_1| > R_0, \tag{6.3.11}$$

where $|Z_P|$ is the support of the divisor Z_P.

Proof. Since $P(z_1,z_2)$ is irreducible and its degree in z_2 differs from 0, the discriminant of $P(z_1,z_2)$, as a polynomial in z_2, is not identically equal to zero. Therefore outside the disc $|z_1| \le R_0$, for some $R_0 > 0$, the equation $P(z_1,z_2) = 0$ (considered with respect to z_2) has precisely m pairwise distinct solutions. Besides, in a sufficiently small neighbourhood of each point z_1^0, $|z_1^0| > R_0$, these solutions can be numbered in such a way that the corresponding functions $z_2^{(j)}(z_1)$, $j = 1, \dots, m$, are holomorphic in this neighbourhood. Therefore the function

$$g(z_1) = \prod_{j=1}^{m} Q(z_1,z_2^{(j)}(z_1)),$$

---

[46] An entire function f(z) is called a factor of an entire function g(z) (or "f(z) divides g(z)") if there is an entire function $h(z) \not\equiv 0$ such that $h\cdot f = g$. A function f is a factor of g if and only if their divisors $Z_g = (|Z_g|,\gamma_g)$ and $Z_f = (|Z_f|,\gamma_f)$ are related by $Z_g \ge Z_f$, i.e. $|Z_f| \subset |Z_g|$ and $\gamma_f(z) \le \gamma_g(z)$, $\forall z\in|Z_f|$.

Recall that $|Z_f| = \{z\in\mathbb{C}^n: f(z) = 0\}$ is the support of the divisor Z_f and $\gamma_f(z)$ is the multiplicity of vanishing of the function f at the point z.

which is single-valued in the domain $|z_1| > R_0$, is holomorphic at every point of this domain and hence holomorphic in the whole domain. Note also that the estimate

$$|z_2^{(j)}(z_1)| \le \frac{1}{|a_0(z_1)|} \sum_{j=1}^{m} |a_j(z_1)|$$

holds. Hence

$$\max_j |z_2^{(j)}(z_1)| \le c_1 |z_1|^{k_1}, \ \forall z_1 \colon |z_1| > 2R_0$$

for some $c_1 > 0$ and $k_1 \in \mathbb{N}$. In turn, this implies follows that

$$\max_j |Q(z_1, z_2^{(j)})| \le c_2 |z_1|^{k_2}, \ \forall z_1 \colon |z_1| > 2R_0 \tag{6.3.12}$$

for some $c_2 > 0$ and $k_2 \in \mathbb{N}$. Therefore,

$$|g(z_1)| \le c_2^m |z_1|^{mk_2}, \ \forall z_1 \colon |z_1| > 2R_0 . \tag{6.3.13}$$

Since $g(z_1)$ is holomorphic for $|z_1| > R_0$, (6.3.13) implies that the following representation holds:

$$g(z_1) = \sum_{\nu=-\infty}^{q} b_\nu z_1^\nu, \ |z_1| > R_0,$$

where $b_q \ne 0$, $-\infty < q \le mk_2 < \infty$. Therefore, for some $c_3 > 0$ and $R_1 > 2R_0$ we have

$$|g(z_1)| \ge c_3 |z_1|^q, \ \forall z_1 \colon |z_1| > R_1. \tag{6.3.14}$$

Now we estimate the quantity $\min_j |Q(z_1, z_2^{(j)}(z_1))|$. By (6.3.12) and (6.3.14) we find that

$$|Q(z_1, z_2^{(j)})| = |g(z_1)| \left(\prod_{i \ne j} |Q(z_1, z_2^{(i)}(z_1))| \right)^{-1} \ge$$

$$\ge \frac{c_3 |z_1|^q}{(\max_j |Q(z_1, z_2^{(j)}(z_1))|)^{m-1}} \ge$$

$$\ge \frac{c_3 |z_1|^q}{(c_2 |z_1|^{k_2})^{m-1}} = \frac{c_4}{|z_1|^\alpha}, \ \forall j, \ |z_1| > R_1. \tag{6.3.15}$$

It is clear that view of (6.3.12) this estimate is equivalent to (6.3.11).

The proof of the lemma is finished.

Lemma 6.3.4. Let $\chi \subset \mathbb{C}^2$ be an analytic set such that $\chi = |Z_P|$, where $P(z_1,z_2)$ is a polynomial, and such that no irreducible component of χ is contained in a plane of the form $z_1 = \gamma$. Let there be polynomials $Q_1(z_1,z_2)$ and $Q_2(z_1,z_2)$ such that the sets $\chi \cap |Z_{Q_1}|$ and $\chi \cap |Z_{Q_2}|$ have dimension 0 or -1[47]. Then the analytic set

$$\tilde{\chi} = \{(z_1,z_2)\in\chi:\ Q_1(z_1,z_2)e^{iz_1} + Q_2(z_1,z_2) = 0\}$$

is such that $dim\tilde{\chi} \le 0$.

Proof. Let χ_1 be an irreducible component of $\tilde{\chi}$ and $dim\chi_1 = 1$. Since $\chi_1 \subset \chi$ and $dim\chi = 1$, χ_1 is an irreducible component of the algebraic set χ; therefore, it is algebraic itself and does not coincide with any plane $z_1 = \gamma$. Thus $\chi_1 = |Z_{P_1}|$, where

$$P_1(z_1,z_2) = a_0(z_1)z_2^m + \dots + a_m(z_1),\ m \ge 1,\ a_0(z_1) \not\equiv 0.$$

is an irreducible polynomial in z_1 and z_2. Therefore we can apply **Lemma 6.3.3**, which implies that for some $q\in\mathbb{N}$, $c_1 > 0$ and $R_0 > 0$,

$$\frac{1}{c_1|z_1|^q} \le \left|\frac{Q_1(z_1,z_2)}{Q_2(z_1,z_2)}\right| \le c_1|z_1|^q, \forall(z_1,z_2)\in\chi_1,\ |z_1| > R_0, \tag{6.3.16}$$

Note also that, as already shown in the proof of **Lemma 6.3.3**,

$$|z_2| \le c_2|z_1|^{q_1},\ \forall(z_1,z_2)\in\chi_1,\ |z_1| > R_0,$$

for some $c_2 > 0$ and $q_1 > 0$. Thus condition (6.3.7) of **Lemma 6.3.2** is valid for the set χ_1. Applying this lemma we find that

$$\overline{lim}_{\substack{|z_1|+|z_2|\to\infty\\(z_1,z_2)\in\chi}} \frac{|Imz_1|}{ln(|z_1|+|z_2|)} = \infty. \tag{6.3.17}$$

At the same time, since the set χ is algebraic, the conditions $dim(\chi \cap$

---

[47] By definition, the equality $dim\chi = -1$ means that $\chi = \varnothing$.

$|Z_{Q_i}|) \leq 0$, $i = 1, 2$, imply that the sets $\chi \cap |Z_{Q_i}|$ are finite. Therefore,

$$|Imz_1| = \left| ln \left| \frac{Q_1(z_1,z_2)}{Q_2(z_1,z_2)} \right| \right| < \infty$$

on χ_1 and outside any ball of sufficiently large radius. This and (6.3.16) imply that

$$|Imz_1| \leq q|ln|z_1|| + |lnc_1|, \ \forall (z_1,z_2) \in \chi_1, \ |z_1| > R_0,$$

contradicting (6.3.17).

The proof of the lemma is finished.

Lemma 6.3.5. Let $f(z)$, $z \in \mathbb{C}^n$, be a *P*-quasipolynomial of the form

$$f(z) = Q_1(z)e^{iz_1} + Q_2(z),$$

where $Q_1 \not\equiv 0$, $Q_2 \not\equiv 0$ are polynomials. Further, let $P(z)$ be an irreducible polynomial which is a factor of the function $f(z)$ and not a factor of the polynomials Q_1 and Q_2[48]. Then $P(z)$ is a linear function in z_1.

Proof. To prove the lemma it suffices to show that the degree of the polynomial $P(z)$ in each of variable $z_2, \ldots, z_n$ is equal to zero. Assume that the opposite is valid. Let, for example, the degree of $P(z)$ in z_2 be equal to $m > 0$. Then the projection of the set $|Z_P|$ to the space of variables $z_1, z_3, \ldots, z_n$ contains a set $\mathbb{C}^{n-1} \setminus E$, where E is the zero set of some polynomial[49] in $z_1, z_3, \ldots, z_n$. In turn, this implies that the projection of the set $|Z_P|$ to the space of variables $z_3, \ldots, z_n$ equals $\mathbb{C}^{n-2} \setminus E_1$, where E_1 is an algebraic[50] set in the space

---

[48] It is obvious that if P divides one of the polynomials Q_i, then it divides the other too.

[49] It is clear that the polynomial $\frac{\partial^m P}{\partial z_2^m}$ will do.

[50] It is easy to see that

$$E_1 = \left\{ (z_3,\ldots,z_n): \left. \frac{\partial^{m+\nu} P}{\partial z_2^m \partial z_1^\nu} \right|_{z_1=0} = 0, \ \forall \nu = 0, 1, 2, \ldots \right\}.$$

of variables $z_3, \ldots, z_n$. We set

$$\chi_j = \chi_j(c_3,\ldots,c_n) =$$

$$= \{(z_1, z_2)\colon (z_1,z_2,c_3,\ldots,c_n)\in|Z_P| \cap |Z_{Q_j}|\}\ j = 1,\ 2.$$

Since, according to the condition of the lemma $P(z)$ is not a factor of the polynomials Q_1 and Q_2, the analytic (more exactly, algebraic) sets $|Z_P|\cap|Z_{Q_j}|$ are such that $dim(|Z_P| \cap |Z_{Q_j}|) \le n-2$, $j = 1, 2$. This implies the set $E_2 = \{(z_3,\ldots,z_n)\colon dim\chi_1(z_3,\ldots,z_n) \ge 1\}$ has Lebesgue measure zero ($mes_{2n-4}E_2 = 0$). Let $(z_3^0,\ldots z_n^0)\notin E_2$. Consider the functions

$$F(z_1,z_2) = f(z_1,z_2,z_3^0,\ldots,z_n^0),$$

$$q_j(z_1,z_2) = Q_j(z_1,z_2,z_3^0,\ldots,z_n^0)\ ,$$

$$p(z_1,z_2) = P(z_1,z_2,z_3^0,\ldots,z_n^0).$$

It is clear that $F(z_1,z_2)$ is a P-quasipolynomial, while $p(z_1,z_2)$, $q_1(z_1,z_2)$, $q_2(z_1,z_2)$ are polynomials. Besides, $p(z_1,z_2)$ divides $F(z_1,z_2)$,

$$F(z_1,z_2) = q_1(z_1,z_2)e^{iz_1} + q_2(z_1,z_2)$$

and $dim(|Z_p| \cap |Z_{q_j}|) = dim\chi_j(z_3^0,\ldots,z_n^0) \le 0$. Thus **Lemma 6.3.4**. can be applied and, therefore, the irreducible components of $|Z_p|$ are of the form $z_1 = \gamma$. Since the degree of $p(z_1,z_2)$ is at most that of $P(z)$, the number of such components is bounded by a constant m_0 independent of $z_3, \ldots, z_n$. Hence

$$p(z_1,z_2) = \alpha_1(z_1-\gamma_1)\ldots(z_1-\gamma_k),\ k\le m_0,$$

where the constants k, α_1, γ_1, ..., γ_k are independent of z_1 and z_2. This it immediately implies that the polynomial $P(z)$, as a polynomial in z_2, is of degree zero for all fixed $(z_1,z_3,\ldots,z_n)\in\mathbb{C}\times(\mathbb{C}^{n-2}\setminus E_2)$. The same is, of course true for all $(z_1,z_3,\ldots,z_n)\in\mathbb{C}^{n-1}$. This contradicts the assumption made.

The proof of the lemma is finished.

As already mentioned, the proof of **Theorem 6.3.4** will be carried out by induction with respect to the number of points in the spectrum $\Lambda(f)$. In the case $m = 2$ the theorem follows from **Lemma 6.3.5**, because any quasipolynomial with two-point spectrum can be reduced to the form $Q_1 e^{iz_1} + Q_2$ by a linear change of variables. We assume that the theorem is true for P-quasipolynomials with spectra consisting of at most m-1 points. Then we consider a P-quasipolynomial

$$f(z) = \sum_{l=1}^{m} a_l(z) e^{\langle \lambda_l, z\rangle}$$

and an irreducible polynomial $P(z)$ that divides f and does not divide all the coefficients $a_l(z)$. We choose a regular point $z^0 \in |Z_P|$ at which $a_m(z^0) \neq 0$. Let, for sake of being specific, $P'_{z_n}(z^0) \neq 0$. Then the equation $P(z) = 0$ is uniquely solvable with respect to z_n in some neighbourhood Ω of the point z^0, and this solution $z_n = \varphi('z)$, $'z = (z_1, \dots, z_{n-1})$, is holomorphic in a neighbourhood of $'z^0$. Note that in this situation the mapping

$$(z_j, z_n) \to (f(z_1^0, \dots, z_{j-1}^0, z_j, z_{j+1}^0, \dots, z_n), P(z_1^0, \dots, z_{j-1}^0, z_j, z_{j+1}^0, \dots, z_n))$$

is degenerate at the point (z_j^0, z_n^0) for any $j = 1, \dots, n-1$. Hence the functions

$$J_j(z) = \begin{vmatrix} \dfrac{\partial P}{\partial z_j} & \dfrac{\partial f}{\partial z_j} \\ \dfrac{\partial P}{\partial z_n} & \dfrac{\partial f}{\partial z_n} \end{vmatrix}$$

vanish at any such point z^0. In view of the irreducibility of $|Z_P|$, it thus follows that $J_j(z) = 0$, $\forall z \in |Z_P|$, $\forall j$. Since the polynomial $P(z)$ is irreducible, the latter means that $P(z)$ divides each function $J_j(z)$.

It can be immediately checked that

$$J_j(z) = \sum_{l=1}^{m} b_{l,j}(z) e^{\langle \lambda_l, z\rangle},$$

where

$$b_{1,j} = \frac{\partial P}{\partial z_j}\left(\frac{\partial a_1}{\partial z_n} - a_1\lambda_{1,n}\right) - \frac{\partial P}{\partial z_n}\left(\frac{\partial a_1}{\partial z_j} - a_1\lambda_{1,j}\right). \tag{6.3.18}$$

We set

$$F_j(z) = a_m J_j(z) - b_{m,j}(z) f(z).$$

It is clear that $F_j(z)$ is divisible by $P(z)$ for any j. Besides, from the form of the functions $J_j(z)$ and $f(z)$ it follows that $F_j(z)$ is a P-quasipolynomial with spectrum consisting of at most m-1 points. More exactly,

$$F_j(z) = \sum_{l=1}^{m-1} c_{l,j}(z)\, e^{\langle\lambda_l, z\rangle},$$

where

$$c_{l,j}(z) = a_m(z) b_{l,j}(z) - b_{m,j}(z) a_l(z).$$

If for some $j = 1, \ldots, m-1$ at least one of the coefficients $c_{l,j}(z)$ is not identically equal to zero on $|Z_P|$, according to the induction assumption the polynomial $P(z)$ is of the form mentioned in the theorem to be proved. Now we consider the case when $c_{l,j}(z) \equiv 0$ on $|Z_P|$, $\forall l, j$. By (6.3.18) we have

$$c_{1,j} = \frac{\partial P}{\partial z_j}\left\{\left(a_m \frac{\partial a_1}{\partial z_n} - a_1 \frac{\partial a_m}{\partial z_1}\right) - a_m a_1(\lambda_{1,n} - \lambda_{m,n})\right\} +$$

$$- \frac{\partial P}{\partial z_n}\left\{\left(a_m \frac{\partial a_1}{\partial z_j} - a_1 \frac{\partial a_m}{\partial z_j}\right) - a_m a_1(\lambda_{1,j} - \lambda_{m,j})\right\} =$$

$$= a_m^2 \frac{\partial P}{\partial z_j}\left\{\frac{\partial}{\partial z_n}\left(\frac{a_1}{a_m}\right) - \frac{a_1}{a_m}(\lambda_{1,j} - \lambda_{m,n})\right\} +$$

$$- a_m^2 \frac{\partial P}{\partial z_n}\left\{\frac{\partial}{\partial z_j}\left(\frac{a_1}{a_m}\right) - \frac{a_1}{a_m}(\lambda_{1,j} - \lambda_{m,j})\right\}. \tag{6.3.19}$$

Note that

$$\varphi'_{z_j}('z) = -\left(\frac{P'_{z_j}}{P'_{z_n}}\right)_{z_n = \varphi('z)}, \tag{6.3.20}$$

and, correspondingly,

$$\frac{\partial}{\partial z_j}\left(\frac{a_1('z,\varphi('z))}{a_m('z,\varphi('z))}\right) = \left\{ \varphi'_{z_j}\frac{\partial}{\partial z_n}\left(\frac{a_1}{a_m}\right) + \frac{\partial}{\partial z_j}\left(\frac{a_1}{a_m}\right)\right\}_{z_n=\varphi('z)} =$$

$$= \left\{ \frac{1}{P'_{z_n}}\left(-P'_{z_j}\frac{\partial}{\partial z_n}\left(\frac{a_1}{a_n}\right) + P'_{z_n}\frac{\partial}{\partial z_j}\left(\frac{a_1}{a_n}\right)\right)\right\}_{z_n=\varphi('z)}.$$

From this and (6.3.19) and (6.3.20) it follows that the equality $c_{1,j}(z) = 0,\ \forall z\in|Z_P|$, leads to the identities

$$\frac{1}{\Psi}\Psi'_{z_j} = (\lambda_{1,n} - \lambda_{m,n})\varphi'_{z_j} + (\lambda_{1,j} - \lambda_{m,j}),\ j = 1,\ \dots,\ m-1,$$

for the function

$$\Psi('z) = \frac{a_1('z,\varphi('z))}{a_m('z,\varphi('z))}.$$

From these identities it follows that

$$\Psi('z) = c\cdot exp\left\{ \sum_{j=1}^{m} (\lambda_{1,j} - \lambda_{m,j})z_j + (\lambda_{1,n} - \lambda_{m,n})\varphi('z)\right\}$$

with some $c\in\mathbf{C}$. The latter is clearly equivalent to the equality

$$a_1(z) - a_m(z)c\cdot e^{\langle\lambda_1-\lambda_m,z\rangle} = 0,\ \forall z\in|Z_P|\cap\Omega. \tag{6.3.21}$$

Since $|Z_P|$ is an irreducible analytic set, (6.3.21) implies that

$$a_1(z) - a_m(z)c\cdot e^{\langle\lambda_1-\lambda_m,z\rangle} = 0,\ \forall z\in|Z_P|.$$

Recall that $P(z)$ is not a factor of any coefficient $a_1(z)$, including the coefficients $a_1(z)$ and $a_m(z)$. Therefore **Lemma 6.3.5** can be applied and we obtain that

$$|Z_P| = \{z\in\mathbf{C}^n:\ (\langle z,\lambda_1-\lambda_m\rangle -\gamma)c_1 = 0\}.$$

This finishes the proof of **Theorem 6.3.4**.

This theorem gives us a possibility to refine **Theorems 6.3.1** and **6.3.3** by noting that the polynomials figuring in these theorems are products of linear functions. Similarly, after suitable reformulation we can refine **Theorem 6.3.2** too. The above-mentioned refinements clearly follow from **Theorem 6.3.4**.

3. Reducibility in the ring of quasipolynomials. The concepts of reducibility and irreducibility in the ring of quasipolynomials are defined in a standard manner. Namely, a $\mathbb{C}$-quasipolynomial (P-quasipolynomial, M_0-quasipolynomial) $f(z)$ is said to be irreducible in the ring of $\mathbb{C}$-quasipolynomials (P-quasipolynomials, M_0-quasipolynomials) if it cannot be represented as the product of two $\mathbb{C}$-quasipolynomials (P-quasipolynomials, M_0-quasipolynomials) with spectra of at least two points. If a quasipolynomial is not irreducible, then it is said to be reducible (in the corresponding ring).

Here we show that the problem of reducibility in the ring of quasipolynomials can be reduced to the problem of reducibility in the ring of polynomials.

A set $M \subset \mathbb{C}^n$ is called an integer (rational) basis in a set $M_1 \subset \mathbb{C}^n$ if the following is valid:

1) the points of M are linearly independent over $\mathbb{Z}$;
2) every point $z \in M_1$ is a linear combination with integer (rational) coefficients of points $\zeta \in M$.

It is easy to see (see, for example, B.Levin [1]) that any set M_1 contains an integer as well as a rational basis. It is also obvious that an integer basis is a rational basis, but a rational basis need not be an integer basis.

Besides, any rational basis of the finite set can be transformed into an integer basis by multiplication by a suitable number $\frac{1}{q}$, $q \in \mathbb{N}$.

Let a set $M = \{\tilde{\lambda}_l\}_{l=1}^{\tilde{m}}$ be an integer basis in the spectrum $\Lambda(f)$ of the $\mathbb{C}$-quasipolynomial

$$f(z) = \sum_{l=1}^{m} c_l e^{\langle \lambda_l, z \rangle} .$$

This means that every exponent λ_l can be represented in the form

$$\lambda_l = \sum_{j=1}^{\tilde{m}} \alpha_{l,j} \tilde{\lambda}_j ,$$

where $\alpha_{l,j} \in \mathbb{Z}$, $\forall l, j$. Hence the quasipolynomial $f(z)$ can be represented in the form

$$f(z) = \sum_{l=1}^{m} c_l \prod_{j=1}^{\tilde{m}} \left(e^{\langle \tilde{\lambda}_j, z \rangle} \right)^{\alpha_{l,j}}.$$

We set

$$\tilde{f}(\zeta) = \sum_{l=1}^{m} c_l \zeta_1^{\alpha_{l,1}} \cdot \ldots \cdot \zeta_{\tilde{m}}^{\alpha_{l,\tilde{m}}}.$$

In this way we associate the polynomial $\tilde{f}(\zeta)$ of the variables $\zeta_1, \ldots, \zeta_{\tilde{m}}, \frac{1}{\zeta_1}, \ldots, \frac{1}{\zeta_{\tilde{m}}}$ to the $\mathbb{C}$-quasipolynomial $f(z)$. We call (for brevity) such polynomials L-polynomials in ζ. Then we define the mapping $\rho_M: \mathbb{C}^n \to \mathbb{C}^{\tilde{m}}$ by the equality

$$\rho_M(z) = \left(e^{\langle \tilde{\lambda}_1, z \rangle}, \ldots, e^{\langle \tilde{\lambda}_{\tilde{m}}, z \rangle} \right).$$

Now we associate to each L-polynomial $Q = Q(\zeta)$, $\zeta \in \mathbb{C}^{\tilde{m}}$, the $\mathbb{C}$-quasipolynomial $\rho_M^* Q = Q \circ \rho_M$. It is obvious that $\rho_M^* \tilde{f} = f$. Note that ρ_M^* is an injection. Indeed, since the $\tilde{\lambda}_l$ are linearly independent over $\mathbb{Z}$, then $\sum_{l=1}^{\tilde{m}} j_l \tilde{\lambda}_l \neq \sum_{l=1}^{\tilde{m}} i_l \tilde{\lambda}_l$ if $J = (j_1, \ldots, j_{\tilde{m}}) \neq I = (i_1, \ldots, i_{\tilde{m}})$, $I \in \mathbb{Z}^{\tilde{m}}$, $J \in \mathbb{Z}^{\tilde{m}}$. Therefore the mapping ρ_M^* maps distinct monomials $\zeta^J = (\zeta_1^{j_1} \cdot \ldots \cdot \zeta_{\tilde{m}}^{j_{\tilde{m}}})$ to distinct exponentials $exp\left\{ \sum_{l=1}^{\tilde{m}} j_l \langle \tilde{\lambda}_l, z \rangle \right\}$, and hence the equality $\rho_M^* Q = 0$ is possible only in the case $Q = 0$.

The above-mentioned relationship between reducibility in the ring of polynomials and that in the ring of $\mathbb{C}$-quasipolynomials is contained in the following theorem.

Theorem 6.3.5. Let a set $M = \{\tilde{\lambda}_l\}_{l=1}^{\tilde{m}}$ be an integer basis in the spectrum $\Lambda(f)$ of the $\mathbb{C}$-quasipolynomial

$$f(z) = \sum_{l=1}^{m} c_l e^{\langle \lambda_l, z \rangle}$$

Then the quasipolynomial f is reducible in the ring of $\mathbb{C}$-quasipolynomials if and only if there is a natural number p such that the polynomial $\tilde{f}(\zeta_1^p, \ldots, \zeta_{\tilde{m}}^p)$, where $\tilde{f} = f \circ \rho_M$, can be represented as the product of two L-polynomials in ζ each of which consists of at least two different monomials.

Proof. First we prove sufficiency. Suppose L-polynomial $\tilde{f}_1(\zeta) = f(\zeta_1^p,\dots,\zeta_{\tilde{m}}^p)$, $p\in\mathbb{N}$, can be represented in the form $\tilde{f}_1(\zeta) = Q_1(\zeta)Q_2(\zeta)$, where $Q_1(\zeta)$ and $Q_2(\zeta)$ are L-polynomials each of which consists of at least two different monomials. From the basis M we construct the basis $\Gamma = \left\{\frac{1}{p}\,\tilde{\lambda}_l\right\}_{l=1}^{\tilde{m}}$ and corresponding mappings ρ_Γ and ρ_Γ^*. Then we have

$$f = \rho_M^*\tilde{f} = \rho_\Gamma^*\tilde{f}_1 = \rho_\Gamma^*Q_1\cdot\rho_\Gamma^*Q_2$$

and each of the spectra of the $\mathbb{C}$-quasipolynomials $\rho_\Gamma^*Q_1$ and $\rho_\Gamma^*Q_2$ consists of at least two points. Thus we have proved the sufficiency part of the theorem.

Now let $f(z) = f_1(z)f_2(z)$, where $f_1(z)$ and $f_2(z)$ are $\mathbb{C}$-quasipolynomials with spectra consisting of at least two points. We choose points $\tilde{\lambda}_{\tilde{m}+1}, \dots, \lambda_{\hat{m}}$, $\hat{m} \le m$, belonging to $\Lambda^* = \Lambda_f \cup \Lambda_{f_1} \cup \Lambda_{f_2}$ in such a way that they complement the set M to a rational basis in Λ^*. Then we choose an integer q such that $\Gamma = \left\{\frac{1}{q}\,\tilde{\lambda}_l\right\}_1^{\hat{m}}$ is an integer basis in Λ^*. Consider the L-polynomials

$$\tilde{f}_3(\zeta) = (f\circ\rho_\Gamma)(\zeta),$$

$$\tilde{f}_1(\zeta) = (f_1\circ\rho_\Gamma)(\zeta),$$

$$\tilde{f}_2(\zeta) = (f_1\circ\rho_\Gamma)(\zeta),$$

where $\zeta = (\zeta_1,\dots,\zeta_{\hat{m}})\in\mathbb{C}^{\hat{m}}$. From the construction of Γ it follows that the L-polynomial $\tilde{f}_3(\zeta)$ is independent of $\tilde{\lambda}_{\tilde{m}+1}, \dots, \lambda_{\hat{m}}$. Therefore it is natural to write $\tilde{f}_3(\zeta^*)$, $\zeta^* = (\zeta_1,\dots,\zeta_{\tilde{m}})$, instead of $\tilde{f}_3(\zeta)$. Note that

$$\rho_\Gamma^*(\tilde{f}_1\cdot\tilde{f}_2) = \rho_\Gamma^*\tilde{f}_1\cdot\rho_\Gamma^*\tilde{f}_2 = f_1\cdot f_2 = f = \rho_\Gamma^*\tilde{f}_3 .$$

From this, taking into account that the mapping ρ_Γ^* is an injection, we conclude that

$$\tilde{f}_1(\zeta_1,\dots,\zeta_{\hat{m}})\tilde{f}_2(\zeta_1,\dots,\zeta_{\hat{m}}) = \tilde{f}_3(\zeta_1,\dots,\zeta_{\tilde{m}}). \tag{6.3.22}$$

This is possible only if

$$\tilde{f}_1(\zeta) = \tilde{f}_4(\zeta^*)\zeta^J,$$

$$\tilde{f}_2(\zeta) = \tilde{f}_5(\zeta^*)\zeta^{-J},$$

where $\tilde{f}_4$ and $\tilde{f}_5$ are some L-polynomials of ζ^* and $J\in\mathbb{Z}^{\hat{m}}$. Thus, (6.3.22) implies that $\tilde{f}_3(\zeta^*) = \tilde{f}_4(\zeta^*)\tilde{f}_5(\zeta^*)$. Note that the connection between the bases Γ and M implies the equality

$$\tilde{f}_3(\zeta_1,\dots,\zeta_{\tilde{m}}) = \tilde{f}(\zeta_1^q,\dots,\zeta_{\tilde{m}}^q).$$

Now, to complete the proof of **Theorem 6.3.5** it remains to note that since the spectra of the quasipolynomials f_1 and f_2 consist of at least two points, each of the corresponding L-polynomials $\tilde{f}_1$, $\tilde{f}$ and $\tilde{f}_2\tilde{f}_5$ contains at least two monomials.

The proof of the theorem is finished.

According to a theorem by Ehrenfeucht and Pełczynski (see J.Cassels [1]), if a polynomial $Q(z)$, $z\in\mathbb{C}^n$, $n \geq 3$, can be represented in the form $Q_1(z_1) + \dots + Q_n(z_n)$, where the $Q_1, \dots, Q_n$ are polynomials in one variable, then such a polynomial is irreducible. This statement, together with **Theorem 6.3.5** leads to the following sufficient condition for irreducibility of $\mathbb{C}$-quasipolynomials.

Theorem 6.3.6. Let the exponents λ_k of a $\mathbb{C}$-quasipolynomial be linearly independent over $\mathbb{Z}$. Then this quasipolynomial is irreducible in the ring of $\mathbb{C}$-quasipolynomials.

The problem of sufficient conditions for irreducibility of M_0-quasipolynomials can be reduced to the corresponding problem for $\mathbb{C}$-quasipolynomials. This follows from the following theorem.

Theorem 6.3.7. Let M_0-quasipolynomials

$$f_i(z) = \sum_{l=1}^{m_i} a_l^{(i)}(z)e^{\langle\lambda_l^{(i)},z\rangle}, \quad i = 1, 2, 3,$$

be such that $f_3(z) = f_1(z)f_2(z)$. Then for any $z^0 \in \mathbb{C}^n$ the following equality holds:

$$\sum_{l=1}^{m_3} a_l^{(3)}(z^0)e^{\langle\lambda_l^{(3)},z\rangle} =$$

$$= \left(\sum_{l=1}^{m_1} a_l^{(1)}(z^0)e^{\langle\lambda_l^{(1)},z\rangle}\right)\left(\sum_{l=1}^{m_2} a_l^{(2)}(z^0)e^{\langle\lambda_l^{(2)},z\rangle}\right), \forall z \in \mathbb{C}^n.$$

Proof. We set $\varphi(z) = f_3(z) - f_1(z)f_2(z)$. It is clear that $\varphi(z)$ is an M_0-quasipolynomial and

$$\varphi(z) = \sum_{l=1}^{m} A_l(z)e^{\langle\mu_l,z\rangle}.$$

where the exponents μ_l are not equal to one another and are contained in the set $\Lambda(f_3) \cup \{\Lambda(f_1) + \Lambda(f_2)\}$[51], and it is easy to see that the coefficients $A_l(z)$ are defined by

$$A_l(z) = \sum_p \delta_l^p a_p^{(3)}(z) + \sum_{j,i} a_j^{(1)}(z)a_i^{(2)}(z)\delta_l^{j,i},$$

where

$$\delta_l^p = \begin{cases} 0 \text{ if } \lambda_p^{(3)} \neq \mu_l \\ 1 \text{ if } \lambda_p^{(3)} = \mu_l \end{cases},$$

$$\delta_l^{j,i} = \begin{cases} 0 & \text{if } \lambda_j^{(1)} + \lambda_i^{(2)} \neq \mu_l \\ 1 & \text{if } \lambda_j^{(1)} + \lambda_i^{(2)} = \mu_l \end{cases}.$$

Thus the functions $A_l(z)$ are polynomials in $a_p^{(3)}$, $a_j^{(2)}$, $a_i^{(1)}$ with coefficients that depend on the spectra of the quasipolynomials f_1, f_2,

---

[51] The arithmetical sum X+Y of the sets X and Y is defined by the equality $X+Y = \{z: z = x+y, x \in Z, y \in Y\}$.

f_3 only. From this it follows that for any z^0 the function

$$\varphi_{z^0}(z) = \sum_{l=1}^{m_3} a_l^{(3)}(z^0)e^{\langle\lambda_l^{(3)},z\rangle} +$$

$$- \left(\sum_{l=1}^{m_1} a_l^{(1)}(z^0)e^{\langle\lambda_l^{(1)},z\rangle}\right)\left(\sum_{l=1}^{m_2} a_l^{(2)}(z^0)e^{\langle\lambda_l^{(2)},z\rangle}\right)$$

can be represented in the form

$$\varphi_{z^0}(z) = \sum_{l=1}^{m} A_l(z^0)e^{\langle\mu_l,z\rangle}.$$

According to the condition of the theorem, $f_3 = f_1f_2$ and, therefore, $\varphi(z) \equiv 0$. From this it follows that $A_l \equiv 0$, $\forall l$, and hence $\varphi_{z^0}(z) \equiv 0$, $\forall z^0$. According to the definition of the function $\varphi_{z^0}(z)$ this is equivalent to the validity of equality (6.3.23).

The proof of the theorem is finished.

From **Theorem 6.3.7** it follows that the irreducibility of the C-quasipolynomial

$$f_{z^0}(z) = \sum_{l=1}^{m} a_l(z^0)e^{\langle\lambda_l,z\rangle}$$

for even one $z^0 \in \mathbb{C}^n$ is a sufficient condition for irreducibility of the M_0-quasipolynomial

$$f(z) = \sum_{l=1}^{m} a_l(z)e^{\langle\lambda_l,z\rangle}$$

(in the ring of M_0-quasipolynomials). Note that for such a z^0, in accordance with the definition of irreducibility, $f_{z^0} \not\equiv 0$ and $\Lambda(f_{z^0}(z))$ consists of at least two points.

Notes

The results of §6.1 concerned with quasipolynomials in several variables were obtained by A.Ronkin [1], [2]. It is difficult to trace the discoverer of **Theorem 6.1.6**. It is contained in A.Ronkin [2] in the form given above, but various variants of it have been met earlier. **Theorems 6.2.1 - 6.2.5** were established by A.Ronkin [3], [4]. **Theorems 6.2.4** and **6.2.5** on the extraction of a root of a quasipolynomial were obtained earlier for functions of one variable by H.Selberg [1]. The part of **Theorem 6.3.1** concerned with C-quasipolynomials was obtained by V.Avanissian and R.Gay [1]. It is an extension to the multidimensional case of J.Ritt's theorem [1] on the division of C-quasipolynomials of one variable. The statement of **Theorem 6.3.1** concerning *P*-quasipolynomials is due to C.Berenstein and A.Dostal [1]. **Theorem 6.3.2** was obtained by A.Ronkin [5]. Statements for functions of one variable close to **Theorem 6.3.2** are contained in the papers of A.Shields [1], and Gordon-Levin [1]. **Theorem 6.3.3** is due to B.Levin and A.Ronkin [1], [2]. In the case of one variable, on the one hand this theorem is a generalization of the corresponding **J.Ritt's Theorem** [2], where the coefficients a_j are assumed to be C-quasipolynomials. On the other hand, it is a partial solution of **H.Selberg's Problem** [1]: find the form of entire functions $w(z)$, $z\in\mathbb{C}$, which are solutions of the equation

$$a_0(z)w^m + \dots a_m(z) = 0$$

with coefficients

$$a_j(z) = \sum_{l=1}^{k_j} c_j e^{\lambda_j z^p},$$

where $p\in\mathbb{N}$, $c_j\in\mathbb{C}$, $\lambda_j\in\mathbb{C}$.

Theorem 6.3.4 is due to C.Berenstein and A.Yger [1]. **Theorems 6.3.5 - 6.3.7** were obtained by A.Ronkin [6].

CHAPTER 7

MAPPINGS

The theory of mappings of completely regular growth is in an initial stage of development, unlike the corresponding theory for functions. There are almost no results and, moreover, there is no clearness in the definitions of basic concepts and in the formulation of concrete problems.

One of the possible directions of development of the theory is connected with so-called mappings of α-regular growth. Here we give an account of a result in this direction and its corollaries. Besides, in this chapter we give a result concerning the distribution of zeros of almost periodic holomorphic mappings. Such mappings are not of α-regular growth, or, in the case of mappings $\mathbb{C}^n \to \mathbb{C}$, are not functions of c.r.g. But the method was used for proving this result is conceptually close to the method by which functions of c.r.g. are studied in this book. That is why we include this result in this book.

§1 Information on the general theory of holomorphic mappings

The theory of currents is an important tool in the study of holomorphic mappings. Here we recall some concepts and facts of this theory. One can find the more detailed information on currents in P.Lelong [1], Lelong-Gruman [1], E.Čirka [1], etc.

We denote by $\mathcal{E}_{p,q}(\Omega) = \mathcal{E}_{p,q}$, where Ω is a domain in $\mathbb{C}^n$, the set of exterior differential forms of type (p,q) with coefficients in $C_0^\infty(\Omega)$. Thus, if $\alpha \in \mathcal{E}_{p,q}(\Omega)$, then

$$\alpha = \sum_{I,J} \alpha_{I,J} dz_I \wedge d\bar{z}_J , \tag{7.1.1}$$

where $I = (i_1, \ldots, i_p)$, $1 \le i_1 < i_2 < \ldots < i_p \le n$, $J = (j_1, \ldots, j_q)$, $1 \le j_1 < j_2 < \ldots < j_q \le n$, $dz_I = dz_{i_1} \wedge \ldots \wedge dz_{i_p}$, $d\bar{z}_J = d\bar{z}_{j_1} \wedge \ldots \wedge d\bar{z}_{j_q}$ and $\alpha_{I,J} \in C_0^\infty(\Omega)$, $\forall I, J$.

Convergence of forms α in the space $\mathcal{E}_{p,q}(\Omega)$ is defined as convergence of the corresponding coefficients $\alpha_{I,J}$, $\forall I, J$, in the space $C_0^\infty(\Omega)$.

Linear continuous functionals on the space $\mathcal{E}_{p,q}(\Omega)$ are called currents of the type (or of degree) $(n-p, n-q)$[52]. We denote by $\mathcal{E}^*_{n-p,n-q}(\Omega) = \mathcal{E}^*_{n-p,n-q}$ the space of currents in Ω of type $(n-p,n-q)$. We can interpret each such current t as an exterior differential form with coefficients in the space of distributions. In other words, t can be represented in the form

$$t = \sum_{I,J} t_{\bar{I},\bar{J}} dz_{\bar{I}} \wedge d\bar{z}_{\bar{J}} , \tag{7.1.2}$$

where $\bar{I}$, $\bar{J}$ are the tuples of numbers from the set $\{1, 2, \ldots, n\}$ complementary to I, J, respectively, and $t_{\bar{I},\bar{J}} \in \mathcal{D}'(\Omega)$. We denote by $\langle t, \alpha \rangle$, or $\int t \wedge \alpha$, the result of applying the current t to the form $\alpha \in \mathcal{E}_{p,q}(\Omega)$, and define it by the equality

$$\langle t, \alpha \rangle = \left(\frac{2}{i} \right)^n \sum_{I,J} \varepsilon_{I,J} \langle t_{\bar{I},\bar{J}}, \alpha_{I,J} \rangle , \tag{7.1.3}$$

where the numbers $\varepsilon_{I,J}$ are equal to +1 or −1 and satisfy the condition

$$\varepsilon_{I,J} dz_{\bar{I}} \wedge d\bar{z}_{\bar{J}} \wedge dz_I \wedge d\bar{z}_J = dz_1 \wedge d\bar{z}_1 \wedge \ldots \wedge dz_n \wedge d\bar{z}_n .$$

The operators d, ∂, $\bar{\partial}$, d^c are defined in the space of currents as well as in the space of forms. Namely, we denote by ∂t, where $t \in \mathcal{E}^*_{n-p,n-q}$,

---

[52] The terminology on "type" and "degree" for currents is not yet standardized. In, e.g., Čirka [1], (n-p,n-q) is called the bedegree of the current, and "degree" is defined differently.

the current from $\mathcal{E}^*_{n-p+1,n-q}$ that acts on the form $\varphi\in\mathcal{E}_{p-1,q}$ by the rule

$$\langle\partial t,\varphi\rangle = (-1)^{p+q+1}\langle t,\partial\varphi\rangle.$$

If a current t is given in the form (7.1.2) then the following representation of ∂t holds:

$$\partial t = \sum_{I,J}\sum_{l=1}^{n}\frac{\partial t_{I,J}}{\partial z_l}\,dz_l\wedge dz_{\bar{I}}\wedge d\bar{z}_{\bar{J}}.$$

The operators $\bar{\partial}$ and $d = \partial+\bar{\partial}$ are defined similarly. As in the case of forms we define the operator d^c by the equality

$$d^c = \frac{i}{4}(\bar{\partial}-\partial).$$

When we study holomorphic mappings, positive currents are the ones most commonly used. A current $t\in\mathcal{E}^*_{n-p,n-p}$ is said to be positive if for any forms

$$\beta_j = \sum_{i=1}^{n} c_{j,i}dz_i\ ,\ c_{j,i}\in\mathbb{C},\ j = 1,\ \dots,\ p,$$

and any non-negative function $\varphi\in C_0^\infty(\Omega)$ the following inequality holds:

$$\langle t,\varphi\cdot(i\beta_1\wedge\bar{\beta}_1)\wedge\dots\wedge(i\beta_p\wedge\bar{\beta}_p)\rangle \ge 0.$$

The set of all positive currents $t\in\mathcal{E}^*_{n-p,n-p}$ is denoted by $\mathcal{E}^{*+}_{n-p,n-p} = \mathcal{E}^{*+}_{n-p,n-p}(\Omega)$. It is useful to introduce the factor $\left(\frac{i}{2}\right)^{n-p}$ in the representation (7.1.2) for such currents, i.e.

$$t = \left(\frac{i}{2}\right)^{n-p}\sum_{I,J} t_{\bar{I},\bar{J}}\,dz_{\bar{I}}\wedge d\bar{z}_{\bar{J}}. \qquad (7.1.4)$$

The coefficients $t_{\bar{I},\bar{J}}$ in this representation of a positive current t are positive measures when $\bar{I} = \bar{J}$ and are complex-valued measures when $\bar{I} \ne \bar{J}$. Therefore the integral

$$\int_E t\wedge\varphi$$

is correctly defined for any Borel set $E \subset\subset \Omega$ and for any form φ of type (p,p) with coefficients that are continuous on Ω. In particular, the quantity

$$\sigma(E) = \sigma(E;t) = \frac{1}{p!}\int_E t\wedge(dd^c|z|^2)^p = \sum_I t_{\bar{I},\bar{I}}(E) \tag{7.1.5}$$

is defined and positive. Thus, we associate each positive current t some non-negative measure σ. This measure is called the Kähler mass (or, simply, the mass) of the current t. Some authors refer to σ as the trace of t. We agree to define the quantity $\sigma(E,t)$ by (7.1.5), or, equivalently, the measure σ is defined also in case t is not a positive current but the integral in (7.1.5) makes sense. This occurs, in particular, if the coefficients $t_{\bar{I},\bar{I}}$ are locally summable functions.

Let t be a positive current of the form (7.1.4), φ an exterior differential form of type (p,p) with continuous coefficients $\varphi_{I,J}$, and E a Borel set as above. Then we put

$$\|t\|_E = \sup_{I,J} |t_{\bar{I},\bar{I}}|(E)$$

and

$$\|\varphi\|_E = \max_{I,J} \sup_{z\in E} |\varphi_{I,J}(z)|.$$

The following estimates for these quantities are often used:

$$\left.\begin{aligned} \left|\int t\wedge\varphi\right| &\le c\|\varphi\|_E\sigma(E;t), \\ \|t\|_E &\le c\sigma(E;t). \end{aligned}\right\} \tag{7.1.6}$$

Here c is a constant which depends on p,n only, $t\in\mathcal{E}^{*+}_{n-p,n-p}$, and φ is a form of type (p,p) with continuous coefficients.

When p = n-1, the second estimate is especially simple. In this case

$$|t_{j,i}| \le t_{j,j} + t_{i,i} \le \sigma$$

and, correspondingly, $\|t\|_E \le \sigma(E;t)$. Note also that a current $t\in\mathcal{E}^*_{1,1}(\Omega)$ is positive if and only if for any $\lambda\in\mathbb{C}^n$ the distribution

$$\sum_{j,i} t_{j,i}\lambda_j\bar{\lambda}_i$$

is positive.

The following theorem plays an important role in various problems of multidimensional complex analysis.

Theorem 7.1.1. Let t be a closed positive current of type (n-p,n-p) in a ball B_R. Then the function

$$\sigma(r) = \frac{\sigma(B_r;t)}{r^{2p}}$$

is monotonically increasing on the interval (0,R).

In particular, from this theorem it follows that if $t\in\mathcal{E}^{*+}_{n-p,n-p}(\Omega)$ is a closed current, then the limit

$$\nu_t(z) = \frac{p!}{\pi^p} \lim_{r\to 0} \frac{\sigma(B_r(z);t)}{r^{2p}}$$

exists at every point $z\in\Omega$. The value $\nu_t(z)$ is called the Lelong number of the current t at the point z.

We also note the following useful formula obtained in the process of proving **Theorem** 7.1.1; it represents the function $r^{-2p}\sigma(B_r;t)$ as the value of the current t on a special form:

$$\frac{\sigma(B_r;t)}{r^{2p}} = \frac{\pi^p}{p!}\nu_t(0) + \frac{1}{p!}\int_{0<|z|<r} t\wedge(dd^c ln|z|^2)^p. \tag{7.1.7}$$

The currents $t = dd^cu$, where $u\in PSH(\Omega)$, form an important class of positive currents.

Note that the product of currents, as well as the product of distributions, is not defined in the general case. But the product of currents of the above form, i.e. of currents $t = dd^cu$, can be defined for continuous and, moreover, for locally bounded functions $u\in PSH(\Omega)$. Besides, such a product is a closed positive current. This definition can be made by induction with respect to the number of factors. Namely, let $v_1\in PSH(\Omega)$, ..., $v_l\in PSH(\Omega)$ be locally bounded functions. We assume that the product $t_{l-1} = dd^cv_1\wedge\ldots\wedge dd^cv_{l-1}$ has been defined and is a closed positive current. The coefficients of the current t_{l-1} are measures and therefore the following equality, defining the current t_l, makes sense:

$$\langle t_l,\varphi\rangle = \langle v_lt_{l-1}, dd^c\varphi\rangle = \int_\Omega v_lt_{l-1}\wedge dd^c\varphi, \quad \forall\varphi\in\mathcal{E}_{n-l+1,n-l+1}(\Omega).$$

It can be shown that this current is closed and positive, and in the case $v_j\in C^\infty(\Omega)$ the current t_l coincides with the current defined by the form $dd^c v_1 \wedge \ldots \wedge dd^c v_1$.

The case $v_1 = v_2 = \ldots = v_n = u$, i.e. the case $t_n = (dd^c u)^n$, plays an essential role in applications. The operator that associates to a plurisubharmonic function $u(z)$ the current $(dd^c u)^n$ is usually called the complex Monge-Ampère operator.

Now we pass directly to the consideration of holomorphic mappings. Let $f: \Omega \to \mathbb{C}^n$ be a holomorphic mapping with Jacobian $\frac{\partial f}{\partial z} \not\equiv 0$. Let also a neighbourhood ω of a point $z^0 \in \Omega$ be such that the point $a = f(z^0)$ has the unique pre-image $f^{-1}(a) = z^0$ in this neighbourhood. If also

$$\left.\frac{\partial f}{\partial z}\right|_{z=z^0} \neq 0,$$

then the point z^0 is called a simple a-point and we say that the multiplicity $\gamma_f(z)$ of the mapping f at the point z^0 is equal to 1. In case

$$\left.\frac{\partial f}{\partial z}\right|_{z=z^0} = 0,$$

the quantity

$$\gamma_f(z^0) = \overline{\lim_{b \to a}} \#(f^{-1}(b) \cap \omega),$$

is called the multiplicity of the mapping f at z^0 (the multiplicity of the a-point z^0). Here $\#(f^{-1}(b) \cap \omega)$ is the number of pre-images of the point b that are contained in ω. Note that $\gamma_f(z^0)$ is independent of ω and, for almost all b close enough to a, every point of the set $f^{-1}(b) \cap \omega$ is simple and the equality

$$\#(f^{-1}(b) \cap \omega) = \gamma_f(z^0)$$

holds.

Besides, if $\Omega_1 \subset\subset \Omega$ and $|f(z)-b| \neq 0$, $\forall z \in \partial\Omega_1$, $|b-a| < r$, then the quantity

$$\sum_{z \in (f^{-1}(b) \cap \Omega_1)} \gamma_f(z),$$

which is the number of pre-images of the point b in Ω_1 counted with multiplicities, is independent of $b \in B_r(a)$ (**Rouché's Theorem** for mappings).

For our purposes it is sufficient to define the multiplicity of a mapping $f: \Omega \to \mathbb{C}^m$, $m < n$, on the set κ^* of regular points of the analytic set $\kappa = f^{-1}(a)$ only. And moreover, it is sufficient to do this in case κ is of pure dimension $\dim\kappa = n-m$.

Let $z^0 \in \kappa^*$, and let κ_1 be a complex manifold in $\mathbb{C}^n$ of dimension m that contains z^0 and is transversal to κ. Then the mapping $f_1 = f|_{\kappa_1}$ has a multiplicity[53] $\gamma_{f_1}(z)$ at the point z^0. This multiplicity is independent of the choice of the manifold κ_1. It is called the multiplicity of the mapping f at the point $z^0 \in f^{-1}(a)$. We denote by $\gamma_f(z^0)$ this multiplicity, just as in the equidimensional case. Note that $\gamma_f(z)$ is constant on the connected components of the set κ^*, and for $m = 1$ it coincides with the above-mentioned multiplicity (order) of vanishing of the holomorphic function $f(z)-a$.

Let κ be an analytic set in $\mathbb{C}^n$ of pure dimension n-m and κ^* the set of its regular points. A pair $Z = (\kappa, \gamma)$, where γ is an integer-valued function on the set κ which is constant on the connected components of κ, is called as an analytic chain with support κ. A chain is called positive if $\gamma(z) \geq 0$, $\forall z \in \kappa^*$.

Let $f: \mathbb{C}^n \to \mathbb{C}^m$, $m \leq n$, be a holomorphic mapping such that the set $f^{-1}(0) = \{z: f(z) = 0\}$ is of pure dimension n-m. Then we associate to f the analytic chain $Z_f = (f^{-1}(0), \gamma_f(z))$, where $\gamma_f(z)$ is the above-defined multiplicity. It is clear that for $m = 1$ this chain is none other than the divisor of the function $f(z)$.

We define the current t_Z of integration over an analytic chain $Z = (\kappa, \gamma)$ by the equality

$$\langle t_Z, \varphi\rangle = \int_{\kappa^*} \gamma(z)\varphi|_{\kappa^*}, \quad \varphi \in \mathcal{E}_{n-m,n-m},$$

where $\varphi|_{\kappa^*}$ is the restriction of the form φ to κ^*.

This current is closed and positive, and at each point $z \in \kappa^*$ its Lelong number coincides with the multiplicity $\gamma(z)$.

---

[53] It is clear that the multiplicity defined above in the equidimensional case is invariant under biholomorphic mappings. Thus the multiplicity is correctly defined in the case of holomorphic mappings of complex manifolds too.

The volume form (element) on κ^* is the restriction to κ^* of the form

$$\alpha_{n-m} = \frac{1}{(n-m)!}(dd^c|z|^2)^{n-m}.$$

Correspondingly we define the volume $V(Z,\Omega)$ of an analytic chain Z in a domain Ω by the equality

$$V(Z,\Omega) = \int_{\Omega\setminus\kappa^*} \gamma\cdot\alpha_{n-m}\Big|_{\kappa^*} = \sigma(\Omega;t_Z).$$

If a chain Z comes from mapping f, i.e. $Z = Z_f$, then the current of integration over Z_f, which is denoted by t_f for brevity, can be defined directly in terms of f. The corresponding equality is called the Poincaré-Lelong formula. It is contained in the following theorem.

Theorem 7.1.2. Let a holomorphic mapping $f\colon \Omega \to \mathbb{C}^m$, $m \le n$, be such that the analytic set $f^{-1}(0)$ is of pure dimension $n-m$. Then the coefficients of the differential form

$$\Lambda_f = \frac{1}{\pi^m}\, ln|f|^2(dd^c(ln|f|^2))^{m-1}$$

are locally summable and

$$t_f = dd^c\Lambda_f,$$

i.e.

$$\langle t_f,\varphi\rangle = \int_\Omega \Lambda_f\wedge dd^c\varphi,\quad \forall\varphi\in\mathcal{E}_{n-m,n-m}(\Omega).$$

As already noted, expressions of the form $(dd^cu)^m$ frequently occur in various branches of complex analysis. They also play an important role in the theory of value distribution of holomorphic mappings, in particular, in the definition of so-called order functions.

Let f be a non-degenerate holomorphic mapping $\mathbb{C}^n \to \mathbb{C}^m$, $m \le n$. The function

$$T_q(r;f) = \int_0^r \frac{1}{s}\,A_q(s;f)ds, \tag{7.1.8}$$

where

$$A_q(s;f) = \frac{1}{\pi^n}\int_{B_s} (dd^c ln(1+|z|^2))^q\wedge(dd^c ln|z|^2)^{n-q},$$

is said to be the q-order function of f. The behaviour of these functions is connected with the values distribution of the considering mappings (see Griffiths-King [1]).

In conclusion of this section we give a statement, to be used in the sequel, concerning the convergence of currents of the form $u(dd^c u)^m$. This statement, unlike the others presented here, will be given with proof, because it is not contained in the sources known to us.

Theorem 7.1.3. Let Ω be a domain in $\mathbb{C}^n$ and let a sequence of holomorphic mappings $f_j: \Omega \to \mathbb{C}^m$, $1 \le m \le n$, be such that f_j, $j \to \infty$, converge uniformly on each compact set $K \subset \Omega$ to a mapping f. Let also the support $|Z_f| = f^{-1}(0)$ of the chain $Z_f = (|Z_f|, \gamma_f(z))$, coming from the mapping f be of pure codimension m. Then the currents (forms) Λ_{f_j} converge to the current (form) Λ_f in the space $\mathcal{E}^*_{m-1,m-1}$.

Proof. We need some notation. Let $\mathfrak{M}_m$ be the set of all arrangements I of m elements taken from the set $\{1, \ldots, n\}$, and let $\bar{I}$ be the corresponding complement to I. Also, set $Z_I = (z_{i_1}, \ldots, z_{i_m})$, $U_I(r_I) = \left\{Z_I: |z_{i_1}| < r_{i_1}, \ldots, |z_{i_m}| < r_{i_m}\right\}$, $R^I = (R^I_{i_1}, \ldots, R^I_{i_m})$.

The following lemma is basic in the proof of **Theorem 7.1.3**.

Lemma 7.1.1. Let the mappings f, f_j and the domain Ω be the same as in Theorem 7.1.3, and let the polydisc[54] $U_r(z^0) \subset\subset \Omega$ satisfy the condition: $\forall I \in \mathfrak{M}_m$, $\exists R^I \in \mathbb{R}^m_+$,

$$\min\left\{|f(z)|: z_{\bar{I}} \in U_{r_{\bar{I}}}(z^0_{\bar{I}}),\ z_I \in \partial U_{R^I}(z^0_I)\right\} > 0.$$

Let also ω be an open set containing $|Z_f| \cap \bar{U}_r(z^0)$, and let $\omega' = \omega \cap U_r(z^0)$. Then

$$\exists \lim_{j \to \infty} \sigma(\omega'; \Lambda_{f_j}) = \sigma(\omega'; \Lambda_f). \tag{7.1.9}$$

Proof. First we consider the case $m = n$. If a set $D \subset\subset \Omega$ is such that $\bar{D} \cap |Z_f| = \varnothing$, then the coefficients of the forms Λ_{f_j} for all sufficiently large j are clearly continuous and converge uniformly on

---

[54] Recall that $U_r(z^0) = \{z \in \mathbb{C}^n: |z_1 - z^0_1| < r_1, \ldots, |z_n - z^0_n| < r_n\}$, $z^0 \in \mathbb{C}^n$, $r \in \mathbb{R}^n_+$.

D, as $j \to \infty$, to the coefficients of Λ_f. Therefore, to prove (7.1.9) it suffices to show that for any $\varepsilon > 0$ there is δ-neighbourhood ω_δ of the set $|Z_f| \cap \bar{U}_r(z^0)$ such that

$$|\sigma(\omega_\delta \cap U_r(z^0);\Lambda_f)| < \varepsilon$$

and for all sufficiently large j the inequalities

$$|\sigma(\omega_\delta \cap U_r(z^0);\Lambda_{f_j})| < \varepsilon$$

hold. Without loss of generality we may assume that the estimates $|f| \le 1$, $|f_j| \le 1$, $j = 1, 2, \ldots$, hold on ω_δ. Then the currents $-\Lambda_{f_j}$ and $-\Lambda_f$ are positive on ω_δ. Hence

$$|\sigma(\omega_\delta \cap U_r(z^0);\Lambda_{f_j})| \le |\sigma(\omega_\delta;\Lambda_{f_j})|$$

and

$$|\sigma(\omega_\delta \cap U_r(z^0);\Lambda_f)| \le |\sigma(\omega_\delta;\Lambda_f)|.$$

Thus it suffices to show the existence of a $\delta > 0$ such that $|\sigma(\omega_\delta;\Lambda_{f_j})| < \varepsilon$, $\forall j \le j_0(\varepsilon)$ and $|\sigma(\omega_\delta;\Lambda_f)| < \varepsilon$. In the present case we have $dim Z_f = 0$. Therefore the set $|Z_f| \cap \bar{U}_r(z^0)$ consists of a finite number of points, and ω_δ is the union of a finite number of non-intersecting balls for δ sufficiently small. Hence it suffices to obtain the above inequality for the case $\omega_\delta = B_\delta$, $|Z_f| \cap \bar{B}_j = \{0\}$. We choose $\delta > 0$ in such a way that $|\sigma(B_\delta;\Lambda_f)| < \varepsilon/2$. Such a choice is possible because the coefficients of the form Λ_f are locally summable functions (see **Theorem 7.1.2**). Now, let a function $\varphi \in C_0^\infty(B_\delta)$ satisfy the condition $\varphi(z) = 1$, $\forall z \in B_{\delta/2}$. Then we represent $\sigma(B_\delta;\Lambda_f)$ in the form

$$\sigma(B_\delta;\Lambda_{f_j}) = \int_{B_\delta} \Lambda_{f_j} \wedge dd^c(\varphi|z|^2) + \int_{B_\delta \setminus B_{\delta/2}} \Lambda_{f_j} \wedge dd^c((1-\varphi)|z|^2). \qquad (7.1.10)$$

Note that the coefficients of the forms Λ_{f_j} converge to uniformly on $B_\delta \setminus B_{\delta/2}$ the coefficients of Λ_f, as $j \to \infty$. Therefore

$$\exists \lim_{j\to\infty} \int_{B_\delta \setminus B_{\delta/2}} \Lambda_{f_j} \wedge dd^c((1-\varphi)|z|^2) = \int_{B_\delta \setminus B_{\delta/2}} \Lambda_f \wedge dd^c((1-\varphi)|z|^2). \quad (7.1.11)$$

At the same time, in accordance with Poincaré-Lelong formula, the first integral in (7.1.10) is the value of the current t_f (of the current of integration over the chain Z_f) on the function $\varphi|z|^2$, i.e.

$$\int_{B_\delta} \Lambda_{f_j} \wedge dd^c(\varphi|z|^2) = \sum_{z\in|Z_f|\cap B_\delta} \gamma_{f_j}(z)\varphi(z)|z|^2. \quad (7.1.12)$$

Correspondingly,

$$\int_{B_\delta} \Lambda_f \wedge dd^c(\varphi|z|^2) = \sum_{z\in|Z_f|\cap B_\delta} \gamma_f(z)\varphi(z)|z|^2. \quad (7.1.13)$$

From the analogue of **Rouché's Theorem** for holomorphic mappings it follows that in this situation the number of points of the set $f^{-1}(0) \cap B_\delta$ counted with multiplicities is independent of j, and these points converge to the point $z = 0$ as $j \to \infty$. Therefore (7.1.12) implies that

$$\lim_{j\to\infty} \int_{B_\delta} \Lambda_{f_j} \wedge dd^c(\varphi|z|^2) = 0.$$

From this and (7.1.10), (7.1.11), (7.1.13) we conclude that

$$\lim_{j\to\infty} \sigma(B_\delta;\Lambda_{f_j}) = \sigma(B_\delta;\Lambda_f).$$

Hence, for the above choice of δ the inequality $|\sigma(B_\delta;\Lambda_{f_j})| < \varepsilon$ is valid for all sufficiently large j. Thus we have proved the lemma in the case $m = n$.

Let now $1 \le m < n$. Elementary calculations show that the equality

$$\Lambda_f \wedge (dd^c|z|^2)^{n-m+1} = \sum_{I\in\mathfrak{M}_m} \Lambda^I_{f,z_{\bar I}} \wedge dd^c|z_I|^2 \wedge (dd^c|z_{\bar I}|^2)^{n-m}, \quad (7.1.14)$$

where $\Lambda^I_{f,z_{\bar I}}$ is the restriction of the form Λ_f to the space $\mathbb{C}^m_{(z_I)}$ for fixed $z_{\bar I}$, is valid. Further we note that

$$\int_{\omega'} \Lambda^I_{f_j,z_{\bar I}} \wedge dd^c|z_I|^2 \wedge (dd^c|z_{\bar I}|^2)^{n-m} =$$

$$= \int_{U_{r_I}(z^0_I)} (dd^c|z_{\bar I}|^2)^{n-m} \int_{\omega'_{z_{\bar I}}} \Lambda^I_{f_j,z_{\bar I}} \wedge dd^c|z_I|^2 =$$

$$= \int_{U_{r_{\bar I}}(z^0_{\bar I})} \sigma(\omega'_{z_{\bar I}}; \Lambda^I_{f_j,z_{\bar I}})(dd^c|z_{\bar I}|^2)^{n-m}, \tag{7.1.15}$$

where $\omega'_{z_{\bar I}} = \{z_I\colon z\in\omega',\ z_{\bar I}\ \text{fixed}\}$. It is clear that if $z_{\bar I}\in U_{r_{\bar I}}(z^0_{\bar I})$ is fixed, then we are in the case n = q, which has already been considered for the variables $z_{i_1}, \ldots, z_{i_q}$. Therefore,

$$\lim_{j\to\infty} \sigma(\omega'_{z_{\bar I}}; \Lambda^I_{f_j,z_{\bar I}}) = \sigma(\omega'_{z_{\bar I}}; \Lambda^I_{f,z_{\bar I}}) . \tag{7.1.16}$$

We us show that

$$sup\{|\sigma(\omega'_{z_{\bar I}}; \Lambda^I_{f_j,z_I})|\colon z_{\bar I}\in U_{r_{\bar I}}(z^0_{\bar I}),\ j = 1,\ 2.\ \ldots\} < \infty. \tag{7.1.17}$$

As before, without loss of generality we may assume that $|f_j| < 1$, $|f| < 1$ on $U_{r_{\bar I}}(z^0_{\bar I})\times U_{R^I}(z^0_I)$. Then the currents under consideration are negative. Therefore, to prove inequality (7.1.17) it suffices to show that

$$sup\left\{|\sigma(U_{R^I}(z^0_I); \Lambda^I_{f_j,z_{\bar I}})|\colon z_{\bar I}\in U_{r_{\bar I}}(z^0_{\bar I}),\ j = 1,\ 2.\ \ldots\right\} < \infty. \tag{7.1.18}$$

Let $\tilde R^I = (\tilde R^I_{i_1},\ldots,\tilde R^I_{i_m})$ be such that $r_{i_k} < \tilde R^I_{i_k} < R^I_{i_k}$, k = 1, ..., m, and

$$|Z_f| \cap \left\{\bar U_{r_{\bar I}}(z^0_{\bar I})\times(\bar U_{R^I}(z^0_I)\setminus U_{\tilde R^I}(z^0_I)\right\} = \varnothing.$$

As in the case n = m we allow a corresponding function φ in the expression $dd^c|z|^2$. Namely, let $\varphi(z_I)\in C_0^\infty(U_R^I(z_I^0))$ satisfy the condition $\varphi(z_I) = 1$, $\forall z_I\in U_{\tilde{R}^I}(z_I^0)$. Then

$$\sigma(U_{R^I}(z_I^0);\Lambda^I_{f_j,z_{\bar{I}}}) =$$

$$= \int_{U_{R^I}(z_I^0)} \Lambda^I_{f_j,z_{\bar{I}}}\wedge dd^c(\varphi|z_I|^2) + \int_{U_{R^I}(z_I^0)\setminus U_{\tilde{R}^I}(z_I^0)} \Lambda^I_{f_j,z_{\bar{I}}}\wedge dd^c((1-\varphi)|z_I|^2).$$

The first integral at the right-hand side of this equality is equal to the sum of the values of the function $\varphi(z_I)|z_I|^2$ at the points of the set $|Z_{f,z_{\bar{I}}}|\cap U_{R^I}(z_I^0)$, where $|Z_{f,z_{\bar{I}}}| = \{z_I\colon\ z\in Z_f,\ z_{\bar{I}}\ \text{fixed}\}$. Here, summation is done counted with multiplicities of the points of the chain $Z_{f,z_{\bar{I}}}$ which comes from by the restriction of the mapping f to the m-dimensional analytic plane $\{z\in\mathbb{C}^n\colon\ z_{\bar{I}}\ \text{fixed}\}$. As before, using the multidimensional generalization of **Rouché's Theorem**, we conclude that for all sufficiently large $j > j_0$ the number of points of the set $|Z_{f,z_{\bar{I}}}|\cap U_{R^I}(z_I^0)$ counted with multiplicities, is independent of j and $z_{\bar{I}}\in U_{r_{\bar{I}}}(z_I^0)$. Therefore,

$$sup\left\{\left|\int_{U_{R^I}(z_I^0)} \Lambda^I_{f_j,z_{\bar{I}}}\wedge dd^c(\varphi|z_I|^2)\right|\colon\ j\ge j_0,\ z_{\bar{I}}\in U_{r_{\bar{I}}}(z^0_{\bar{I}})\right\} < \infty. \qquad (7.1.19)$$

At the same time, since

$$|Z_f|\cap\left\{supp((1-\varphi)|z_I|^2)\times\bar{U}_{r_{\bar{I}}}(z^0_{\bar{I}})\right\} = \emptyset,$$

the coefficients of the form $\Lambda^I_{f_j,z_{\bar{I}}}$ in the second integral are uniformly bounded; consequently,

$$sup\left\{\left|\int_{U_{R^I}(z_I^0)} \Lambda^I_{f_j,z_{\bar{I}}} \wedge dd^c((1-\varphi)|z_I|^2)\right| : j \geq j_0,\ z_{\bar{I}} \in U_{r_{\bar{I}}}(z^0_{\bar{I}})\right\} < \infty.$$

From this and (7.1.18), (7.1.19) we conclude that the inequality (7.1.17) is true and therefore (7.1.16) is true too. Then, taking into account (7.1.13), (7.1.14) and (7.1.15), we obtain

$$\lim_{j\to\infty} \sigma(\omega';\Lambda_{f_j}) = \sigma(w';\Lambda_f) .$$

The proof of the lemma is finished.

Now we continue directly with the proof of the theorem. It is clearly sufficient to show the convergence of the currents Λ_{f_j} in a sufficiently small neighbourhood of an arbitrary point $z^0 \in |Z_f|$. Since $codim|Z_f| = m$ it is possible to substitute $z-z^0 = T\zeta$, where T is unitary operator, such that $\forall I \in \mathfrak{M}_m$ the intersection $B_\varepsilon \cap |Z_{f,\zeta_{\bar{I}}}|$ for $\zeta_{\bar{I}} = 0$ consists of the single point $\zeta_I = 0$ for some sufficiently small $\varepsilon > 0$. For simplicity of notation we assume that $z^0 = 0$ and $T = Id$, is the identity operator. Then for some $S^{\bar{I}}$ and R^I the condition

$$min\left\{|f(z)|: z_{\bar{I}} \in U_{r_{\bar{I}}},\ z_I \in \partial U_{R^I}\right\} > 0$$

is satisfied. We set $r_i = min\{S_i^{\bar{I}}: \bar{I} \in \mathfrak{M}_{n-m}\}$, $r = (r_1,\dots,r_n)$ and show that the statement of the theorem is true for the domain $\Omega = U_r$. For this it is clearly suffices to prove the following statement. If $\varphi \in \mathcal{E}_{n-m+1,n-m+1}$, then for any $\varepsilon > 0$ there is an open set ω in $\mathbb{C}^n$ such that $\omega \supset (|Z_f| \cap \bar{U}_r)$ and for all sufficient large j the inequality

$$\left|\int_{\omega \cap U_r} \Lambda_{f_j} \wedge \varphi\right| < \varepsilon$$

holds.

Let ω be an open set such that $(|Z_f| \cap \bar{U}_r) \subset \omega$, $|f(z)| < \frac{1}{2}$, $\forall z \in \omega$, and $|\sigma(\omega;\Lambda_f)| < \frac{\varepsilon}{2}$. Note that **Lemma 7.1.1** can be applied, because of the choice of the polydisc U_r. Therefore for all sufficiently large j >

j_0, the inequalities $|\sigma(\omega \cap U_r; \Lambda_{f_j})| < \varepsilon$ hold. Then, using the estimate (7.1.7), we conclude that

$$\left| \int_{\omega \cap U_r} \Lambda_{f_j} \wedge \varphi \right| \le c\varepsilon \|\varphi\|_{\mathbb{C}^n}, \quad \forall j \ge j_0.$$

Thus we have proved that the theorem is true in the case $\Omega = U_r$, and hence it is true for an arbitrary domain Ω.

We also note the following statement, which follows from the theorem just proved.

Corollary. Let f and f_j, $j = 1, 2, \ldots$, be as in **Theorem 7.1.3**. Then for any domain $\Omega' \subset\subset \Omega$ the equality

$$\lim_{j \to \infty} \sigma(\Omega'; \Lambda_{f_j}) = \sigma(\Omega'; \Lambda_f)$$

holds.

§2 Plurisubharmonic functions of α-regular growth and asymptotic behaviour of order functions of holomorphic mappings

A function $u \in PSH(\mathbb{C}^n, \rho]$ is said to be a function of α-regular growth if there is a set $E \subset \mathbb{C}^n$ such that

1) E can be covered by balls $B^{(j)} = B_{r_j}(a_j)$ in such a way that

$$\lim_{R \to \infty} \frac{1}{R^\alpha} \sum_{j: B^{(j)} \cap B_R \ne \emptyset} r_j^\alpha = 0 ;$$

2) for any $\varepsilon > 0$ there exists an $r' = r'(\varepsilon) > 0$ such that

$$|u(z) - \mathcal{L}_u^*(z)| < \varepsilon |z|^\rho, \quad \forall z \notin (E \cup B_{r'}).$$

It follows from the statement in §2 of **Chapter 2** concerning the convergence with respect to Carleson's α-measure that this definition is equivalent to the following one.

A function $u \in PSH(\mathbb{C}^n, \rho]$ is said to be a function of α-regular growth if the functions[55] $u^{[r]}(z)$ converge with respect to Carleson's

---

[55] Recall that $u^{[r]}(z) = r^{-\rho} u(rz)$.

α-measure to the function $\mathcal{L}_u^*(z)$ on every compact set in $\mathbb{C}^n$ as $r \to \infty$.

It is clear that the functions of c.r.g. considered in the previous chapters are functions of α-regular growth[56) (with $\alpha \in (2n-2,2n]$ for entire functions in $\mathbb{C}^n$ and with $\alpha \in (n-2,n]$ for functions belonging to $SH(\mathbb{R}^n,\rho]$).

A holomorphic mapping $f: \mathbb{C}^n \to \mathbb{C}^m$, $m \le n$, is said to be a mapping of at most normal type with respect to the order $\rho > 0$ if the plurisubharmonic function $ln|f(z)| \in PSH(\mathbb{C}^n,\rho]$. As in the case of entire functions, the indicator $\mathcal{L}^*_{ln|f|}(z)$ of the function $ln|f(z)|$ is called the (regularized) indicator of f. For brevity we denote it by $\mathcal{L}_f^*(z)$.

Let $f: \mathbb{C}^n \to \mathbb{C}^m$, $m \le n$, be a holomorphic mapping of at most normal type with respect to the order ρ. Then it is said to be of α-regular growth if the function $ln|f(z)|$ is of α-regular growth.

As already mentioned, the order functions $T_q(r;f)$ (see (7.1.8)) play an important role in the study of the value distribution of holomorphic mappings. The following theorems state some properties of such functions for α-regular mappings. Namely, the asymptotic behaviour of the order functions and their " logarithmic " derivatives (i.e. the functions $A_q(r;f)$) are established.

Theorem 7.2.1. Let $\{u_r\}_{r>1}$ be a uniformly bounded family of continuous plurisubharmonic functions in a domain $\Omega \subset \mathbb{C}^n$. Let also the $u_r(z)$ converge with respect to Carleson's α-measure to the function $u(z)$ on every compact subset of Ω as $r \to \infty$.

Then, as $r \to \infty$, the currents $(dd^c u_r)^q$ converge to the current $(dd^c u)^q$ in the space $\mathcal{E}^*_{q,q}$ for $q \le n+1-\alpha/2$.

Proof. We prove this by the induction with respect to the number q. Note that if the functions $u_r(z)$ converge (as $r \to \infty$) with respect to Carleson's α-measure to the function $u(z)$, then they converge to $u(z)$ with respect to any α_1-measure with $\alpha_1 \ge \alpha$ (this follows immediately from the definition of Carleson's α-measure). Applying **Lemma 2.1.8**, we find that the functions $u_r(z)$ converge to $u(z)$ as elements of the space

---

[56) In accordance with **Theorems 2.2.1** and **2.2.2**, convergence of the functions $u^{[r]}(z)$ with respect to Carleson's α-measure for some $\alpha \in (n-2,n]$ implies the corresponding convergence for every $\alpha \in (n-2,n]$.

$\mathcal{D}'(\Omega)$. Besides, $u\in PSH(\Omega)$. Convergence in $\mathcal{D}'$ of the functions u_r implies convergence of the derivatives, and hence convergence in $\mathcal{E}^*_{1,1}$ of the currents $dd^c u_r$ to the current $dd^c u$. Thus we have proved the the theorem in the case $q = 1$. Therefore we have the base of the induction. To prove the possibility of passing from $q-1$ to q, i.e. the induction step, we need the following lemma.

Lemma 7.2.1. Let $u_r(z)$ and $u(z)$ be as in **Theorem 7.2.1**. Let also $\{t_s\}_{s>1}$ be a family of closed positive currents of type (q,q), $q \le n-\alpha/2$, satisfying the condition

$$\sup_S \sigma(\Omega_1; t_s) = c_1(\Omega_1) < \infty, \ \forall \Omega_1 \subset\subset \Omega.$$

Then, as $r \to \infty$, the quantity[57] $\langle (u_r-u)t_s, \varphi\rangle$ tends to zero uniformly with respect to $s\in(1,\infty)$ and $\varphi\in\{\psi\in\mathcal{E}_{n-q,n-q}(\Omega'): \|\psi\|_{\Omega'} < 1\}$, $\Omega' \subset\subset \Omega$.

Proof. We set

$$M = \sup\{|u_r(z)|: r > 1, z\in\Omega\}$$

and

$$E_{r,\varepsilon} = \{z\in\Omega: |u_r(z)-u(z)| > \varepsilon\}.$$

It is clearly sufficient to prove the lemma in the case when Ω' is an arbitrary ball such that $\bar{\Omega}' \subset \Omega$. For simplicity of exposition we assume that $\Omega' = B_1$.

Let $\varphi\in\mathcal{E}_{n-q,n-q}(B_1)$. Then we have

$$\langle (u_r-u)t_s, \varphi\rangle = \int_{B_1} (u_r-u)t_s\wedge\varphi =$$

$$= \int_{B_1\setminus E_{r,\varepsilon}} (u_r-u)t_s\wedge\varphi + \int_{B_1\cap E_{r,\varepsilon}} (u_r-u)t_s\wedge\varphi. \qquad (7.2.1)$$

We estimate separately both terms at the right-hand side of this equality. Using inequality (7.1.6) we obtain

$$\left| \int_{B_1\setminus E_{r,\varepsilon}} (u_r-u)t_s\wedge\varphi \right| \le c\|\varphi\|_{B_1}\sigma(B_1\setminus E_{r,\varepsilon}; (u_r-u)t_s) \le$$

---

[57] This quantity makes sense because the coefficients of the current t_s are measures.

$$\le c\varepsilon\|\varphi\|_{B_1}\sigma(B_1\setminus E_{r,\varepsilon};t_s) \le c\varepsilon\|\varphi\|_{B_1}\sigma(B_1;t_s) \le c\varepsilon\|\varphi\|_{B_1}c_1(B_1). \quad (7.2.2)$$

To estimate the second term we note that according to the lemma, $mes_c^{\alpha}(B_1\cap E_{r,\varepsilon}) \to 0$ as $r \to \infty$. This means that for all $r \ge 1$ the set $B_1\cap E_{r,\varepsilon}$ can be covered by balls $B^{(r,j)}$, $j = 1, 2, \ldots$, with $R_{j,r}$ satisfying the condition

$$\lim_{r\to\infty}\sum_{j=1}^{\infty}(R_{j,r})^{\alpha} = 0. \quad (7.2.3)$$

Besides, without loss of generality we may assume that for some $\delta > 0$ and all r and j the inclusions $B^{(r,j)} \subset B_j \subset\subset \Omega$ hold.

Estimating the mass of the current in the balls $B^{(r,j)}$ with the aid of **Theorem 7.1.1** we obtain

$$\left|\int_{B_1\cap E_{r,\varepsilon}}(u_r-u)t_s\wedge\varphi\right| \le c\|\varphi\|_{B_1}\sigma(B_1\cap E_{r,\varepsilon};t_s) \le$$

$$\le c\|\varphi\|_{B_1}\sum_{j=1}^{\infty}\sigma(B^{(r,j)};t_s) \le c\|\varphi\|_{B_1}\sum_{j=1}^{\infty}\frac{\sigma(B_\delta;t_s)}{(\delta-1)^{2n-2q}}(R_{j,r})^{2n-2q} \le$$

$$\le c\|\varphi\|_{B_1}c_1(B_\delta)(\delta-1)^{2q-2n}\sum_{j=1}^{\infty}(R_{j,r})^{2n-2q}.$$

From this and (7.2.3) it follows that if $q \le n - \frac{\alpha}{2}$ (i.e. if $2n-2q \ge \alpha$), then

$$\int_{B_1\cap E_{r,\varepsilon}}(u_r-u)t_s\wedge\varphi \to 0, \quad (7.2.4)$$

as $r \to \infty$, uniformly with respect $s > 1$ and $\varphi\in\{\psi\in\mathcal{E}_{n-q,n-q}(B_1): \|\psi\|_{B_1} \le 1\}$. Taking into account that ε is arbitrary and comparing (7.2.4), (7.2.2) and (7.2.1), we find that $\langle(u_r-u),t_s\rangle \to 0$ (with required uniformity) as $r \to \infty$.

The proof of the lemma is finished.

Lemma 7.2.2. Let $u_r(z)$ and $u(z)$ be as in **Theorem 7.2.1**, and let $\{t_s\}_{s>1}$ be a family of closed positive currents of type (q,q), $q \le$

$n - \frac{\alpha}{2}$, that converges to the current t in $\mathcal{E}^*_{q,q}$ as $s \to \infty$. Then

$$\lim_{s \to \infty} \langle u(t_s - t), \varphi \rangle = 0, \ \forall \varphi \in \mathcal{E}_{n-q,n-q}(\Omega).$$

Proof. We us use the equality

$$\langle u(t_s - t), \varphi \rangle = \langle (u - u_r)(t_s - t), \varphi \rangle + \langle u_r(t_s - t), \varphi \rangle. \qquad (7.2.5)$$

We estimate the first term at the right-hand side. Since the currents t_s converge to the current t, for any domain $\Omega_1 \subset\subset \Omega$ there is a number $s_0 = s_0(\Omega_1)$ such that

$$\sup\{\sigma(\Omega_1; t_s) : s \geq s_0\} \leq 2\sigma(\Omega_1; t).$$

Thus we can apply **Lemma 7.1.1**, and hence the quantity $\langle (u_r - u)t_s, \varphi \rangle$ converges to zero uniformly with respect to $s \geq s_0$ as $r \to \infty$. Similarly,

$$\lim_{r \to \infty} \langle (u_r - u)t, \varphi \rangle = 0$$

and, therefore,

$$\lim_{r \to \infty} \langle (u_r - u)(t_s - t), \varphi \rangle = 0$$

uniformly with respect to $s \geq s_0$.

To estimate the second term at the right-hand side of equality (7.2.5), we note that since the functions u_r are continuous and the currents t_s converge, then for every fixed r the equality

$$\lim_{s \to \infty} \langle u_r(t_s - t), \varphi \rangle = 0$$

holds.

Summing up what has been said above, we conclude that for every $\varepsilon > 0$ we can first choose r such that

$$|\langle (u_r - u)(t_s - t), \varphi \rangle| < \frac{\varepsilon}{2}$$

for all $s \geq s_0$, and then chose $s_1 \geq s_0$ in such a way that

$$|\langle u_r(t_s - t), \varphi \rangle| < \frac{\varepsilon}{2}$$

for $s \geq s_1$ and our fixed r. Hence we obtain

$$|\langle u(t_s - t), \varphi \rangle| \leq \frac{\varepsilon}{2} + \frac{\varepsilon}{2} = \varepsilon, \ \forall s \geq s_1.$$

Thus

$$\lim_{s\to\infty} \langle u(t_s - t), \varphi\rangle = 0.$$

This finishes the proof of lemma.

Now we assume that for some q, $1 \le q \le n+1-\alpha/2$, the currents $(dd^c u_r)^{q-1}$ converge to the current $(dd^c u)^{q-1}$ as $r \to \infty$. We show that in this situation the currents $(dd^c u_r)^q$ converge to $(dd^c u)^q$. From the definition of $(dd^c v)^q$ we have

$$\langle (dd^c v)^n, \varphi\rangle = \langle v(dd^c v)^{n-1}, dd^c \varphi\rangle.$$

Thus, to prove the relationship

$$\lim_{r\to\infty} (dd^c u_r)^q = (dd^c u)^q$$

it suffices to show that

$$\lim_{r\to\infty} u_r (dd^c u_r)^{q-1} = u(dd^c u)^{q-1}.$$

We set $t_r = (dd^c u_r)^{q-1}$ and $t = (dd^c u)^{q-1}$. By the assumption, $t_r \to t$ as $r \to \infty$. Therefore,

$$sup\{|\langle t_r, \varphi\rangle|:\ r \ge 1,\ \varphi \in \mathcal{E}_{n-q+1,n-q+1}(\Omega'),\ \|\varphi\|_{\Omega'} < 1\} < \infty,\ \forall \Omega' \subset\subset \Omega.$$

Since $q-1 < n - \frac{\alpha}{2}$, then we can use **Lemmas 7.2.1** and **7.2.2** to obtain

$$\lim_{r\to\infty} |\langle (u_r - u)t_r, \varphi\rangle| = 0,$$

$$\lim_{r\to\infty} |\langle u(t_r - t), \varphi\rangle| = 0.$$

It follows from these equalities that

$$\lim_{r\to\infty} |\langle u_r t_r - ut, \varphi\rangle| = 0,\ \forall \varphi \in \mathcal{E}_{n-q+1,n-q+1}(\Omega),$$

i.e.

$$\lim_{r\to\infty} u_r (dd^c u_r)^{q-1} = u(dd^c u)^{q-1}.$$

As mentioned above, this implies the equality sought for:

$$\lim_{r\to\infty} (dd^c u_r)^q = (dd^c u)^q .$$

The proof of the theorem is finished.

Remark. At the same time we have proved that if the conditions of **Theorem** 7.2.1 are satisfied, then the relationship

$$\lim_{r \to \infty} u_r(dd^c u_r)^{q-1} = u(dd^c u)^{q-1}$$

holds.

The second definition (which is is equivalent to the first one) of functions of α-regular growth is based on the concept of convergence with respect to Carleson's α-measure. We use this definition and apply **Theorem** 7.2.1 to the functions $u^{[r]}(z)$ which come from a function $u(z)$ that is of α-regular growth. Then we obtain the following theorem.

Theorem 7.2.2. Let $u(z) \in PSH(\mathbb{C}^n \rho] \cap C(\mathbb{C}^n)$ be a function of α-regular growth. Then for any $q \le n+1-\frac{\alpha}{2}$ the currents $(dd^c u^{[r]})^q$ converge (in the topology of the space $\mathcal{E}^*_{q,q}(\mathbb{C}^n)$) to the current $(dd^c \mathcal{L}^*_u)^q$ as $r \to \infty$.

Note that $\mathcal{L}^*_u(z)$, the current $(dd^c \mathcal{L}^*_u)^q$ and its mass are positively homogeneous (in a natural sense). Therefore $\sigma(S_r; (dd^c \mathcal{L}^*_u)^q) = 0$. From this using **Lemma** 2.1.1 we obtain the following corollary to **Theorem** 7.2.2.

Corollary. Let $u(z) \in PSH(\mathbb{C}^n \rho] \cap C(\mathbb{C}^n)$ be a function of α-regular growth. Let K be an open cone in $\mathbb{C}^n$ with the vertex at the origin which that for some $q \le n+1-\frac{\alpha}{2}$ satisfies the condition

$$\sigma(\partial K; (dd^c \mathcal{L}^*_u)^q) = 0.$$

Then

$$\exists \lim_{r \to \infty} \frac{1}{r^{\rho q+2n-2q}} \sigma(K \cap B_r; (dd^c u)^q) = \sigma(K \cap B_1; (dd^c \mathcal{L}^*_u)^q).$$

Note that this **Corollary** is similar to the statement on the existence of the cone density of a measure that is Riesz associated with a subharmonic function of c.r.g.

Using **Theorem** 7.2.2, and more exactly, using the **Corollary** above, we can easily obtain the asymptotic behaviour of the order functions of α-regular mappings.

Theorem 7.2.3. Let $f: \mathbb{C}^n \to \mathbb{C}^m$, $m \le n$, be a holomorphic mapping of α-regular growth (with respect to the order $\rho > 0$). Then for $q \le n+1-\frac{\alpha}{2}$ the order function $T_q(r;f)$ of this mapping satisfies the condition

$$\lim_{r\to\infty} \frac{1}{r^{\rho q}} T_q(r;f) = 2^q \int_0^1 \frac{ds}{s} \int_{B_s} (dd^c((\pounds_f^*)^+))^q \wedge (dd^c \ln|z|^2)^{n-q},$$

and the function[58] $A_q(r;f)$ satisfies the condition

$$\lim_{r\to\infty} \frac{1}{r^{\rho q}} A_q(r;f) = 2^q \int_{B_1} (dd^c((\pounds_f^*)^+))^q \wedge (dd^c \ln|z|^2)^{n-q}.$$

Proof. It is obvious that α-regularity of growth of the mapping f or, equivalently, of the function $\ln|f(z)|$, implies α-regularity of growth of the function $\ln(1+|f(z)|^2)$. The regularized indicator of the last function is equal to $2(\pounds_f^*)^+$. Applying the above **Corollary** to **Theorem** 7.2.2 (with $K = \mathbb{C}^n$) to the function $\ln(1+|f|^2)$, we find that

$$\lim_{r\to\infty} \frac{1}{r^{\rho q+2n-2q}} \sigma(B_r;(dd^c \ln(1+|z|^2))^q) =$$

$$= 2^q \sigma(B_1;(dd^c((\pounds_f^*)^+))^q). \tag{7.2.6}$$

It follows from (7.1.7) that

$$A_q(r;f) = \text{const} + \frac{(n-q)!}{\pi^{2n-q}} \frac{1}{r^{2n-2q}} \sigma(B_r;(dd^c \ln(1+|z|^2))^q).$$

This and (7.2.6) imply that

$$\exists \lim_{r\to\infty} \frac{1}{r^{\rho q}} A_q(r;f) = 2^q \frac{(n-q)!}{\pi^{2n-q}} \sigma(B_1;(dd^c((\pounds_f^*)^+))^q).$$

Since the function $(\pounds_f^*)^+$ is positively homogeneous the mass of the current $(dd^c((\pounds_f^*)^+))^q$ is also positively homogeneous and it is easy to see that its Lelong number vanishes identically. Therefore (7.1.7) implies

$$\frac{(n-q)!}{\pi^{2n-q}} \sigma(B_1;(dd^c((\pounds_f^*)^+))^q) = \frac{1}{\pi^n} \int_{B_1} (dd^c(\pounds_f^*)^+)^q \wedge (dd^c \ln|z|^2)^{n-q}.$$

Hence

---

[58] Recall that $A_q(r;f) = \frac{1}{\pi^n} \int_{B_r} (dd^c \ln(1+|f|^2))^q \wedge (dd^c \ln|z|^2)^{n-q}$.

$$\lim_{r\to\infty} \frac{1}{r^{\rho q}} A_q(r;f) = \frac{2^q}{\pi^n} \int_{B_1} (dd^c(\pounds_f^*)^+)^q \wedge (dd^c \ln|z|^2)^{n-q}. \qquad (7.2.7)$$

Thus we have obtained the required asymptotics of the function $A_q(r;f)$.

Now, to prove a similar statement for the function $T_q(r;f)$ it suffices to note that

$$T_q(r;f) = \int_0^r \frac{A_q(s;f)}{s} ds = \int_0^1 \frac{A_q(rs;f)}{s} ds,$$

and to use (7.2.7). As a result we obtain

$$\lim_{r\to\infty} \frac{1}{r^{\rho q}} T_q(r;f) = \frac{2^q}{\pi^n} \int_0^1 \frac{ds}{s} \int_{B_s} (dd^c((\pounds_f^*)^+))^q \wedge (dd^c \ln|z|^2)^{n-q}.$$

This finishes the proof of the theorem.

§3 Jessen's Theorem for almost periodic holomorphic mappings

Recall that a continuous function $f(x)$ on the real axis is said to be almost periodic[59] (a.p.) if for any $\varepsilon > 0$ and some $l = l(\varepsilon) > 0$ each interval $(a,a+l)$, $a\in\mathbb{R}$, contains a number τ such that $sup\{|f(x+\tau)-f(x)|: -\infty < x < \infty\} < \varepsilon$.

Proceeding from this definition a function $f(z)$, $z\in\mathbb{C}$, which is holomorphic in the strip $T_{(a,b)} = \{z = x+iy\in\mathbb{C}: -\infty < x < \infty,\ a < y < b\}$, is called almost periodic (in $T_{(a,b)}$) if for any $\varepsilon > 0$ and any strip $T_{(a_1,b_1)}$, $a < a_1 < b_1 < b$, there exists $l = l(\varepsilon,a_1,b_1) > 0$ such that for any $c\in\mathbb{R}$ and some $\tau\in(c,c+l)$ the inequality

$$sup\{|f(z+\tau)-f(z)|: z\in T_{(a_1,b_1)}\} < \varepsilon$$

holds.

The following definitions are equivalent to the ones above (see, for

---

[59] A detailed account of the theory of almost periodic functions can be found in B.Levitan [1].

example, B.Levitan [1]).

A function $f \in C(\mathbb{R})$ (respectively, $f \in H(T_{(a,b)})$) is called almost periodic if for every sequence $h_j \in \mathbb{R}$, $j = 1, 2, \ldots$, we can choose a subsequence h_{j_k}, $k = 1, 2, \ldots$, such that the sequence of functions $f(x+h_{j_k})$ (respectively, $f(z+h_{j_k})$) converges uniformly on the whole axis (respectively, converge with respect to any norm $sup\{|:| -\infty < x < \infty, a_1 < y < b_1\}$, $a < a_1 < b_1 < b$) as $k \to \infty$.

The following well-known theorem describes the distribution of the zeros of almost periodic holomorphic functions.

Jessen's Theorem. Let $f(z)$, $z \in \mathbb{C}$, be an a.p. holomorphic function in a strip $T_{(a,b)}$. Then

1) the following limit exists

$$\lim_{r \to \infty} \frac{1}{2r} \int_{-r}^{r} \ln|f(x+iy)|dx \overset{def}{=} A_f(y), \ \forall y \in (a,b);$$

2) the function $A_f(y)$ (the Jessen function) is convex;

3) if the derivatives of the function $A_f(y)$ at the points a_1 and b_1, $a < a_1 < b_1 < b$, exist, then

$$\lim_{r \to \infty} \frac{1}{2r} V_f(r;a_1,b_1) = \frac{1}{2\pi} (A_f(b_1)-A_f(a_1)),$$

where $V_f(r;a_1,b_1)$ is the number of zeros of the function $f(z)$ in the domain $\{z \in \mathbb{C}: |x| < r, a_1 < y < b_1\}$ counted with multiplicities.

The idea of using weak convergence (as in this book) for the study of functions of c.r.g., turned out to be fruitful in the study of a.p.f. too. Namely, it can be used for extending **Jessen's Theorem** to the multidimensional case.

We give the necessary definitions and notation.

Let G be a domain in $\mathbb{R}^n$. We denote by T_G the tube domain with base G. In other words we set $T_G = \{z = x+iy \in \mathbb{C}^n: x \in \mathbb{R}^n, y \in G\}$. We also set $\|f\|_G = sup\{|f(z)|: z \in T_G\}$.

A holomorphic mapping $f: T_G \to \mathbb{C}^m$, $m \le n$, is said to be almost periodic if for any $\varepsilon > 0$ and any domain $G' \subset\subset G$ there exists $l =$

$l(\varepsilon,G') > 0$ such that in every domain[60] $\{z\in\mathbb{C}^n: \|x-a\|_\infty < l,\ y\in G'\}$, $a\in\mathbb{R}^n$, there is a point τ such that $\|f(z+\tau)-f(z)\|_{G'} < \varepsilon$.

The following definition is equivalent to the previous one.

A holomorphic mapping $f: T_G \to \mathbb{C}^m$, $m \le n$, is said to be almost periodic if for every sequence $h_j\in\mathbb{R}^n$, $j = 1, 2, \ldots$, there is a subsequence h_{j_k}, $k = 1, 2, \ldots$, such the sequence of mappings $f(z+h_{j_k})$ converges with respect to any norm $\|:\|_{G'}$, $G' \subset\subset G$.

Here we consider not arbitrary almost periodic holomorphic mappings f, but only those that satisfy the following condition: for every function $F(z)$ that can be represented in the form $F(z) = \lim_{j\to\infty} f(z+h_j)$, the set $|Z_F|$ has pure dimension n-m. We call such mappings regular. Note that in the case m = 1 every almost periodic mapping (function) is regular. In the case m > 1 there exist non-regular almost periodic mappings. It is easy to see that the following mapping is of the above type: $f: \mathbb{C}^2 \to \mathbb{C}^2$ with $f_1(z_1,z_2) = = exp(iz_1-iz_2)$, $f_2(z_1,z_2) = exp(2iz_1+2iz_2)$. At the same time the following statement can be shown. Let a mapping $f: \mathbb{C}^n \to \mathbb{C}^m$, $1 < m \le n$, be such that its components are $\mathbb{C}$-quasipolynomials f_j, $j = 1, \ldots, m$, with purely imaginary spectra and with stars Z_{f_j} (see §1 of **Chapter 6**) which satisfy the condition $dim\{Z_{f_1}\cap \cdots \cap Z_{f_m}\} = 2n-m$. Then f is a regular almost periodic mapping. Whether the requirement of regularity is necessary for the validity of the facts below is presently unknown.

Let f be a holomorphic mapping from $T_G \subset \mathbb{C}^n$ into $\mathbb{C}^m$ and, as in §1 of this chapter, let Λ_f be the current defined by the form $\frac{1}{\pi^m}ln|f|^2(dd^c|f|^2)^{m-1}$, which is also denoted by Λ_f. The coefficients of this current (form), which we denote by $\alpha_{I,J} = \alpha_{I,J}(z;f)$, $I, J\in\mathfrak{M}_{m-1}$, are locally summable (if $codim_z|Z_f| = m$, $\forall z\in|Z_f|$). Thus,

$$\Lambda_f = \left(\frac{i}{2}\right)^{m-1} \sum_{I,J\in\mathfrak{M}_{m-1}} \alpha_{I,J}dz^I\wedge d\bar{z}^J.$$

For the sake of brevity we denote by $V_f(G)$ the volume $V(Z_f;G)$ of the

---

[60] Recall that $\|(a_1,\ldots,a_n)\|_\infty = \max_{1\le j\le n} |a_j|$.

chain Z_f in the domain G. Besides, everywhere below we set:

$$\nu\in\mathbb{R}^n_+,\ z = x+iy,\ y\in\mathbb{R}^n,\ x\in\mathbb{R}^n,\ \nu\cdot x = (\nu_1x_1,\dots,\nu_nx_n),\ \mathbf{1} = (1,\dots,1),$$

$$\nu^{\mathbf{1}} = \nu_1\cdot\nu_2\cdot\dots\cdot\nu_n,\ \Pi_\nu = \{x\colon\ |x_1| < \nu_1,\ \dots,\ |x_n| < \nu_n\},$$

$$\underline{\nu} = \min_{1\le i\le n}\{\nu_i\},\ \Pi_\nu(x^0,G) = \{\ z\colon\ x\in x^0+\Pi_\nu,\ y\in G\}.$$

The following theorem is the above-mentioned multidimensional analogue of **Jessen's Theorem**.

Theorem 7.3.1. Let f be a regular almost periodic holomorphic mapping from a tube domain $T_G \subset \mathbb{C}^n$ into the space $\mathbb{C}^q$, $1 \le q \le n$. Then

1) the currents (forms)

$$\Lambda^\nu_f \overset{def}{=} \left(\frac{i}{2}\right)^{q-1} \sum_{I,J\in\mathfrak{M}_{q-1}} \alpha_{I,J}(x\cdot\nu+iy;f)dz^I\wedge d\bar{z}^J$$

converge in the space $\mathcal{D}'_{q-1,q-1}(T_G)$ as $\nu \to \infty$ to the current (Jessen's current)

$$\tilde{\Lambda}_f = \left(\frac{i}{2}\right)^{q-1} \sum_{I,J\in\mathfrak{M}_{q-1}} \tilde{\alpha}_{I,J}dz^I\wedge d\bar{z}^J$$

with locally summable coefficients $\tilde{\alpha}_{I,J}$ that depend on y only;

2) the current $\tilde{\Delta}_f \overset{def}{=} dd^c\tilde{\Lambda}_f$ is positive;

3) if a domain $G' \subset\subset G$ is such that $\mu_f(\partial G') = 0$, where the positive measure μ_f is defined in G by the equality $\mu_f(K) = \sigma(\Pi_\nu\times K;\tilde{\Lambda}_f)$, then $\lim\limits_{\underline{\nu}\to\infty}\frac{1}{\nu^{\mathbf{1}}} V_f(\Pi_\nu\times G')$ exists and

$$\lim_{\underline{\nu}\to\infty}\frac{1}{\nu^{\mathbf{1}}} V_f(\Pi_\nu\times G') = \mu_f(G').$$

Proof. First we prove the following lemma.

Lemma 7.3.1. Let the mapping f be as in the theorem. Let also $E_\theta = \{z\in T_G\colon\ |f(z)| < \theta\}$, $0 < \theta < \frac{1}{2}$.

Then for any $\varepsilon > 0$ and $G_0 \subset\subset G$ there exists $\theta_0 = \theta_0(\varepsilon,G_0) > 0$ such that the inequality

$$\int_{\Pi_1(x^0,G_0)\cap E_\theta} |\alpha_{I,J}(z;f)|dxdy < \varepsilon \tag{7.3.1}$$

holds for $x^0 \in \mathbb{R}^n$, $I,J\in\mathfrak{M}_{q-1}$, $0 < \theta < \theta_0$.

Proof. It follows from (7.1.6) that in view of the positivity of the current (form) $(dd^c ln|f|^2)^{m-1}$, to prove the inequality (7.3.1) it is sufficient to show the existence of θ_0 such that

$$|\sigma(\Pi_1(x^0,G_0) \cap E_\theta;\Lambda_f)| < \varepsilon,\ \forall x^0\in\mathbb{R}^n,\ 0 < \theta < \theta_0.$$

Assume that there is no such θ_0. Then there exist a number $\varepsilon > 0$ and sequences $\{\theta_j\}$, $\theta_j\downarrow 0$, and $\{x^{(j)}\}$, $x^{(j)}\in\mathbb{R}^n$, exist such that

$$|\sigma(\Pi_1(x^{(j)},G_0) \cap E_\theta;\Lambda_f)| \geq \varepsilon,\ \forall j. \tag{7.3.2}$$

Setting $f_j(z) = f(x+x^{(j)})$ and $\Gamma_j = \{z\colon x\in\Pi_1,\ y\in G_0,\ |f(z+x^{(j)})| < \theta_j\}$ we rewrite (7.3.2) as

$$|\sigma(\Gamma_j;\Lambda_{f_j})| \geq \varepsilon,\ \forall j.$$

Now we choose a subsequence $\{j'\}$ of the sequence $\{j\}$ in such a way that the mappings $f_{j'}(z)$ converge to some mapping F with respect to every norm $\|\colon\|_{G_0}$, $G_0 \subset\subset G$. This is possible because of the definition of almost periodic mappings. We set $E_{\theta,F} = \{z\in T_G\colon\ |F(z)| < \theta\}$. Since $f_{j'} \rightrightarrows F$, for some $j_0 = j_0(\theta)$ the inclusion $\bigcup_{j'\leq j_0} \Gamma_{j'} \subset E_{2\theta,F}$ is valid. Hence

$$|\sigma(\Pi_1(0,G_0) \cap E_{2\theta,F};\Lambda_{f_{j'}})| \geq |\sigma(\Gamma_{j'};\Lambda_{f_{j'}})| \geq \varepsilon,\ \forall j' \geq j_0. \tag{7.3.3}$$

In accordance with the **Corollary** to **Theorem 7.1.3**, the following relationship is true:

$$\lim_{j'\to\infty} \sigma(\Pi_1(0,G_0) \cap E_{2\theta,F};\Lambda_{f_{j'}}) = \sigma(\Pi_1(0,G_0) \cap E_{2\theta,F};\Lambda_F).$$

From this and (7.3.3) it follows that

$$|\sigma(\Pi_1(0,G_0) \cap E_{2\theta,F};\Lambda_F)| \geq \varepsilon,\ \forall\theta > 0.$$

This is impossible, because the coefficients of the form Λ_F are locally summable and $\bigcap_{\theta>0} E_{2\theta,F} = |Z_F|$. Hence we can choose $\theta > 0$ such that the Lebesgue measure of the set $\sigma(\Pi_1(0,G_0) \cap E_{2\theta}$ is arbitrary small.

The proof of the lemma is finished.

Now we continue with the proof of the theorem.

It follows from the definition of the functions $\alpha_{I,J}(z;f)$ (as coefficients of the form Λ_f) that

$$\alpha_{I,J}(z;f) = \frac{1}{\pi^q} \frac{\ln|f|^2}{|f|^{4(q-1)}} \det\left\{|f|^2 \frac{\partial^2|f|^2}{\partial z_{i_k} \partial \bar{z}_{j_l}} - \frac{\partial^2|f|^2}{\partial z_{i_k}} \cdot \frac{\partial^2|f|^2}{\partial \bar{z}_{j_l}}\right\}_{k,l}$$

We set

$$|f|_\theta = \max\{\theta, |f|\}$$

and

$$\alpha^\theta_{I,J} = \alpha^\theta_{I,J}(z;f) = \frac{1}{\pi^q} \frac{\ln|f|^2_\theta}{|f|^{4(q-1)}_\theta} \det\left\{|f|^2 \frac{\partial^2|f|^2}{\partial z_{i_k} \partial \bar{z}_{j_l}} - \frac{\partial^2|f|^2}{\partial z_{i_k}} \cdot \frac{\partial^2|f|^2}{\partial \bar{z}_{j_l}}\right\}_{k,l}$$

From the elementary properties of almost periodic functions in tube domains (these properties are similar to the corresponding properties of functions of one variable) it follows that the functions $|f|_\theta$ and $\alpha^\theta_{I,J}$ are almost periodic in T_G for any $\theta > 0$.

Then it follows from the general theory of a.p.f. that the averages of the functions $\alpha^\theta_{I,J}$ exist, i.e. the following limits exist:

$$\lim_{R\to\infty} \frac{1}{(2R)^m} \int_{-R}^{R} \cdots \int_{-R}^{R} \alpha^\theta_{I,J}(x+iy)dx =$$

$$= \lim_{R\to\infty} \frac{1}{2^n} \int_{\Pi_1} \alpha^\theta_{I,J}(Rx+iy)dx \overset{\text{def}}{=} a^\theta_{I,J}(y;f).$$

Moreover, for any $\varepsilon > 0$, any $y \in G_0 \subset\subset G$, and any $x^0 \in \mathbb{R}^n$ the inequality

$$\left|\frac{1}{2^n} \int_{\Pi_1} \alpha^\theta_{I,J}(x\cdot\nu+iy+x^0;f)dx - a^\theta_{I,J}(y;f)\right| < \varepsilon$$

is valid for every $\nu \in \mathbb{R}^n_+$ such that $\underline{\nu}$ is larger than some $l_0 = l_0(\varepsilon, G_0, \theta)$.

It is obvious that the functions $a^{\theta}_{I,J}$ are continuous in G. We show that the limit

$$\lim_{\theta \to 0} a^{\theta}_{I,J}(y) \overset{\text{def}}{=} \tilde{\alpha}_{I,J}(y)$$

exists in $L^1_{loc}(G)$ and that for any $\varepsilon > 0$ and any $G_0 \subset\subset G$ there are numbers $\hat{\theta} = \hat{\theta}(\varepsilon, G_0)$ and $\hat{l} = \hat{l}(\varepsilon, \theta, G_0)$ such that the inequality

$$\int_{G_0} \left| \frac{1}{2^n} \int_{\Pi_1} \alpha^{\theta}_{I,J}(x \cdot \nu + iy + x^0)dx - \tilde{\alpha}_{I,J}(y) \right| dy < \varepsilon \qquad (7.3.4)$$

holds for $x^0 \in \mathbb{R}^n$, $\theta < \hat{\theta}$, $\underline{\nu} > \hat{l}$. For this we choose positive numbers θ' and θ'' such that

$$\tilde{\theta} \overset{\text{def}}{=} max\{\theta', \theta''\} < \theta_0(\varepsilon, G_0),$$

where $\theta_0(\varepsilon, G_0)$ is the same as in **Lemma 7.3.1**. Let also l be a number such that

$$l > max\{l_0(\varepsilon, G_0, \theta'), l_0(\varepsilon, G_0, \theta'')\}.$$

Then

$$\int_{G_0} \left| a^{\theta'}_{I,J} - a^{\theta''}_{I,J} \right| dy \le \int_{G_0} \left| a^{\theta'}_{I,J}(y) - \frac{1}{2^n} \int_{\Pi_1} \alpha^{\theta'}_{I,J}(x \cdot \nu + iy + x^0) \right| dy +$$

$$+ \int_{G} \frac{1}{2^n} \left| \int_{\Pi} \left\{ \alpha^{\theta'}_{I,J}(x \cdot \nu + iy + x^0) - \alpha^{\theta''}_{I,J}(x \cdot \nu + iy + x^0) \right\} dx \right| dy +$$

$$+ \int_{G_0} \left| \frac{1}{2^n} \int_{\Pi_1} \alpha^{\theta''}_{I,J}(x \cdot \nu + iy + x^0)dx - a^{\theta''}_{I,J}(y) \right| dy \le$$

$$\le \varepsilon mes_n G_0 + \varepsilon mes_n G_0 + \frac{2}{2^n t^1} \int_{\Pi_1(0;G_0) \cap E_{\tilde{\theta}}} |\alpha_{I,J}(x + iy + x^0)| dxdy \le$$

$$\le 2\varepsilon mes_n G_0 + 2\varepsilon.$$

Hence there exist functions $\tilde{\alpha}_{I,J}(y) \in L^1_{loc}$ such that

$$\int_{G_0} |a^{\theta}_{I,J}(y) - \tilde{\alpha}_{I,J}(y)|dy \to 0 \quad \text{as } \theta\downarrow 0,\ \forall G_0 \subset\subset G,\ I\in\mathfrak{M}_{q-1},\ J\in\mathfrak{M}_{q-1}.$$

The existence of numbers $\hat{\theta}$ and $\hat{1}$ for which inequality (7.3.4) is valid, can be proved similarly. Then we note that (7.3.4) implies the weak convergence (as $\underline{\nu} \to \infty$) of the functions $\alpha_{I,J}(x\cdot\nu+iy)$ to the function $\tilde{\alpha}_{I,J}(y)$ on any "step", i.e. on any function φ that satisfies the conditions: 1) $\varphi = \text{const} \neq 0$ for $(x,y)\in Q = (x^0+\Pi_{\nu'})\times(y^0+\Pi_{\nu''})$; 2) $\varphi = 0$ for $(x,y)\notin Q$. The finite linear combinations of these "steps" are dense in the space of continuous functions of compact support. Therefore the functions $\alpha_{I,J}(x\cdot\nu+iy)$ converge weakly in T_G as $\underline{\nu} \to \infty$ to the function $\tilde{\alpha}_{I,J}(y)$ on the functions which are continuous in T_G and have compact support in T_G. Moreover, they converge to $\tilde{\alpha}_{I,J}(y)$ as elements of the space $\mathcal{D}'(T_G)$. Hence statement 1) of the theorem is true.

To prove statements 2) and 3), we consider the current of integration over the chain Z_f; we denote it by Δ_f.

From the current Δ_f we construct a family of currents Δ^{ν}_f. First we associate each form

$$\varphi = \left(\frac{i}{2}\right)^{n-q} \sum_{I,J\in\mathfrak{M}_{n-q}} \varphi_{I,J} dz^I \wedge d\bar{z}^J$$

in the space $D_{n-q,n-q}(T_G)$ and each $\nu\in\mathbb{R}^n_+$ the form

$$\varphi^{\nu} = \left(\frac{i}{2}\right)^{n-q} \sum_{I,J} \varphi_{I,J}\left(\frac{x}{\nu} + iy\right) dz^I \wedge d\bar{z}^J,$$

where $\frac{x}{\nu} = \left(\frac{x_1}{\nu_1}, \dots, \frac{x_n}{\nu_n}\right)$. Then we define the current Δ^{ν}_f by the equality

$$\langle\Delta^{\nu}_f, \varphi\rangle = \frac{1}{2^n \nu^1} \langle\Delta_f, \varphi^{\nu}\rangle.$$

As has been mentioned in §1 of this chapter, $\Delta_f = dd^c\Lambda_f$. Therefore,

$$\langle\Delta^{\nu}_f, \varphi\rangle = \frac{1}{\nu^1} \langle\Lambda_f, dd^c\varphi^{\nu}\rangle.$$

Now we define the operators ∂_y and $\bar{\partial}_y$ to act on forms $\psi = \sum \psi_{I,J} dz^I \wedge d\bar{z}^J$ as follows:

$$\partial_y \psi = \frac{1}{2} \sum_{I,J} \sum_j \frac{\partial \psi_{I,J}}{\partial y_j} dz_j \wedge dz^I \wedge d\bar{z}^J,$$

$$\bar{\partial}_y \psi = \frac{1}{2} \sum_{I,J} \sum_j \frac{\partial \psi_{I,J}}{\partial y_j} d\bar{z}_j \wedge dz^I \wedge d\bar{z}^J.$$

The operators ∂_x and $\bar{\partial}_x$ are defined similarly. Taking these notations into account, we have

$$\langle \Delta_f^\nu, \varphi \rangle = \frac{1}{\nu^1} \langle \Lambda_f, \frac{i}{2} \partial_y \bar{\partial}_y \varphi^\nu \rangle + \frac{1}{\nu^1} \langle \Lambda_f, \frac{i}{2} (i\partial_y \bar{\partial}_x - i\partial_x \bar{\partial}_y + \partial_x \bar{\partial}_x) \varphi^\nu \rangle =$$

$$= \frac{1}{\nu^1} \left(\frac{i}{2}\right)^n \int_{T_G} \sum_{I,J \in \mathfrak{M}_{q-1}} \alpha_{I,J} \beta^\nu_{\bar{I},\bar{J}} \, dz^I \wedge d\bar{z}^J \wedge dz^{\bar{I}} \wedge d\bar{z}^{\bar{J}} +$$

$$+ = \frac{1}{\nu^1} \left(\frac{i}{2}\right)^n \int_{T_G} \sum_{I,J \in \mathfrak{M}_{q-1}} \alpha_{I,J} \gamma^\nu_{\bar{I},\bar{J}} \, dz^I \wedge d\bar{z}^J \wedge dz^{\bar{I}} \wedge d\bar{z}^{\bar{J}},$$

where the $\beta^\nu_{\bar{I},\bar{J}}$ are the coefficients of the form $\frac{i}{2}\partial_y \bar{\partial}_y \varphi^\nu$ and the $\gamma^\nu_{\bar{I},\bar{J}}$ are the coefficients of the form $\frac{i}{2}(i\partial_y \bar{\partial}_x - i\partial_x \bar{\partial}_y + \partial_x \bar{\partial}_x)\varphi^\nu$.

We substitute $\frac{x}{\nu} \to x$ in the above integrals. Then we obtain

$$\langle \Delta_f^\nu, \varphi \rangle = \left(\frac{i}{2}\right)^n \int_{T_G} \sum_{I,J} \alpha_{I,J}(x \cdot \nu + iy) \left\{ \beta_{\bar{I},\bar{J}} + \gamma_{\bar{I},\bar{J},\nu} \right\} dz^I \wedge d\bar{z}^J \wedge dz^{\bar{I}} \wedge d\bar{z}^{\bar{J}},$$

where the $\beta_{\bar{I},\bar{J}}$ are the coefficients of the form $\frac{i}{2}\partial_y \bar{\partial}_y \varphi^\nu$, and the functions $\gamma_{\bar{I},\bar{J},\nu}$ are algebraic sums of terms of one of the following forms:

$$\frac{1}{\nu_j} \frac{\partial^2 \varphi_{I',J'}}{\partial y_i \partial x_j}, \quad \frac{1}{\nu_i \nu_j} \frac{\partial^2 \varphi_{I',J'}}{\partial x_i \partial x_j}, \quad I', J' \in \mathfrak{M}_{n-q+1}.$$

According to the statement 1), which has already been proved, the functions $\alpha_{I,J}(x \cdot \nu + iy)$ converge in the space $\mathcal{D}'(T_G)$ to the functions $\tilde{\alpha}_{I,J}(y)$ as $\underline{\nu} \to \infty$. Therefore,

$$\lim_{\nu \to \infty} \left(\frac{i}{2}\right)^n \int_{T_G} \sum_{I,J} \alpha_{I,J}(x\cdot\nu+iy)\beta_{\bar{I},\bar{J}} dz^I \wedge d\bar{z}^J \wedge dz^{\bar{I}} \wedge d\bar{z}^{\bar{J}} =$$

$$= \left(\frac{i}{2}\right)^n \int_{T_G} \sum_{I,J} \tilde{\alpha}_{I,J}(y)\beta_{\bar{I},\bar{J},} dz^I \wedge d\bar{z}^J \wedge dz^{\bar{I}} \wedge d\bar{z}^{\bar{J}} = \langle \tilde{\Lambda}_f \cdot \tfrac{i}{2}\partial_y \bar{\partial}_y \varphi \rangle$$

and

$$\lim_{\nu \to \infty} \int_{T_G} \sum_{I,J} \alpha_{I,J}(x\cdot\nu+iy)\gamma_{\bar{I},\bar{J},\nu}(z) dz^I \wedge d\bar{z}^J \wedge dz^{\bar{I}} \wedge d\bar{z}^{\bar{J}} = 0.$$

Hence

$$\lim_{\nu \to \infty} \langle \Delta_f^\nu, \varphi \rangle = \langle \tilde{\Lambda}_f \cdot \tfrac{i}{2}\partial_y \bar{\partial}_y \varphi \rangle, \quad \forall \varphi \in D_{n-q,n-q}(T_G).$$

Since the current $\tilde{\Lambda}_f$ is independent of x,

$$\langle \tilde{\Lambda}_f \cdot \tfrac{i}{2}\partial_y \bar{\partial}_y \varphi \rangle = \langle dd^c\tilde{\Lambda}_f \cdot \varphi \rangle, \quad \forall \varphi \in D_{n-q,n-q}(T_G).$$

Thus, in the space $\mathcal{D}'_{q,q}(T_G)$ the limit $\lim_{\nu \to \infty} \Delta_f^\nu$ exists and equals $dd^c\tilde{\Lambda}_f \overset{\text{def}}{=} \tilde{\Delta}_f$. In view of the positivity of the currents Δ_f^ν statement 2) on the positivity of the current $dd^c\tilde{\Lambda}_f$ follows. Moreover, the convergence of the currents Δ_f^ν and their positivity imply that for each domain $\Omega \subset\subset T_G$ satisfying the condition $\sigma(\partial\Omega;\tilde{\Delta}_f) = 0$, the equality

$$\lim_{\nu \to \infty} \sigma(\Omega;\Delta_f^\nu) = \sigma(\Omega;\tilde{\Delta}_f)$$

holds. We apply this property to the domain $\Omega = \Pi_1(0;G_0)$, where $G_0 \subset\subset G$. Taking into account the definition of the current Δ_f^ν, the x-independence of the current $\tilde{\Delta}_f$ and equality (7.1.7), find that

$$\lim_{\nu \to \infty} \frac{1}{\nu^1} V_f(\Pi_\nu(0;G_0)) = \lim_{\nu \to \infty} \frac{1}{\nu^1} \sigma(\Pi_\nu(0;G_0);\Delta_f) =$$

$$= \lim_{\nu \to \infty} \sigma(\Pi_1(0;G_0);\Delta_f^\nu) = \sigma(\Pi_1(0;G_0);\tilde{\Delta}_f).$$

The proof of the theorem is finished.

In the case q = 1 the current $\tilde{\Lambda}_f$ is a locally summable function. The positivity of the current $\tilde{\Delta}_f = dd^c\tilde{\Lambda}_f$ means that this function is plurisubharmonic in T_G. This in view of the x-independence of $\tilde{\Lambda}_f$ is equivalent to its convexity. Finally, the measure μ_f is defined by the equality

$$\mu_f(E) = \mu_{\tilde{\Lambda}_f}(E\times\Pi_1),$$

where $\mu_{\tilde{\Lambda}_f}$ is the Riesz associated measure of the function $\tilde{\Lambda}_f$. Thus, in the case $n = 1$ and when the conditions of **Jessen's Theorem** are satisfied, we have

$$\mu_f((\alpha,\beta)) = \frac{1}{\pi}\int_\alpha^\beta \frac{\partial^2\tilde{\Lambda}_f(y)}{\partial y^2}\,dy = \frac{1}{\pi}\,(\tilde{\Lambda}'_f(\beta)-\tilde{\Lambda}'_f(\alpha)).$$

Nevertheless, formally Jessen's theorem is not contained in **Theorem 7.3.1**, because different kinds of convergence figure in the definitions of the functions $A_f(y)$ and $\tilde{\Lambda}_f(y)$. However, we note that a small and even simplifying, change in the proof of **Theorem 7.3.1**, leads to the following theorem, which coincides with **Jessen's Theorem** in the case $n = 1$.

Theorem 7.3.2. Let $f(z)$ be an almost periodic holomorphic function in a tube domain $T_G \subset \mathbb{C}^n$. Then

1) $\forall y\in G$ the following limit exists

$$\varliminf_{\nu\to\infty} \frac{1}{2^n\nu^1}\int_{\Pi_\nu} ln|f(x+iy)|dx \overset{\text{def}}{=} A_f(y);$$

2) the function $A_f(y)$ (the Jessen function) is convex;

3) for any domain $G_0 \subset\subset G$ such that $\mu_f(\partial G_0) = 0$, where $\mu_f = \frac{1}{\theta_n}\Delta A_f(y)$, the equality

$$\varliminf_{\nu\to\infty} \frac{1}{2^n\nu^1} V_f(\Pi_\nu(0;G_0)) = \frac{\theta_n}{2\pi}\,\mu_f(\partial G_0)$$

holds.

Notes

The main facts of the theory of currents, which have been briefly presented in §1, were obtained by P.Lelong [3]. In particular, **Theorem 7.1.1** and formula (7.1.7) are due to him. The definition of the product of the currents $dd^c v$ was given in the paper of E.Bedford and B.Taylor [1]. **Theorem 7.1.2**, known as the Poincaré-Lelong formula, was obtained by P.Griffiths and J.King [1]. **Theorem 7.1.3** was obtained by L.Ronkin [14]. All results in §2 are due to L.Ronkin [15]. These results were preceded by papers by P.Degtjar [1], in which the asymptotics of the order functions was obtained for holomorphic mappings similar to the ones considered in §2. The results in §3 are due to L.Ronkin [13], [14]. Some other problems concerning the zero distributions of almost periodic entire mappings were considering by O.Gel'fond [1] and B.Kazarnovskiǐ [1].

BIBLIOGRAPHY

Many Russian journals are translated into English cover-to-cover. Below, only the (russian) source is given.

Agranovič, P.Z.

1. Functions of several variables of completely regular growth (Russian), Teor. Funkciĭ Funkcional. Anal. i Priložen, Vyp. 30(1978), 3-13.

Agranovič, P.Z., Logvinenko, V.N.

1. The analogue of the Valiron-Titchmarsh Theorem for binomial asymptotics of subharmonic functions with masses on a finite system of rays (Russian), Sibirsk. Mat. Z., 26(1985), No 5, 2-19.

2. Polynomial asymptotic representation of functions subharmonic in the plane (Russian), Sibirsk. Mat. Z., 32(1991), No 1, 3-22.

Agranovič, P.Z., Ronkin, L.I.

1. Functions of several variables of completely regular growth (Russian), Preprint FTINT AN USSR (1976), Kharkov.

2. Functions of completely regular growth of several variables, Ann. Polon. Math. 39(1981), 239-254.

3. Functions of completely regular growth (Russian), Mat. Anal. i Teor. Ver., Naukova Dumka (1977), Kiev, 1-4.

4. Conditions of pluriharmonicity of the indicator of holomorphic function of several variables (Russian), Mat. Sb., 98 (1975), No 2, 319-332.

Ahiezer, N.I., Ronkin, L.I.

1. Separatly-analytic functions of several variables and theorems "edge-of-the-wedge", (Russian), Uspehi Mat. Nauk, 27(1973), No 3, 27-42.

2. Separatly-analytic functions of several variables (Russian), Probl. Mat. Fiz. and Funk. Anal., FTINT AN USSR, Kharkov, 3-10.

Avanissian, V., Gay, R.

1. Sur une transformation des fonctionelles analytiques et ses applications aus fonctions entieres de plusieurs variables, Bull. Soc.

Math. France. 103(1975), 341-384.

Azarin, V.S.

1. On rays of completely regular growth of entire function (Russian), Mat. Sb., 79(1969), 460-476.

2. Asymptotic behaviour of subharmonic and entire functions (Russian), Dokl. Akad. Nauk SSSR, 229(1976), No 6, 1289-1290.

3. Asymptotic behaviour of subharmonic functions of finite order (Russian), Mat. Sb., 108(1979), No 2, 147-167.

4. A characteristic property of functions of completely regular growth in the interior of an sector (Russian), Teor. Funkciǐ Funkcional Anal. i Priložen., Kharkov, 2(1966), 55-66.

5. Subharmonic functions of completely regular growth in multidimensional space (Russian), Dokl. Akad. Nauk SSSR, 146(1962), No 4, 743-746.

6. Functions of completely regular growth that are subharmonic in the whole space (Russian), Zapiski Meh.-Mat. Fak. KhGU, 28(1961), Ser.4, 128-148.

7. Generalization of a theorem of Hayman on subharmonic functions in ann-dimensional cone (Russian), Mat. Sb., 66(1965), No 2, 248-264.

Bedford, E., Taylor, B.A.

1. A new capacity for plurisubharmonic functions, Acta Math. 149(1982), No. 1-2, 1-40.

2. The Dirichlet problem for a complex Monge-Ampere equation, Invent. Math. 37(1976), 1-44.

Berenstein, C.A., Dostal, M.A.

1. The Ritt's theorem in several variables, Ark. Mat. 12(1974), No. 2, 267-280.

Berenstein, C.A., Yger, A.

1. Ideals generated by exponential polynomials, Prepr. TR 83, - 46, VI-1983, 152 pp.

Boas, R.P.

1. Entire functions, Academic Press(1954), New York.

Brelot, M.

1. Etude des fonctions sousharmoniques au voisinage d'un point singulier, Ann. Inst. Fourier. 1(1950), 121-156.

Cassels, J.W.S.

1. Factorisation of polynomials in several variables, Proc. 15th Scand. congr. Oslo 1968. Lecture Notes in Math. 118, Springer - Verlag, New York. 1-16.

Čirka, E.M.

1. Complex analytic sets (Russian), Nauka(1985), Moscow. English translation: Kluwer Academic Publishers(1989), Dordrecht.

Degtjar, P.B.

1. Some problems on the distribution of values of holomorphic mappings and complex variations (Russian), Mat. Sb., 115(1981), No 2, 307-318.

Džrbasjan, M.M.

1. Integral transforms and representations of functions in the complex domain (Russian), Nauka(1966), Moscow. MR 35 #370.

Favorov, S.Ju.

1. Reduction sets for subharmonic functions of completely regular growth (Russian), Sibirsk. Mat. Z., 20(1979), No 6, 1294-1302.

2. Sets of growth reduction for entire and subharmonic functions (Russian), Mat. Zametki, 40(1986), No 4, 460-467.

3. Addition of indicators of entire and subharmonic functions of several variables (Russian), Mat. Sb., 105(1976), No 1, 129-140.

4. Growth of plurisubharmonic functions (Russian), Sibirsk. Mat. Z., 24(1983), No 1, 168-174.

5. Entire functions of several variables with completely regular growth (Russian), Teor. Funkciǐ, Funkcional. Anal. i Priložen., Kharkov, 1982, Vyp. 38, 103-111.

Gel'fond, O.A.

1. Average number of zeros of systems of holomorphic almost periodic functions (Russian), Uspehi Mat. Nauk, 39(1984), Vyp. 1, 123-124.

Gol'dberg, A.A., Ostrovskii, I.V.

1. Distribution of values of meromorphic functions (Russian), Nauka(1970), Moscow.

Gordon, A.J., Levin, B.J.

1. On the division of quasipolynomials (Russian), Funk. Analiz, 5(1971), No 1, 22-29.

Govorov, N.

1. Functions of completely regular growth in the half-plane (Russian), Theses, Rostov-na-Donu (1966).

2. On the indicator of functions of nonintegral order that are analytic and of completely regular growth in a half plane (Russian), Dokl. Akad. Nauk SSSR. 162(1965), No 3, 495-498.
3. Riemann boundary value problems with infinite index (Russian), Nauka (1986), Moscow.
Griffiths, P.A., King, J.
1. Nevanlinna theory and holomorphic mappings between algebraic varieties, Acta Math. 130(1973), 145-220.
Grishin, A.F.
1. Regularity of the growth of subharmonic functions: III (Russian), Teor. Funkciĭ, Funkcional. Anal. i Priložen. 1968, Vyp.7, 59-84.
Gruman, L.
1. Les zeros des fonctions entieres d'ordre fini de croissance reguliere dans C^n, C.R. Acad. Sci. Paris. 282(1976), 363-365.
2. Entire holomorphic mappings in one and several complex variables and their asymptotic growth, Ark. Mat. 9(1971), No. 1, 23- 71.
Guzman, M.
1. Differentiation of integrals in R^n, Springer-Verlag(1975), Berlin, Heidelberg, New York.
Hayman, W.K., Kennedy, P.B.
1. Subharmonic functions, Vol.1, Academic Press(1976), London New York San Francisco.
Heĭfic, A.I.
1. On subharmonic functions of completely regular growth in half-plane, Dokl. Akad. Nauk SSSR, V.239(1978), No 2, 282-285.
Hoffman, K.
1. Banach spaces of analytic functions, Prentice-Hall Inc.(1962), Englewood Cliffs, N.Y.
Hörmander, L.
1. The analysis of linear partitial differential operators I, Distribution theory and Fourier analysis, Springer-Verlag(1983), Berlin, Heidelberg, New York, Tokyo.
2. An introduction to complex analysis in several variables, North-Holland, Amsterdam (1973).
Kac'nelson, V.E.
1. Equivalent norms in spaces of entire functions (Russian), Mat. Sb.,

92(1973), No 1, 34-54.
Kac'nelson, V.E., Ronkin, L.I.
1. Minimal volume of analytical sets (Russian), Sibirsk. Mat. Z., 15(1974), 516-528.
Kazarnovskiĭ, B.Ja.
1. Newton polyhedra and zeros of system of exponential sums (Russian), Funk. Anal. i Priložen., 18(1984), Vyp. 4, 40-49.
Kiselman, C.O.
1. On entire functions of exponential type and indicators of analytic functionals, Acta Math. 117(1967), No. 1-2, 1-35.
Korevaar, J., Hellerstein, S.
1. Discrete sets of uniqueness for bounded holomorphic functions f(z,w), in: Entire functions and related parts of analysis. Proc. Symp. Pure Math, XI, AMS Providence, R. I. (1968), 273-284.
Lelong, P.
1. Fonctions entieres (n variables) et fonctions plurisousharmoniques d'ordre fini dans C^n, J. d'Analyse Math. 12(1965), 365-407.
2. Fonctionelles analytiques et fonctions entieres (n variables), Univ. press (1968), Montreal.
3. Fonctions plurisousharmoniques et formes differentielles positives, Gordon and Breach (1968), New York.
Lelong, P., Gruman, L.
1. Entire functions of several complex variables, Springer - Verlag (1986), Berlin, Heidelberg.
Lelong-Ferrand, J.
1. Etude des fonctions subharmoniques positives dans un cylindre ou dans un cone, C.R. Acad. Sci. Paris. 229(1949), No. 5, 340- 341.
Leont'ev, A.F.
1. Series of exponentials (Russian), Nauka(1976), Moscow.
Levin, B. Ja.
1. Distribution of zeros of entire functions, GITTL (1956), Moscow; English transl.: Translations of Math. Monographs, vol. 5, AMS Providence, R.I. (1964). MR **19**, 402; **28** #217.
2. The growth of an entire function along a ray and the distribution of its zeros in arguments (Russian), Mat. Sb., 2(44)(1937), No 6, 1097-1142.

3. Applications of Lagrange interpolation series in the theory of entire functions (Russian), Mat. Sb., 8(50)(1940), No 1, 437-454.
Levin, B.Ja., Logvinenko, V.N.
1. About classes of subharmonic functions that are bounded on some subsets of $\mathbb{R}^n$, Issled. po lin. operat. i teor. funk., XVII (Zapiski seminarov LOMI 170), Nauka(1989), Leningrad, 157-175.
Levin, B.Ja., Ronkin, A.L.
1. Asymptotic series and algebroid functions (Russian), Dokl. Akad. Nauk, 280(1985), No 2, 258-291.
2. Asymptotic series of exponentials and solutions of some algebraic equations (Russian), in: Issl. po Kompleksn. Anal., Ufa(1987), 71-120.
Levinson, N.
1. Gap and density theorems, AMS, Providence, R. I. (1940).
Levitan, B.M.
1. Almost periodic functions (Russian), GITTL (1953), Moscow.
Logvinenko, V.N.
1. Binomial asymptotics for a class of entire functions, Dokl. Akad. Nauk SSSR. 205(1972), No.5. 1037-1040.
2. Multidimensional analogues of M. Cartwright's theorem (Russian), Dokl. Akad. Nauk SSSR, 12(1974), No 4, 17-24.
3. Theorems of the type of M. Cartwright's theorems and real sets of uniqueness for entire functions of several complex variables (Russian), Teor. Funkciĭ in Funkcional. Anal. i Priložen., Vyp. 22(1975), 85-100.
Logvinenko, V.N., Sereda, J.F.
1. Equivalent norms in spaces of entire functions of exponential type (Russian), Teor. Funkciĭ Funkcional. Anal. i Priloženen., 22(1973), 68-78.
Martineau, A.
1. Indicatrices de N variables, Invent. Math. Berlin. 2(1966), 81-85; 3(1967), 16-19 (Correction et compliment).
Napalkov, V.V.
1. Convolution equation in multidimensional spaces (Russian), Nauka(1982),Moscow.
Panejah, B.P.
1. Some problems of harmonic analysis (Russian), Dokl. Akad. Nauk SSSR, 142(1962), No 5, 1026-1029.

2. Some inequalities for entire functions of exponential type and a priori estimates for common differential operators (Russian), Uspehi Mat. Nauk, 21(1966), No 3, 75-114.
Pfluger, A.
1. Die Wertverteilung und das Verhalten von Betrag und Argument einer speziellen Klasse analytischer Funktionen. I, II, Comm. Math. Helv. 11(1938), 180-213; 12(1939), 25-69.
2. Über Interpolation ganzer Funktionen, Comm. Math. Helv. 14(1941/1942), 314-349.
3. Über ganze Funktionen ganzer Ordnung, Comm. Math. Helv. 18(1946), 177-203.
Plancherel, M., Polya, G.
1. Fonctions entieres et integrales Fourier multiples, Comm. Math. Helv. 9(1937), 224-248.
Rashkovskiǐ, A.Ju.
1. Growth and decrease of subharmonic functions in a cone (Russian), Ukr. Mat. Ž., 41(1989), No 9, 1252-1258.
2. Integral representation of subharmonic functions of finite order in a cone (Russian), Sibirsk. Mat. Z., 30(1989), No 3, 109-123.
Rashkovskiǐ, A.Ju., Ronkin, L.I.
1. Subharmonic functions of finite order in a cone (Russian), Dokl.Akad. Nauk SSSR, 287(1987), No 2, 298-302.
2. Subharmonic functions of completely regular growth in a cone (Russian), I, II, Preprint FTINT AN USSR, Kharkov, 1985, 27-85; 1986, 34-86.
Ritt, J.F.
1. On the zeros of exponential polynomials, Trans. Amer. Math. Soc. 31(1929), 680-686.
2. Algebraic combinations of exponentials, Trans. Amer. Math. Soc. 31(1929), 654-679.
Ronkin, A.L.
1. Distribution of zeros of quasipolynomials of several variables (Russian), Funk. Analiz i Priložen., 14(1980), No 3, 91-92.
2. Distribution of zeros of quasipolynomials of several variables (Russian), Dep. BINITI AN, No 375-80 (Dep. RZMat 5v 232 Dep).
3. On quasipolynomials (Russian), Funk. Analiz i Priložen., 12(1978),

No 4, 93-94.
4. On quasipolynomials (Russian), Funkcional. Anal. and Prikl. Mat., Naukova Dumka (1982), Kiev, 131-158.
5. A theorem on division of quasipolynomials (Russian), Teor. Funkciĭ Funkcional. Anal. i Priložen., 1980, Vyp. 34, 104-111.
6. On irreducibility of quasipolynomials (Russian), Teor. Funkciĭ Funkcional. Anal. i Priložen., 1983, Vyp. 39, 106-109.
Ronkin, L.I.
1. Introduction to the theory of entire functions of several variables, Nauka (1971), Moscow; English Transl.: Translations of Math. Monographs, Vol. 44, AMS, Providence, R. I. (1974).
2. Entire functions of finite degree and functions of completely regular growth (of several variables) (Russian), Dokl. Akad. Nauk SSSR (N.S.), 119(1958), 221-214.
3. Functions of completely regular growth in the half-plane (Russian), Dokl. Akad. Nauk USSR, 239(1978), No 2, 282-285.
4. Regularity of growth and $\mathcal{D}'$-asymptotic of functions holomorphic in C^{+} (Russian), Izv. VUSov Matem.,1990, No 2, 16-28.
5. Subharmonic functions of completely regular growth in a closed cone (Russian), Dokl. Akad. Nauk SSSR, 314(1990), No 6, 1331-1335.
6. Certain problems on completeness and uniqueness for functions of several variables (Russian), Funk. Analiz i Priložen., 7(1973), No 1, 45-55.
7. The completeness of systems of functions and real uniqueness sets of entire functions of several variables (Russian), Funk. Analiz i Priložen., 5(1971), No 4, 86-87.
8. Real sets of uniqueness for entire functions of several variables and the completeness of systems of functions (Russian), Sibirsk. Mat. Z., 13(1972), No 3, 638-644.
9. Discrete uniqueness sets for entire functions of several variables (Russian), Dokl. Akad. Nauk SSSR, 218(1974), No 4, 764-767.
10. Discrete uniqueness sets for entire functions of exponential type in several variables (Russian), Sibirsk. Mat. Z., 15(1978), No 1, 142-152.
11. Discrete uniqueness sets for entire functions of exponential type bounded for real values of variables (Russian), Matem., Fiz.,

Funkcional. Anal., Kharkov, FTINT Akad. Nauk USSR (1974), No. 5, 11-26.
12. Elements of the theory of analytic functions of several variables (Russian), Naukova Dumka (1977), Kiev.
13. Jessen's theorem for holomorphic almost periodic functions in tube domains (Russian), Sibirsk. Mat. Z., 28(1987), No 3, 199-204.
14. Jessen's theorems for holomorphic almost mappings (Russian), Dokl. Akad. Nauk USSR, Ser A, 1987, No 6, 10-12.
15. Weak convergence of currents $(dd^c u)^q$ and asymptotics of order functions for holomorphic mappings of regular growth (Russian), Sibirsk. Mat. Z., 25(1984), No 4, 167-173.
Schwartz, L.
1. Analyse Mathematique, V. 1, Hermann (1967), Paris.
Selberg, H.
1. Einige Darstellungsatze aus der Theorie der ganzer Funktionen endlicher Ordnung, J. Matem. Naturvid. Vilasse. 1934, No. 10, 91- 102.
Sigurdsson, R.
1. Growth properties of analytic and plurisubharmonic functions of finite order, Math. Scand. 59(1986), 235-304.
Sire, O.
1. Sur les fonctions de deux variables d'ordre aparrent total fini, Rend. Circolo Mat. Palermo. 31(1911), 1-91.
Stein, E.M., Weiss, G.
1. Introduction to Fourier analysis on Euclidean spaces, Princeton University Press(1971), Princton New Jersey.
Titchmarsh, E.C.
1. Theory of functions, 2nd ed., Oxford University Press (1939), Oxford.
Vauthier, J.
1. Comportement asymptotique des fonctions entieres de type exponentiel dans $\mathbb{C}^n$ et bornees dans le domaine reel, J. Funct. Anal. 12(1973), No. 3, 290-306.
2. Comportement asymptotique des fonctions entieres de type exponentiel dans C^n bornees dans le domaine reel, C. R. Acad. Sci. Paris. 275(1973), 1061-1063.
Vladimirov, V.
1. Distributions in mathematical physics (Russian), Nauka(1979),

Moscow.
2. Methods of the theory of functions of several variables, Nauka (1964), Moscow; English Transl.: M.I.T. Press (1966), Cambridge, Mass. MR 30 #2163; **34** #1551.

Wiegerinck, J.
1. Entire functions of Paley - Wiener type in $\mathbb{C}^n$, Radon transforms and problems of holomorphic extensions, Centrum voor Wiskunde en Informatica (1975), Amsterdam. Thesis.

SUBJECT INDEX